Electronic Materials: Science & Technology

Series Editor: Harry L. Tuller
Professor of Materials Science and Engineering
Massachusetts Institute of Technology
Cambridge, Massachusetts
tuller@mit.edu

For further volumes:
http://www.springer.com/series/5915

Roel van de Krol • Michael Grätzel
Editors

Photoelectrochemical Hydrogen Production

Springer

Editors
Roel van de Krol
Department of Chemical Engineering/
Materials for Energy Conversion
and Storage
Faculty of Applied Sciences
Delft University of Technology
P.O. Box 5045, 2600 GA Delft
The Netherlands
r.vandekrol@tudelft.nl

Michael Grätzel
Laboratory for Photonics and Interfaces
Ecole Polytechnique Fédérale de Lausanne
CH-1015 Lausanne, Switzerland
michael.graetzel@epfl.ch

ISSN 1386-3290
ISBN 978-1-4614-1379-0 e-ISBN 978-1-4614-1380-6
DOI 10.1007/978-1-4614-1380-6
Springer New York Dordrecht Heidelberg London

Library of Congress Control Number: 2011939087

Printed on acid-free paper

Springer is part of Springer Science+Business Media (www.springer.com)

Preface

Hydrogen is a highly versatile fuel that may become one of the key pillars to support our future energy infrastructure. It can be efficiently converted into electricity using a fuel cell, or it can directly drive an internal combustion engine. Using hydrogen is clean; the only reaction product upon oxidation is pure water, with little or no exhaust of greenhouse gases. It can even be converted into more convenient form of fuel, a liquid hydrocarbon, using excess CO_2 and well-established Fischer–Tropsch technology. However, hydrogen does not occur freely in nature, and producing hydrogen in a clean, sustainable, and economic way is a major challenge.

This book is about tackling that challenge with semiconductors, using water and sunlight as the only ingredients. The ultimate aim is to make a monolithic photoelectrode that evolves hydrogen and oxygen at opposite sides of the electrode, so that they can be easily separated. Finding semiconductors that can do this efficiently, at low cost, and without suffering from corrosion is far from trivial. The emphasis in this book is on transition metal oxides, a low-cost and generally very stable class of semiconductors. There is a darker side to these materials, though. The bandgap of metal oxide semiconductors is often a bit too large, and the optical absorption coefficient is usually small. In addition, the catalytic activity for water oxidation or reduction at the surface is generally poor, and the electronic charge transport properties can be downright horrible. This issues have thwarted many earlier efforts in the late 1970s and early 1980s to reach the "Holy Grail" of solar water splitting. In the past few years, however, exciting breakthroughs in nanotechnology have stimulated a huge amount of renewed interest in this field. This book attempts to summarize both the basic principles and some of the important recent developments in photoelectrochemical water splitting. While we cannot even hope to approach completeness in a single volume, we nevertheless hope that both experts and newcomers in this field find something useful here that helps their research.

The book is organized into four parts. The first part covers basic principles and is specifically aimed at undergraduate and graduate students, as well as colleagues who are new to the field. Chapter 1 provides a brief motivation for our interest in solar hydrogen production. The properties of semiconductors, the semiconductor/

electrolyte interface, and basic PEC device operation are covered in Chap. 2, while an overview of photoelectrochemical measurement techniques is given in Chap. 3. The second part of the book is on materials properties and synthesis. In Chap. 4, Kevin Sivula discusses the intrinsic properties of α-Fe_2O_3 (hematite) that limit its performance as a photoanode, and how these limitations can be overcome by nanostructuring. Kazuhiro Sayama outlines the properties of ternary and mixed metal oxide photoelectrodes in Chap. 5, showing recent results on $BiVO_4$ and a high-throughput screening method. In Chap. 6, Bruce Parkinson takes the high-throughput concept to the next level by discussing combinatorial approaches to discover new candidate materials and to screen thousands of compositions in a quick and systematic fashion. The third part of the book is on devices and device characterization. This part consists of a single, extensive chapter by Eric Miller, Alex DeAngelis, and Stewart Mallory on multijunction approaches and devices for solar water splitting (Chap. 7). They analyze the merits of various tandem configurations and materials combinations, and give an overview of key aspects to be considered in future research efforts. The fourth and final part of the book gives an overview of some of the future perspectives for photoelectrochemical water splitting. In Chap. 8, Julian Keable and Brian Holcroft take a closer look at the economic and business perspectives, and set the device performance targets that need to be met in order to commercialize the technology. In the final chapter, Scott Warren describes how some of the recent developments in nanotechnology and nanophotonics can be leveraged in solar water splitting materials, offering an exciting glimpse at future performance breakthroughs (Chap. 9).

Putting together a volume like this is a big undertaking in which many people are involved. First and foremost, the editors express their sincere thanks to all the contributors. We hope they are pleased with the fruits of our collective labor, and greatly appreciate their patience during the lengthy course of this project. We thank the people of Springer for their encouragement and support throughout the project: Elaine Tham, Lauren Danahy, Merry Stuber, and especially Michael Luby. A final and special thanks goes to the series editor, Prof. Harry Tuller, for inviting us to edit a volume on the exciting subject of solar water splitting.

Delft, The Netherlands — Roel van de Krol
Lausanne, Switzerland — Michael Grätzel

Contents

Part I
Basic Principles

Chapter 1
Introduction

Roel van de Krol and Michael Grätzel

1.1 The Energy Challenge

One of the main challenges facing mankind in the twenty-first century is to supply the world's population of sufficient energy to meet the desired living standards. The power consumption of the current (2011) global population of nearly 7 billion people is 15 TW, and these numbers are estimated to increase to ~9 billion and 30 TW by 2050. Fossil fuels, which currently provide about 85% of our energy supply, will be unable to keep up with this increase in demand. In the long run, this is simply a matter of the available reserves. Based on the current consumption rate, estimated reserves range from 150 to 400 years for coal, 40–80 years for oil, and 60–160 years for natural gas. The effect of dwindling reserves, however, will be felt on a much shorter time scale. This is because the available reserves become increasingly hard to recover, and the peak in the production will occur long before the supplies run out. According to the International Energy Agency, the production of conventional (easily recoverable) oil has already peaked in 2006. The oil price has in fact sharply risen since 2001, and it is unlikely that it will ever return to its pre-2001 (inflation-corrected) level.

A perhaps far more serious concern associated with the use of fossil fuels is the impact on the environment. The main concern in this regard is the emission of greenhouse gases, in particular CO_2, and their contribution to global warming. Since the beginning of the industrial revolution, the CO_2 level in the atmosphere

R. van de Krol (✉)
Faculty of Applied Sciences, Department of Chemical Engineering/Materials for Energy Conversion and Storage, Delft University of Technology, P.O. Box 5045, 2600 GA Delft, The Netherlands
e-mail: r.vandekrol@tudelft.nl

M. Grätzel
Laboratory for Photonics and Interfaces, Ecole Polytechnique Fédérale de Lausanne, CH-1015 Lausanne, Switzerland

R. van de Krol and M. Grätzel (eds.), *Photoelectrochemical Hydrogen Production*, Electronic Materials: Science & Technology 102, DOI 10.1007/978-1-4614-1380-6_1,

has risen from 280 to 394 ppm,[1] and it is currently rising by about 2 ppm/year. According to the International Panel on Climate Change (IPCC), a CO_2 level above 450 ppm carries a high risk[2] of causing global warming by more than 2°C. Such a rise is likely to have a severe adverse impact on ecosystems and human society, with effects that will be felt throughout the century. If the temperature change can be limited to less than 2°C, there is a good chance that society can adapt. Several studies agree that the current decade, between 2010 and 2020, is a critical one. Unless we are able to sharply reduce CO_2 emissions within the next 10 years, exceeding the 450 ppm level seems unavoidable [1, 2].

To reduce our dependence on fossil fuels and curb the exhaust of CO_2, we need to make a large-scale transition toward new, sustainable sources of energy. While most scientists and politicians nowadays agree that such a transition is unavoidable, there is much uncertainty about the route to follow, and the speed at which this can and should be done. More often than not, the viability of a certain route is determined by economical considerations, rather than technological impediments. As we see later in this chapter and in Chaps. 7 and 8, cost is indeed a crucially important factor for the photoelectrochemical water splitting route that is the topic of this book.

1.2 Sustainable Energy Sources

While an in-depth review of the various sustainable energy sources and options is far beyond the scope of this chapter, it is instructive to briefly consider their estimated global power generation capacities (Table 1.1). Any future energy infrastructure will

Table 1.1 Overview of global power generating capacities of sustainable energy sources [5, 6]

Energy source	Power (TW)	Remarks
Wind	4	Represents 10–15% of global technical potential for on- and off-shore installations
Hydroelectric	1–2	Remaining untapped potential is 0.5 TW
Tidal and ocean currents	<2	
Geothermal	12	Only a small fraction of this can be exploited
Biomass	10	Requires 10% of earth's land surface to be covered with switchgrass
Nuclear	10	Requires construction of a 1-GW_{peak} power plant every 35 h for the next 40 years. Finite uranium supplies imply need for fast breeder or thorium reactors
Solar	>20	Requires 0.16% of the earth's surface to be covered with 10% efficient solar cells. Total solar power reaching the earth's surface is 120,000 TW

[1] CO_2 level in May 2011.

[2] Estimates vary between 30 and 80%.

Fig. 1.1 Area of land that needs to be covered with 10% efficient solar cells in order to generate 20 TW of electrical power

almost certainly be composed of a mixture of these – and other – technologies, with local circumstances (geography, climate, population density) determining the optimal mix for a particular region. Due to the massive efforts involved in implementing any of these options on a Terawatt scale [3, 4], it is more than likely that fossil fuels will continue to play an important role for the next few decades. In principle, the estimated total fossil fuel reserves can sustain a 25–30-TW energy consumption for at least a few more centuries [5]. This, however, requires efficient capture and storage of CO_2 on an enormous scale, using yet unproven technology.

Of the sources shown in Table 1.1, solar energy is the only source that has the potential to meet all our energy needs. To generate 20 TW of power from the sun, the area to be covered with 10% efficient solar cells is about 816.000 km^2, which corresponds to an area of about 900 $\times$ 900 km^2. While this appears relatively small when projected onto Africa (Fig. 1.1), it should be realized that this is equal to the total surface area of France and Germany combined. Covering such a large area with solar cells presents a daunting task, even when this is undertaken on a (de-centralized) global scale. To illustrate this one would need to produce on average 650 m^2 of solar cell panels per second, 24/7 for 365 days per year, for the next 40 years in order to reach 20 TW of peak power.

The solar cell market is currently growing by 35–40% per year, and is one of today's fastest growing markets. In 2010, the global production of solar cells was 16.6 GW peak, resulting in a total installed capacity of ~40 GW_{peak}. While it is difficult to predict how this market will develop over the next few decades, estimates range from 3 to 7 TW of installed PV capacity in 2050. Irrespective of the uncertainties involved, there can be little doubt that solar energy will become an important component of the energy mix in the decades to come.

1.3 From Solar to Fuel

As the contribution of solar energy to the total energy mix increases, it will become difficult for electricity network operators to cope with the intermittent nature of solar power (day/night cycle, clouds). At a certain point, grid-based storage capacities will be exceeded and large-scale energy storage solutions need to be implemented. One of the more attractive possibilities is to store solar energy in the form of a chemical fuel. The energy of a visible-light photon ranges between 1 and 3 eV, or 100–300 kJ/mol, which is more than sufficient for many chemical synthesis routes. Compared to, e.g., batteries and mechanical or gravity-based storage systems such as flywheels and pumped water reservoirs, chemical fuels combine the advantages of high energy storage densities and ease of transportation. Examples of chemical fuels include hydrogen, methane, methanol, gasoline, diesel, etc.

Except for hydrogen, all of these examples require a source of carbon. While CO_2 is an obvious candidate in view of the environmental concerns discussed in Sect. 1.1, capturing CO_2 from the atmosphere comes at a huge entropic cost because of its dilute nature. Fossil fuel-based power plants seem attractive point-sources of highly concentrated CO_2, but the goal of the exercise was to avoid the use of fossil fuels in the first place. A conceptually more attractive route would be to capture the CO_2 emitted by, e.g., cars, and to reuse it by synthesizing fuels with sunlight as the energy source. This would close the CO_2 loop. One of the challenges that would have to be addressed is to minimize the energy penalty involved in capturing the CO_2. Direct photo(electro)chemical conversion of CO_2 to a fuel seems to be even more challenging, as the electrochemical half-reactions for the conversion of CO_2 to, e.g., methanol or methane involve complex six- and eight-electron transfer steps, respectively.

Based on these considerations, the conversion of solar energy into hydrogen appears to be a much more attractive route. Water is a convenient and abundant source of hydrogen, and there is more than enough water available. A back-of-the-envelope calculation shows that ~3.5×10^{13} L of water is needed to store the energy the world uses in 1 year (4.7×10^{20} J) in the form of hydrogen. This corresponds to 0.01% of the annual rain fall, or 0.000002% of the amount of water in the world's oceans. The water splitting reaction can be written as follows:

$$2H_2O + \text{sunlight} \rightleftarrows 2H_2 + O_2 \quad \Delta G = 237\,\text{kJ/mol}. \tag{3.1}$$

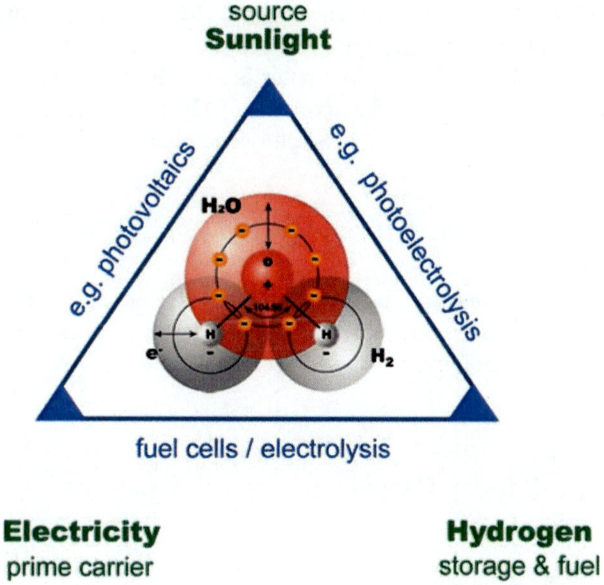

Fig. 1.2 Possible future energy triangle (courtesy of Dr. Andreas Luzzi)

Table 1.2 Gravimetric and volumetric energy densities of several fuels (at 1 bar)

	Energy density	
Fuel	Gravimetric (MJ/kg)	Volumetric (MJ/L)
Coal	24	–
Wood	16	–
Gasoline (petrol)	44	35
Diesel	46	37
Methanol	20	18
Natural gas	54	0.036
Hydrogen	143	0.011

The reduction half-reaction is an easy two-electron transfer reaction, while four electrons are involved in the oxidation of water to form oxygen. While by no means trivial, it is considerably easier to photo-oxidize water than it is to photoreduce CO_2. Moreover, hydrogen can be readily converted into electricity – and back again – with fuel cells and electrolyzers. This offers the prospect of a future energy infrastructure based on sunlight, hydrogen, and electricity, as illustrated in Fig. 1.2.

One of the main concerns associated with hydrogen is the difficulty in storing it. While hydrogen has a very high gravimetric energy density, the volumetric energy density is rather low (Table 1.2). Solutions can be found in the form of high pressure storage containers (up to 700 bar), liquid cryo-storage, or physisorption at high-surface area metal organic frameworks (MOFs) or in clathrate hydrate cages.

Another solution is to store hydrogen by forming chemical bonds. This can be in the form of metal hydrides, such as MgH_2, $LaNi_5H_6$, and $LiBH_4$, or by using hydrogen and CO_2 to make chemical fuels. The latter is a much easier route than the direct photochemical or electrochemical activation of CO_2. For example, CO_2 and hydrogen can be converted into CO via the slightly endothermic reverse water–gas shift reaction:

$$CO_2 + H_2 \rightleftarrows H_2O + CO \quad \Delta H = +\,42\,\text{kJ/mol}. \tag{3.2}$$

The CO and H_2 (*syngas*) can be separated out with a membrane and converted to liquid hydrocarbon fuels, such as methanol and diesel, using well-established Fischer–Tropsch technology.

1.4 Routes to Solar Hydrogen

Many pathways exist for the conversion of water and sunlight into hydrogen:

- Photoelectrochemical water splitting
- Photocatalytic water splitting
- Coupled photovoltaic – electrolysis systems
- Thermochemical conversion
- Photobiological methods
- Molecular artificial photosynthesis
- Plasma-chemical conversion
- Mechano-catalytic, magnetolysis, radiolysis, etc.

Some of these methods are described in more detail in the recent books by Rajeshwar and Grimes [7, 8]. This book focuses exclusively on the first method in the list: photoelectrochemical water splitting with semiconductor photoelectrodes. There are several reasons why this approach is appealing. One of the main advantages is that hydrogen and oxygen are produced at separate electrodes. This avoids serious safety concerns[3] and allows easy separation of these gases without having to pay a heavy energy penalty for postseparation. A second advantage is that it can be carried out at room temperature, i.e., there is no need for large-scale solar concentrators that would limit its application to large central facilities in sunny regions of the world. A third advantage is that a photoelectrochemical water splitting device can be constructed entirely from inorganic materials. This offers a degree of chemical robustness and durability that is difficult to achieve for organic or biological systems.

[3] The explosion limits of hydrogen are between 18 and 59%, and the flammability limits are between 4 and 74% (in air).

1.5 Benchmark for Photoelectrochemical Water Splitting

It should be noted that the advantages mentioned above for photoelectrochemical (PEC) systems are equally applicable to coupled photovoltaic–electrolysis systems. All the necessary components of such a system (solar cell, electrolyzer, dc–dc converter) are commercially available, and solar-to-hydrogen efficiencies around 8% have already been demonstrated. This approach can therefore be considered as a benchmark for PEC water splitting. The PEC approach offers two potential advantages over PV + electrolysis. The first advantage lies in the fact that commercial electrolyzers require cell voltages of ~1.9 V in order to reach their optimal operating current densities of ~1 A/cm^2. Since the thermodynamically required potential for water splitting is 1.23 V, this places an upper limit of 65% (1.23/1.9) on the overall energy conversion efficiency [9]. In contrast, the current density at a semiconductor photoelectrode immersed in water is much smaller (10–20 mA/cm^2 at most) and the required overpotential is therefore substantially lower.[4] The second advantage is that a PEC system can be constructed as a single, monolithic device. This requires fewer packaging components (frame, glass, connections, etc.) and may lead to significantly lower costs. The cost per kg of hydrogen is in fact the key benchmark figure. Estimated costs for hydrogen produced with PV + electrolysis exceed \$8/kg, well above the \$2–4/kg target set by the US Department of Energy for future hydrogen production pathways. As discussed in more detail in Chaps. 7 and 8, photoelectrochemical water splitting may offer a route toward hydrogen production costs of \$3–5/kg, which is competitive with existing energy sources.

1.6 Materials for PEC Devices

The key component for PEC systems is the semiconductor photoelectrode. The ideal photoelectrode fulfills several tasks at once: light absorption, charge separation, charge transport, and H_2 or O_2 evolution at its surface. Moreover, it needs to be stable in an aqueous solution, and have the potential to be made at low cost. No semiconducting material has yet been found that comes even close to meeting these contradictory demands. This means that trade-offs have to be made, leading to the development of composite photoelectrodes in which different materials fulfill different functionalities.

Figure 1.3 illustrates some of the approaches currently being studied in the field. The left-hand figure shows a monolithic PEC device developed by John Turner's group at NREL [9]. It is based on a p-type $GaInP_2$ photocathode that is biased by an

[4] This depends on the catalytic activity of the semiconductor surface, which can be quite low. By attaching suitable co-catalysts, overpotentials as low as 0.3–0.4 V can be achieved.

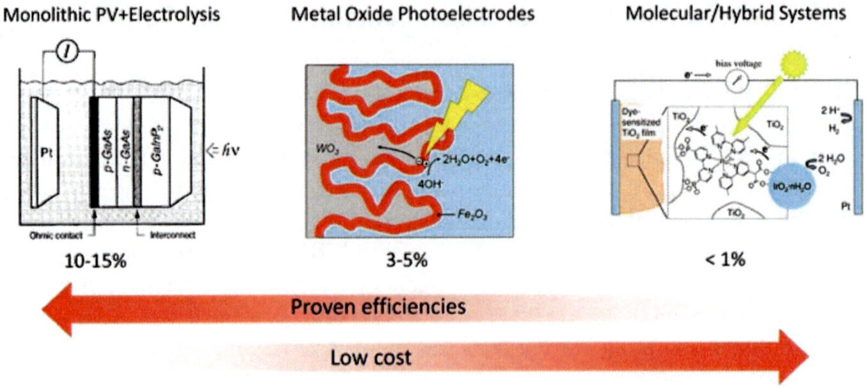

Fig. 1.3 Different approaches toward photoelectrochemical water splitting [9–11]. *Left-hand figure* from ref. [9], reprinted with permission from AAAS. *Right-hand figure* reprinted with permission from ref. [11], copyright 2009 American Chemical Society

integrated GaAs pn junction. With an impressive solar-to-hydrogen efficiency of 12.4%, this example is often quoted as the efficiency benchmark for PEC devices. Challenges to be addressed for this approach are the inherently poor stability of the employed semiconductor materials in water (lifetime only a few hours) and the prohibitively high cost of the GaAs and $GaInP_2$ compounds. The approach shown in the right-hand side of Fig. 1.3 represents a much cheaper solution. It was reported by the group of Thomas Mallouk at Penn State University, and consists of a dye-sensitized nanoporous TiO_2 system in which water oxidation was catalyzed by attaching an IrO_2 nanoparticle to the ruthenium-based dye molecule. The dye molecule and the TiO_2 function as optical absorption centers and charge separators, respectively. The challenges for this system are to improve the performance (currently less than 1% quantum yield) and the degradation of the Ru-based dye by optimizing the kinetic pathways for the reactions of the photogenerated charge carriers.

The middle part of Fig. 1.3 shows an approach in which metal oxide semiconductors, such as Fe_2O_3 and WO_3 are used as light absorbers and charge separators. Inorganic catalysts such as IrO_2 or cobalt-based compounds are generally necessary to catalyze the oxygen evolution reaction. These systems are studied by the authors' groups at EPFL [10, 12] and TU Delft [13, 14], and many other groups around the world. The main advantage of using metal oxides is their low cost and excellent stability against (photo)corrosion in aqueous solutions – although this does depend on the pH of the solution and the choice of metal oxide. One of the main challenges for metal oxide-based absorbers is the combination of modest light absorption and poor charge transport properties. Solutions for these and other challenges have been proposed in the form of mesoporous materials, guest–host nanostructures, tandem junctions, plasmonics, and combinatorial search methods for new metal oxide semiconductors. Chapters 4, 5, 6, 7, and 9 discuss the exciting developments that have been reported in this field over the past few years.

Acknowledgment Both authors thank the European Commission's Framework 7 program (NanoPEC, Project 227179) for support.

References

1. Allison, I., Bindoff, N.L., Bindschadler, R.A., Cox, P.M., de Noblet, N., England, M.H., Francis, J.E., Gruber, N., Haywood, A.M., Karoly, D.J., Kaser, G., Le Quéré, C., Lenton, T.M., Mann, M.E., McNeil, B.I., Pitman, A.J., Rahmstorf, S., Rignot, E., Schellnhuber, H.J., Schneider, S.H., Sherwood, S.C., Somerville, R.C.J., Steffen, K., Steig, E.J., Visbeck, M., Weaver, A.J.: The Copenhagen Diagnosis: Updating the World on the Latest Climate Science. The University of New South Wales Climate Change Research Centre (CCRC), Sydney, Australia (2009)
2. Oppenheim, J., Beinhocker, E.D.: Climate change and the economy - myths versus realities. Davos, Switzerland. McKinsey & Company, Inc. (2009)
3. Pacala, S., Socolow, R.: Stabilization wedges: solving the climate problem for the next 50 years with current technologies. Science **305**, 968–972 (2004)
4. Hoffert, M.I.: Farewell to fossil fuels? Science **329**, 1292–1294 (2010)
5. Lewis, N.S., Nocera, D.G.: Powering the planet: chemical challenges in solar energy utilization. Proc. Nat. Acad. Sci. U S A **103**, 15729–15735 (2006)
6. Lewis, N.S., Crabtree, G.: Basic Research Needs for Solar Energy Utilization: report of the Basic Energy Sciences Workshop on Solar Energy Utilization, April 18-21, 2005. US Department of Energy, Office of Basic Energy Science, Washington, DC (2005)
7. Rajeshwar, K., McConnell, R., Licht, S.: Solar Hydrogen Generation – Toward a Renewable Energy Future. Springer, New York (2008)
8. Grimes, C.A., Varghese, O.K., Ranjan, S.: Light, Water, Hydrogen – The Solar Generation of Hydrogen by Water Photoelectrolysis. Springer, New York (2008)
9. Khaselev, O., Turner, J.A.: A monolithic photovoltaic-photoelectrochemical device for hydrogen production via water splitting. Science **280**, 425–427 (1998)
10. Sivula, K., Le Formal, F., Grätzel, M.: WO_3–Fe_2O_3 photoanodes for water splitting: a host scaffold, guest absorber approach. Chem. Mater. **21**, 2862–2867 (2009)
11. Youngblood, W.J., Lee, S.H.A., Maeda, K., Mallouk, T.E.: Visible light water splitting using dye-sensitized oxide semiconductors. Acc. Chem. Res. **42**, 1966–1973 (2009)
12. Kay, A., Cesar, I., Grätzel, M.: New benchmark for water photooxidation by nanostructured alpha-Fe_2O_3 films. J. Am. Chem. Soc. **128**, 15714–15721 (2006)
13. Van de Krol, R., Liang, Y.Q., Schoonman, J.: Solar hydrogen production with nanostructured metal oxides. J. Mater. Chem. **18**, 2311–2320 (2008)
14. Enache, C.S., Lloyd, D., Damen, M.R., Schoonman, J., van de Krol, R.: Photo-electrochemical properties of thin-film $InVO_4$ photoanodes: the role of deep donor states. J. Phys. Chem. C **113**, 19351–19360 (2009)

Chapter 2
Principles of Photoelectrochemical Cells

Roel van de Krol

2.1 The Photoelectrochemical Cell

Figure 2.1 shows a simplified energy diagram of a photoelectrochemical (PEC) cell based on a single photoanode and a metal counter electrode. More complicated configurations that involve photocathodes and/or more than one photoelectrode are discussed at the end of this chapter. The main component of the PEC cell is the semiconductor, which converts incident photons to electron–hole pairs. These electrons and holes are spatially separated from each other due to the presence of an electric field inside the semiconductor, the origin of which is discussed in Sect. 2.5. The photogenerated electrons are swept toward the conducting back-contact, and are transported to the metal counter-electrode via an external wire. At the metal, the electrons reduce water to form hydrogen gas. The photogenerated holes are swept toward the semiconductor/electrolyte interface, where they oxidize water to form oxygen gas.

For an alkaline electrolyte, the reduction and oxidation reactions can be written as[1]

$$4H_2O + 4e^- \rightleftarrows 2H_2 + 4OH^- \quad E^0_{red} = -0.828 \text{ V vs. NHE} \tag{2.1}$$

$$4OH^- + 4h^+ \rightleftarrows 2H_2O + O_2 \quad E^0_{ox} = -0.401 \text{ V vs. NHE.} \tag{2.2}$$

[1] Note that the sign of the potential for the oxidation half-reactions is opposite from that usually encountered in the literature, which usually lists these reactions as *reduction* reactions.

R. van de Krol (✉)
Faculty of Applied Sciences, Department Chemical Engineering/Materials for Energy Conversion and Storage, Delft University of Technology, P.O. Box 5045, 2600 GA Delft, The Netherlands
e-mail: r.vandekrol@tudelft.nl

R. van de Krol and M. Grätzel (eds.), *Photoelectrochemical Hydrogen Production*, Electronic Materials: Science & Technology 102, DOI 10.1007/978-1-4614-1380-6_2,

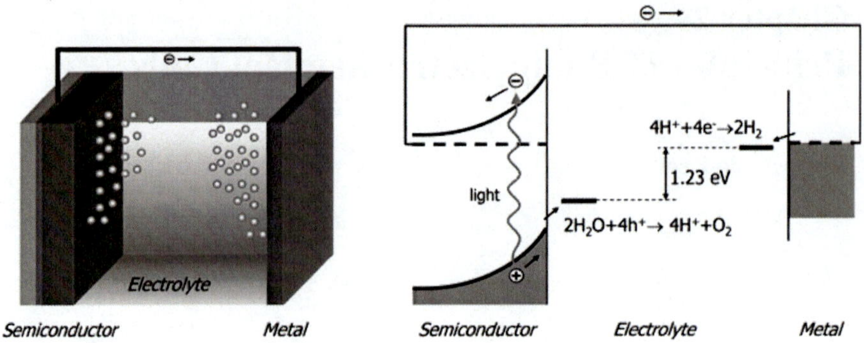

Fig. 2.1 Illustration of a photoelectrochemical cell that consists of a semiconducting photoanode and a metal cathode. The corresponding energy diagram is shown in the *right*

For an acidic environment, the appropriate reactions can be obtained from (2.1) and (2.2) by subtracting or adding the dissociation reaction of water into protons and hydroxyl ions:

$$4\mathrm{H}^+ + 4\mathrm{e}^- \rightleftarrows 2\mathrm{H}_2 \quad E^0_{\mathrm{red}} = +0.000 \text{ V vs. NHE} \tag{2.3}$$

$$2\mathrm{H}_2\mathrm{O} + 4\mathrm{h}^+ \rightleftarrows 4\mathrm{H}^+ + \mathrm{O}_2 \quad E^0_{\mathrm{ox}} = -1.229 \text{ V vs. NHE.} \tag{2.4}$$

The Gibbs free energy change for the overall water splitting reaction is given by the expression:

$$\Delta G = -nF\Delta E. \tag{2.5}$$

At standard temperature (298 K) and concentrations (1 mol/L, 1 bar), the electrochemical cell voltage ΔE of -1.229 V corresponds to a Gibbs free energy change of +237 kJ/mol H_2. This shows that the water-splitting reaction is thermodynamically *uphill*. This is markedly different from the photo*catalysis* reactions that one encounters in, e.g., photo-assisted degradation of organic pollutants, for which the Gibbs free energy change is negative.

2.2 Semiconducting Photoelectrode Materials

Some of the key requirements for a semiconductor photoelectrode are efficient absorption of visible light and good charge transport. It is often – though not always – easy to determine these parameters from an experiment on a particular material. Clearly, this approach becomes impractical if one wants to screen an entire class of candidate photoelectrode materials. For such cases a more fruitful

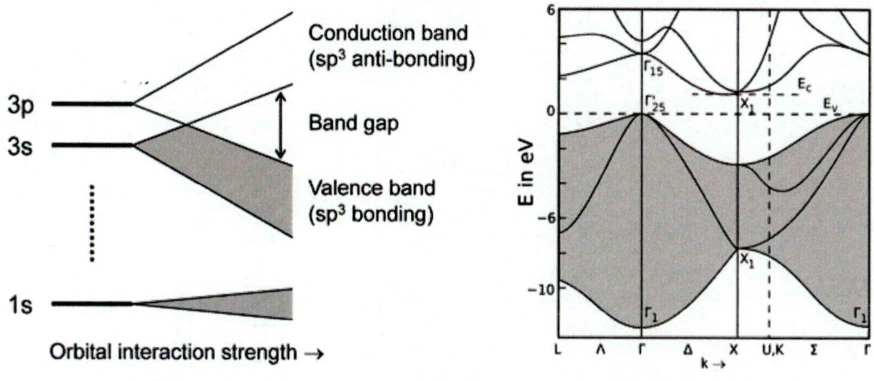

Fig. 2.2 *Left*: Formation of the valence and conduction bands in covalent semiconductors from bonding and antibonding sp^3 orbitals, respectively. *Right*: Calculated electronic band structure of silicon [1]. The *gray area* indicates occupied states in the valence band of the material

approach may be to calculate the electronic structure of a base material, and use this to predict how the properties depend on, e.g., composition. While still far from trivial, electronic structure calculations are now becoming more and more routine. The required computing power and software are readily available, and the number of electronic structure calculations reported in the literature, even by experimentally oriented groups, increases rapidly. However, in order to use these spectra to predict certain photoelectrode properties, one first needs to understand how chemical bonding between the atoms affects the electronic structure. Some of the main principles are discussed below. In contrast to most standard textbooks, we emphasize the properties of metal oxide semiconductors.

In most conventional semiconductors, such as Si and Ge, covalent bonding dominates. In silicon, for example, the outer 3s and 3p orbitals combine to form hybrid sp^3 orbitals. Neighboring sp^3 orbitals interact to form bonding and antibonding combinations that form the valence and conduction bands of the material, respectively. This is schematically illustrated in Fig. 2.2, which also shows the electronic band structure of silicon.

The bonding in metal oxide semiconductors is very different in nature. Since oxygen has a much higher electronegativity than any metal, the valence electrons are either fully or partially transferred from the oxygen to the metal ion. The bonding character of metal oxides is therefore highly polar or even ionic. A qualitative band picture can be obtained by constructing a molecular orbital (MO) diagram from the individual atomic energy levels. Figure 2.3 shows an example for rutile TiO_2, the very first and most extensively investigated photoanode material for water splitting [2, 3]. The main features of the MO diagram correspond quite well to the calculated band structure for rutile TiO_2, which is shown in Fig. 2.4. The valence band is mainly composed of O-2p orbitals, whereas the conduction band is primarily Ti-3d in character. One could think of the valence band as being occupied with the electrons that originally resided on the titanium

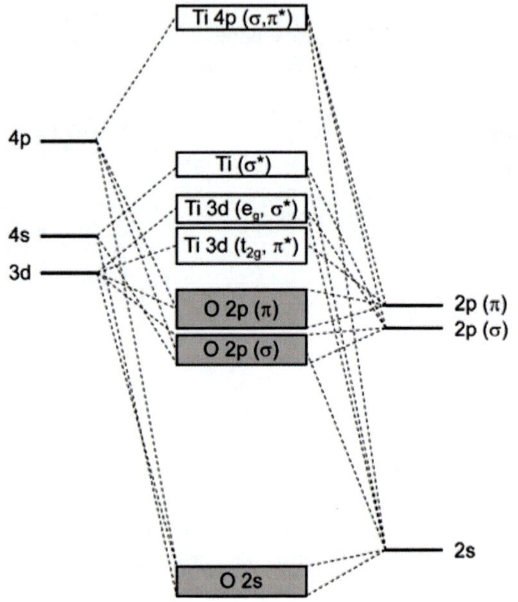

Fig. 2.3 Molecular orbital diagram of rutile TiO_2 (after Stoyanov [4] and Fisher [5])

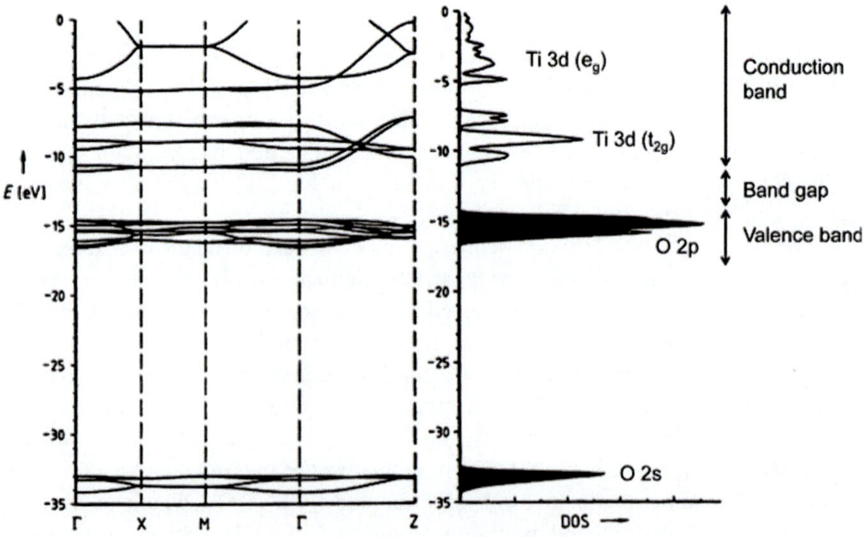

Fig. 2.4 Electronic band structure and density-of-states (DOS) of rutile TiO_2. The black parts of the DOS indicate completely filled bands. Adapted from Hoffmann [6], pp. 31, copyright Wiley-VCH Verlag GmbH & Co. KGaA. Reproduced with permission

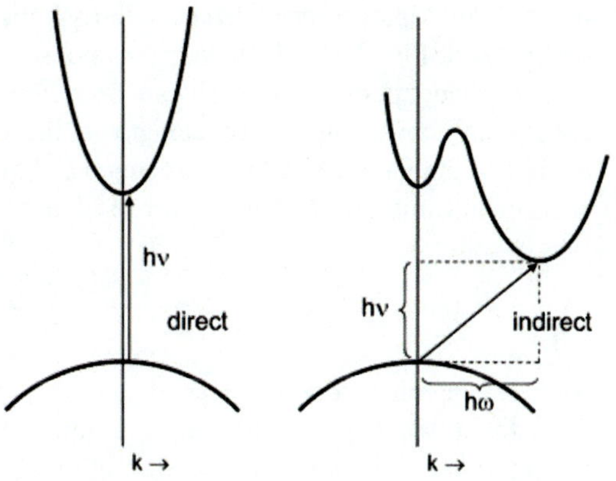

Fig. 2.5 Optical transitions in semiconductors with a direct and an indirect bandgap. The indirect transition requires assistance of a phonon with energy $\hbar\omega$

atoms, before they were transferred to the more electronegative oxygen during the formation of the bond.[2]

It should be noted that such a "local" view is entirely inappropriate for covalent semiconductors such as silicon, in which the electrons are completely delocalized over the material. This local vs. delocalized view is a key feature that distinguishes metal oxide semiconductors from their covalently bonded counterparts. The local character of the energy bands in metal oxides sometimes leads to ambiguous interpretations. For example, "free" electrons in TiO_2 are sometimes distinguished from Ti^{3+} species. While such a distinction may be realistic at the surface of a material, where the electronic structure is distorted due to a disruption of the lattice symmetry, it is not obvious in the bulk in which the conduction band is mainly composed of Ti 3d orbitals.

While seemingly complicated, the electronic band structure can directly lead to several important insights for photoelectrode materials. The first insight is the nature of the optical transition, illustrated in more detail in Fig. 2.5. If the highest point in the valence band is located at the same *k*-vector as the lowest point of the conduction band, the optical transition does not require a change in crystal momentum. Such a transition is called *direct*, and is indicated by a vertical line in the *E*–*k* diagram. In contrast, *indirect* transitions do require a change in crystal momentum, as illustrated in Fig. 2.5. Since photons carry very little momentum, indirect transitions require absorption or emission of a phonon (i.e., a lattice vibration) and are much less likely to occur. As a consequence, the absorption coefficient of indirect semiconductors is much smaller than that of direct semiconductors – the difference can be two orders of magnitude. This is why direct semiconductors, such as amorphous Si or $CuIn_xGa_{1-x}Se_2$ (CIGS) can absorb all incident light in just a few micrometers, whereas crystalline (indirect) Si solar cells typically have a thickness

---

[2] The valence band of a semiconductor is analogous to the highest occupied molecular orbital (HOMO) in a molecule, whereas the conduction band is the solid state analogue of the lowest unoccupied molecular orbital (LUMO).

of ~300 μm. Metal oxides often have fairly horizontal-running *E*-*k* lines in their band structure (cf. Fig. 2.4), which sometimes obscures the direct vs. indirect nature.

The bandgap of a material can be determined from a measurement of the absorption coefficient vs. wavelength. If the bottom of the conduction band and the top of the valence band are assumed to have a parabolic shape, the absorption coefficient (units: m^{-1}) can be expressed as follows:

$$\alpha = \frac{A\left(h\nu - E_g\right)^m}{h\nu}. \tag{2.6}$$

Here, A is a constant and m depends on the nature of the optical transition: $m = ½$ for a direct bandgap, and $m = 2$ for an indirect gap. From (2.6), extrapolation of a plot of $(\alpha h\nu)^{1/2}$ vs. $h\nu$ plot gives the indirect bandgap, while a plot of $(\alpha h\nu)^2$ vs. $h\nu$ yields the direct bandgap of the material. Such a plot is called a "Tauc plot" [7] and is often encountered in the photoelectrochemistry and photocatalysis literature.

The second insight that can be obtained from the electronic band structure is the mobility of charge carriers, which is related to the width of the conduction and valence bands. For Si the bands are rather broad, spanning more than 10 eV. This is a direct consequence of the extensive overlap of the sp^3 orbitals on neighboring atoms. More overlap between atomic wavefunctions results in broader bands and easier transport of free charge carriers through the material. This can be quantified via the curvature of the individual bands, which is directly related to the effective mass and mobility of the charge carriers:

$$m^* = \frac{h^2}{\left(\frac{\partial^2 E}{\partial k^2}\right)} \quad \text{and} \quad \mu = \frac{q\tau}{m^*}. \tag{2.7}$$

Wide bands that are strongly curved indicate a high charge carrier mobility. A high mobility is especially desirable in photoelectrodes with an indirect bandgap. This is because these materials require a large thickness to absorb all the incident light, which means that photogenerated electrons and holes have to travel large distances before reaching the interface.

Table 2.1 gives an overview of the room temperature charge carrier mobilities of several semiconductors. What stands out are the small charge carrier mobilities for the transition metal oxides. One reason for this is the fact that the 3d orbitals in transition metals are fairly contracted, and show less overlap with neighboring atoms than the s- and p-orbitals. One can also look at this from an electrostatic point of view. The electrostatic interaction of a free electron with the cations surrounding it creates a local lattice distortion. This distortion accompanies the electron when it moves through the lattice, and is called a *polaron*. If the interaction is very strong, the polaron can even become trapped at a particular lattice site. The electron can then only move by thermally activated hopping, and its mobility will be very low. Similar considerations apply to "free" holes in the valence band, which can form polarons through electrostatic interaction with the surrounding oxygen ions. Charge transport via polarons is believed to play an important role in, e.g., hematite (α-Fe_2O_3), which will be discussed in more detail in Chap. 4.

Table 2.1 Charge carrier mobilities and effective masses at room temperature

	Mobility (cm^2/V/s)		Effective carrier masses ($\times m_e$)		
Material	Electrons	Holes	Electrons	Holes	References
Si	1,500	450			[8]
GaAs	8,500	400	0.07	0.68	[8, 9]
CdS	340	50	0.19	0.8	[8, 9]
ZnO	200	180	0.24	0.45	[8, 9]
TiO_2 (anatase)		2×10^{-3} [a]	>10	0.8	[10, 11]
WO_3	~10		~0.9		[12]
α-Fe_2O_3	0.1[b]	0.2[c]			[13]
p-Cu_2O		90			[14]
p-$CuAlO_2$		10			[15]
p-$SrCu_2O_2$		0.46			[16]
p-$CuMnO_2$		3.5×10^{-6}			[17]

[a]Theoretical estimate
[b]At 1,200 K
[c]At 1,400 K

The third piece of information that can be obtained from the electronic band structure is the density-of-states (DOS). This is the number of allowed electronic states per unit of energy interval. In *E–k* diagrams the *k*-values are equally spaced along the *k*-axis, so the DOS can be readily estimated from the total length of the black lines that run through a certain interval ΔE. As illustrated in Fig. 2.4 for TiO_2, the highest DOS occur at energies where the *E–k* curves are flat (horizontal). The DOS representation not only allows one to quickly and easily identify the various bands and band widths, but it also gives a rough impression of the optical transition probabilities between various bands. This is because the total number of excitations per second is proportional to the DOS at the occupied ground state (the initial state), multiplied by the transition probability to the final state, λ_{if}. The latter is given by Fermi's Golden Rule, (2.8), and depends linearly on the DOS at the final (excited) state, g_{f}:

$$\lambda_{\text{if}} = \frac{2\pi}{h} |M_{\text{if}}|^2 g_{\text{f}}. \tag{2.8}$$

In some cases, one can actually see certain features of the DOS reflected in the optical absorption spectrum of the material.

2.3 Charge Carriers and Doping

Under equilibrium conditions (i.e., no illumination and no net current flow), the concentration of free electrons in the conduction band and free holes in the valence band is given by the following expressions:

$$n = N_{\text{C}}\,\mathrm{e}^{-(E_{\text{C}}-E_{\text{F}})/kT} \quad \text{with} \quad N_{\text{C}} = 2\left(\frac{2\pi m_{\text{e}}^{*}kT}{h^2}\right)^{3/2}, \tag{2.9}$$

$$p = N_V\,e^{-(E_F - E_V)/kT} \quad \text{with} \quad N_V = 2\left(\frac{2\pi m_h^* kT}{h^2}\right)^{3/2}. \tag{2.10}$$

Equations (2.9) and (2.10) are valid for both undoped and doped semiconductors. They are, however, not valid when the Fermi level is less than ~$3kT$ away from either one of the band edges. Under these conditions, the semiconductor is *degenerate*, and exhibits near-metallic behavior. The relationships for the effective densities of states were derived from the (nearly) free electron model, and may not be entirely accurate for transition metal oxides. Despite these limitations, (2.9) and (2.10) are exceedingly useful for describing the behavior of semiconducting photoelectrodes.

Semiconductors for practical applications are often doped, mainly with the aim to improve the conductivity. In metal oxide photoelectrodes, shallow donors and acceptors are almost always necessary because of the low intrinsic charge carrier mobilities. The conductivity of the material is given by $\sigma = ne\mu_e + pe\mu_h$, so increasing n or p will compensate for a small value of μ_e or μ_h. Examples of donor-type dopants are Ti^{4+} on a Fe^{3+} site in Fe_2O_3, or phosphorus in silicon. The extra valence electron introduced by the donor atom is loosely bound to the donor nucleus, and can be excited to the conduction band where it then contributes to the conductivity. Conversely, holes in acceptor-type dopants can be excited to the valence band. Since a hole is equivalent to a missing electron, one can also picture this as an electron being excited from the valence band into the energy level of the acceptor species.

If the dopant level is within ~$2kT$ of E_C or E_V, it will be (almost) fully ionized at room temperature – this is referred to as a *shallow* dopant. For *deep* donors and acceptors, the degree of ionization can be calculated with the following equations:

$$\frac{N_D^+}{N_D} = \frac{1}{1 + g_D\,e^{(E_F - E_D)/kT}} \quad \text{and} \quad \frac{N_A^-}{N_A} = \frac{1}{1 + g_A\,e^{(E_A - E_F)/kT}}. \tag{2.11}$$

Here, E_D and E_A are the donor and acceptor energies and g_A and g_D are the corresponding degeneracy factors that reflect the multiplicity of the energy state [18]. Under the assumption that $n = N_D^+$, the following useful expressions relate the free electron concentration directly to the position of the donor level in the bandgap [18]:

$$n = \frac{N_\zeta}{2}\left(\sqrt{1 + \frac{4N_D}{N_\zeta}} - 1\right) \quad \text{with} \quad N_\zeta = \left(\frac{N_c}{g_D}\right) e^{-(E_C - E_D)/kT} \tag{2.12}$$

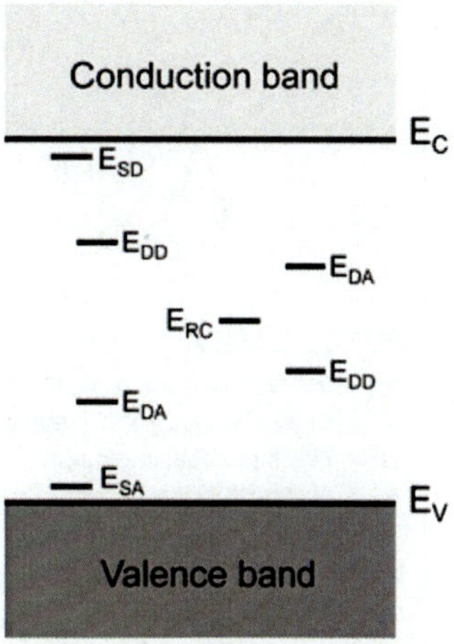

Fig. 2.6 Energy levels of shallow and deep donors (SD, DD) and acceptors (SA, DA) in a semiconductor. Deep donor or acceptor states can also occur below or above midgap, respectively. Midgap states (RC) are often very efficient recombination centers and can be either donor- or acceptor-like in nature

An analogous expression can be obtained for acceptor-doped materials. The energy levels for the various dopants are illustrated in Fig. 2.6. Deep dopants can act as optically active centers or as catalytically active surface sites, as given in the following paragraphs.

2.4 Defect Chemistry

Semiconducting photoelectrodes are almost always doped to improve their properties. In most cases, the aim is to enhance the n- or p-type conductivity, as described in Sect. 2.3. Certain dopants may enhance the optical absorption of wide bandgap semiconductors [19], increase the minority carrier diffusion length [20, 21], or enhance the catalytic activity at the surface of the semiconductor [22]. Other dopants adversely affect the properties, for example, by introducing midgap bulk or surface states that act as recombination centers [23, 24].

In addition to foreign cation and anion dopants, native point defects are also usually present in the material. Examples are vacancies, interstitials, or substituents.[3] These defects can have a similar influence on the optical, electrical, and catalytic properties as dopants. They are formed by intrinsic defect-chemical reactions, or by a change in the lattice stoichiometry due to exchange of, e.g., oxygen with the gas phase. Since virtually every defect affects the performance of the material in some way, the ability to understand and predict the relationship between dopants and defect concentrations is of paramount importance for designing efficient photoelectrodes.

[3] An example of a native substituent is a site exchange of A and B cations in a ternary compound such as AB_xO_y.

Kröger-Vink notation

$$A_{site}^{charge}$$

A : atom/ion, or "V" for vacancy
site : atom/ion, or "i" for interstitial
charge : ′ (negative), • (positive), or × (neutral)

Defect examples:

$V_{Fe}^{///}$: Iron vacancy in e.g. Fe_2O_3
$V_O^{\bullet\bullet}$: Oxygen vacancy in a metal oxide
$Zn_i^{\bullet\bullet}$: Zinc interstitial in e.g. ZnO
$Al_{Ti}^{/}$: Al substitutional dopant in e.g. $SrTiO_3$

Defect reaction requirements:

1. Conservation of **mass**
2. Conservation of **lattice site stoichiometry**
3. Conservation of **charge**

Defect reaction examples:

$0 \rightleftharpoons 2V_{Fe}^{///} + 3V_O^{\bullet\bullet}$	$K_S = [V_{Fe}^{///}]^2[V_O^{\bullet\bullet}]^3 = e^{-\Delta G_S/kT}$	Schottky defect in Fe_2O_3
$0 \rightleftharpoons e^{/} + h^{\bullet}$	$K_e = np = N_C N_V e^{E_g/kT}$	Thermal electron-hole generation
$0 \rightleftharpoons V_O^{\bullet\bullet} + \frac{1}{2}O_2(g) + 2e^{/}$	$K_R = [V_O^{\bullet\bullet}]n^2 p(O_2)^{1/2} = K_R^0 e^{\Delta G_R/kT}$	Reduction of a metal oxide
$2TiO_2 \;\; 2Ti_{Fe}^{\bullet} + 3O_O^{x} + \frac{1}{2}O_2(g) + 2e^{/}$ (irreversible)		Dissolution reaction for Ti-doped Fe_2O_3

Fig. 2.7 Diagram summarizing the key elements of the Kröger–Vink notation for point defects in ionic solids. The formation of defects can be described with defect-chemical reactions and corresponding equilibrium constants

2.4.1 Doping Binary Metal Oxides

In covalently bonded semiconductors, the introduction of dopants is fairly straightforward since all lattice sites are similar to each other. This is even true for a polar compound such as GaAs. In this material, Si can be incorporated on both Ga and As lattice sites [25]. For ionically bonded materials the situation is quite different, since the large electrostatic penalty prevents any exchange of anions and cations. This puts certain restrictions on the incorporation of dopants, as given below.

The incorporation of dopants in metal oxides and other ionic materials can be conveniently described with the so-called *Kröger–Vink* notation [26], which is summarized in Fig. 2.7. For a more extensive description the reader is referred to several excellent textbooks on this subject [27, 28]. When describing defect-chemical reactions, one has to observe the conservation of mass and charge, just as one would for any other chemical (half)reaction. What is different in the defect chemistry of ionic solids is the *conservation of lattice site stoichiometry*. For example, if we dope Fe_2O_3 with zinc oxide, two zinc ions substitute for Fe, and the two oxygen ions occupy two of the three oxygen sites in the Fe_2O_3 unit. This means that the third oxygen site remains empty, resulting in an oxygen vacancy:

$$2ZnO \xrightarrow{Fe_2O_3} 2Zn_{Fe}^{/} + 2O_O^{x} + V_O^{\bullet\bullet}. \tag{2.13}$$

Note that this is an *irreversible* reaction, since spontaneous demixing will not occur. *This means that one cannot define an equilibrium constant for a dissolution reaction.* Closer inspection shows that reaction (2.13) indeed fulfills the required conservation of mass, charge, and lattice site stoichiometry.

One may try to get rid of the oxygen vacancy in reaction (2.13) by performing the synthesis in an oxygen-rich atmosphere. In this case, both mass and lattice site stoichiometry are conserved when adding the oxygen gas, which means that the charge has to be balanced by the addition of a hole:

$$2\text{ZnO} + 1/2\text{O}_2(\text{g}) \xrightarrow{\text{Fe}_2\text{O}_3} 2\text{Zn}_{\text{Fe}}^{/} + 3\text{O}_{\text{O}}^{\text{x}} + \text{h}^{\bullet}. \quad (2.14)$$

Note that the presence of a hole suggests p-type conductivity in Fe_2O_3. In many cases, however, metal oxides remain n-type when acceptor-doped. If this is the case, it is more appropriate to replace the hole with an electron on the left-hand side of (2.14).

In (2.13) and (2.14) the charge of the aliovalent[4] Zn dopant is compensated by an ionic defect ($V_O^{\bullet\bullet}$) and an electronic defect ($h^{\bullet}$), respectively. To illustrate the difference between these ionic and electronic compensation mechanisms in more detail, consider Ti-doped Fe_2O_3. When subtracting the ionic compensation reaction from the electronic one, we obtain

$$6\text{TiO}_2 \xrightarrow{3\text{Fe}_2\text{O}_3} 6\text{Ti}_{\text{Fe}}^{\bullet} + 9\text{O}_{\text{O}}^{\text{x}} + 3/2\text{O}_2(\text{g}) + 6\text{e}^{/} \quad (2.15)$$

$$6\text{TiO}_2 \xrightarrow{4\text{Fe}_2\text{O}_3} 6\text{Ti}_{\text{Fe}}^{\bullet} + 12\text{O}_{\text{O}}^{\text{x}} + 2\text{V}_{\text{Fe}}^{///} \quad (2.16)$$

$$3\text{O}_{\text{O}}^{\text{x}} + 2\text{V}_{\text{Fe}}^{///} \rightleftarrows 3/2\text{O}_2(\text{g}) + 6\text{e}^{/}. \quad (2.17)$$

This shows that the difference between two *irreversible* dissolution reactions can be written as a *reversible* oxidation/reduction reaction. An important implication of this is that the addition of a dopant is, by itself, not sufficient to enhance the conductivity of a metal oxide photoelectrode: one also needs to ensure that the equilibrium of reaction (2.17) lies at the right-hand side. The factors that affect the equilibrium position are oxygen partial pressure, dopant concentration, and temperature. The equilibrium constant of (2.17) is given by

$$K = \frac{n^6 p_{\text{O}_2}^{3/2}}{[\text{V}_{\text{Fe}}^{///}]^2}. \quad (2.18)$$

[4] An "aliovalent" dopant has different charge than the ion that it replaces.

From this, it is immediately clear that a low $p(O_2)$ favors electronic compensation. Moreover, K is constant, so at constant $p(O_2)$ the factor $n^6/[V_{Fe}^{///}]^2$ also remains constant. This implies that at high dopant concentrations, ionic compensation will dominate over electronic compensation. In other words, a plot of the conductivity as a function of the Ti concentration will have its highest slope at low dopant concentrations. Finally, we note that the entropy (ΔS) of (2.17) strongly increases from left to right. Since $\Delta G = \Delta H - T\Delta S$, electronic compensation is favored at high temperatures.

It is interesting to note that after subtracting the Fe_2O_3 Schottky reaction (see Fig. 2.7) from (2.17), the standard oxygen reduction reaction for n-type oxides is obtained:

$$O_O^x \rightleftarrows V_O^{\bullet\bullet} + 1/2O_2(g) + 2e^{/}. \tag{2.19}$$

This reaction is of course valid for both doped and undoped metal oxides.

For doped metal oxides, the aliovalent metal ions influence the overall defect concentrations via what is sometimes called the "First Law of Doping." This law simply states that adding an aliovalent dopant increases the concentration of defects with opposite charges, and decreases the concentration of defects with charges of the same sign.

The concentration of dopants in metal oxides is usually limited to 1–2% at most, which corresponds to a concentration of $\sim 10^{21}$ cm^{-3}. Higher concentrations are not likely to be effective, and may even lead to segregation of the dopant phase. If the conductivity of a doped metal oxide photoelectrode is found to be too low, it may be more effective to anneal it at high temperatures and under low $p(O_2)$ than to increase the dopant concentration.

In (2.13)–(2.16), only one type of ionic defect was considered for each ionic compensation reaction. In the case of ZnO-doped Fe_2O_3, the Zn acceptor could also be compensated by a cation interstitial ($Fe_i^{\bullet\bullet\bullet}$) instead of an oxygen vacancy. However, the higher charge of the interstitial Fe makes this energetically less favorable. Similarly, oxygen interstitials instead of iron vacancies could compensate the Ti donor in reaction (2.16), but in this case the large size of interstitial anions prohibits this.[5] In general, donor dopants are compensated by cation vacancies or electrons, while acceptor dopants are compensated by anion vacancies, cation interstitials, or holes. Which ionic defect is more likely to occur depends on the crystal structure, the ionic radius, and the charge of the ion.

[5] Anion interstitials normally only occur in oxides with the fluorite structure, which can be viewed as an fcc base lattice of cations in which the interstitial sites are occupied by anions.

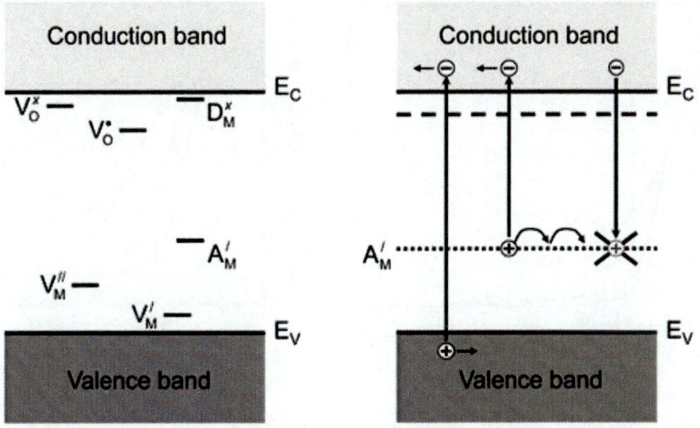

Fig. 2.8 *Left*: Energy diagram for a hypothetical a metal oxide, indicating several shallow and deep donor and acceptor levels. *Right*: n-Type oxide with optically active deep acceptor levels. After excitation, the holes have to hop to the interface via neighboring acceptor levels

2.4.2 Dopant Energy Levels

Most defects in metal oxides can have multiple oxidation states. For example, neutral oxygen vacancies can be ionized to divalent species in two steps:

$$V_O^x \rightleftarrows V_O^\bullet + e^{/} \tag{2.20}$$

$$V_O^\bullet \rightleftarrows V_O^{\bullet\bullet} + e^{/}. \tag{2.21}$$

The corresponding ionization energies are depicted in the energy diagram in the left-hand panel of Fig. 2.8. The common convention is to label the donor states by their charge *before* ionization, whereas acceptor states are labeled by their charge *after* ionization. Hence, D_M^x represents a shallow donor, and $A_M^{/}$ a deep acceptor.

As outlined above, shallow donors are used to increase the conductivity of an n-type oxide. Examples are Ti and Si in Fe_2O_3 [29–31], W in $BiVO_4$ [32], and Nb in TiO_2 [33]. Oxygen vacancies are often considered to be shallow donors as well, based on the observation that the conductivity of the oxide increases when reducing it at low $p(O_2)$. However, in many oxides – especially those with a fairly wide bandgap – the ionization energy of a singly ionized oxygen vacancy is simply too high (>3 kT) to be considered as a shallow donor at room temperature. This implies that the conductivity is mostly determined by ionization of neutral oxygen vacancies, i.e., reaction (2.20). Note that this is somewhat different from the mechanism implied by reaction (2.19), in which reactions (2.20) and (2.21) are *both* assumed to occur. When present as deep

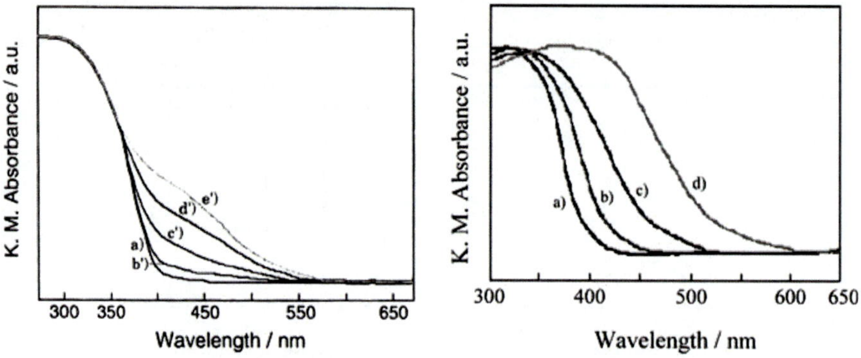

Fig. 2.9 *Left*: Optical absorption spectrum of TiO_2 doped with Cr via chemical impregnation. The amount of Cr increases from 0 wt% (**a**) to 1 wt% (**e′**). *Right*: Optical absorption of TiO_2 doped with Cr by physical implantation, with the Cr concentration increasing from 0 wt% (**a**) to 1.3 wt% (**d**). Reprinted from [34], with permission from Elsevier

donor species, ionized oxygen vacancies may adversely affect the performance of photoanodes by acting as recombination centers. If this is the case, their concentration can be kept low by doping with shallow donors, i.e., by using the First Law of Doping.

Cr and Fe are well-known examples of deep acceptors in TiO_2. As indicated in Fig. 2.8 (right), the electron in these acceptor levels can be optically excited to the conduction band. This requires less energy than the bandgap of the material, and Fe- and Cr-doped TiO_2 actually absorb visible light. An example of this is shown in Fig. 2.9 (left) for Cr-doped TiO_2. This visible light sensitization effect has attracted much interest in the field of photoelectrochemistry, especially in the late 1970s and early 1980s. A problem with this approach is that hole transport toward the interface is slow, as it takes place via hopping along neighboring acceptor levels (cf. Fig. 2.8). Moreover, the optical absorption cross section of these defects is small, requiring relatively thick films and long hole diffusion path lengths. Because of this, most holes recombine before reaching the interface. The photocurrents for Fe- or Cr-doped TiO_2 are indeed found to be low, and use of cation doping to improve the visible light absorption of wide-bandgap semiconductors has all but been abandoned in photoelectrochemistry. However, it still attracts interest in photocatalysis based on doped TiO_2 nanoparticles, where the holes only have to travel a short distance and recombination is less severe.

Interestingly, TiO_2 doped with Cr by physical ion implantation shows less recombination, and the shape of the absorption spectrum suggests an actual decrease of the bandgap instead of a local absorption center [34]. An example is shown in the right-hand side of Fig. 2.9. Clearly, the different incorporation methods for Cr give rise to different defects and a different electronic structure of the material. The physical reason for these differences remains to be established.

2.4.3 Defect Equilibria

If all the defect formation energies and/or equilibrium constants are known, the concentration of all the defects in a material can be calculated. As an example, we briefly consider the case for a hypothetical M_2O_3 photoanode doped with X^{4+}. Let us assume that X substitutes for the metal (as opposed to occupying an interstitial site), that Schottky disorder dominates in M_2O_3, and that the material is n-type ($n \gg p$). We furthermore assume that fully ionized oxygen vacancies are deep donors and can easily trap an electron, whereas singly ionized oxygen vacancies are shallow donors. We can then write the following defect equations and equilibrium constants:

$$2XO_2 \xrightarrow{M_2O_3} 2X_M^{\bullet} + 3O_O^x + 1/2O_2(g) + 2e^{/} \tag{2.22}$$

$$0 \rightleftarrows 3V_M^{///} + 2\,V_O^{\bullet\bullet} \quad K_S = [V_M^{///}]^3[V_O^{\bullet\bullet}]^2 \tag{2.23}$$

$$0 \rightleftarrows e^{/} + h^{\bullet} \quad K_e = np \tag{2.24}$$

$$O_O^x \rightleftarrows V_O^{\bullet\bullet} + 1/2O_2(g) + 2e^{/} \; K_R = n^2[V_O^{\bullet\bullet}]p_{O_2}^{1/2} \tag{2.25}$$

$$V_O^{\bullet} \rightleftarrows V_O^{\bullet\bullet} + e^{/} \; K_i = \frac{[V_O^{\bullet\bullet}]n}{[V_O^{\bullet}]}. \tag{2.26}$$

No equilibrium constant is defined for the dissolution reaction, since it is irreversible. Note that we have chosen to write the electronic compensation form of the dissolution reaction; we could have just as easily have chosen the ionic compensation reaction. As we saw before, they are related through reduction reaction (2.25) and are therefore not independent of each other.

We now have four equilibrium equations and five unknown variables ($[V_M^{///}]$, $[V_O^{\bullet\bullet}]$, $[V_O^{\bullet}]$, n, p). The missing equation is the electroneutrality equation:

$$3[V_M^{///}] + n = 2[V_O^{\bullet\bullet}] + [V_O^{\bullet}] + p \tag{2.27}$$

This system of nonlinear algebraic equations can be numerically solved with a computer, or one can find analytical solutions by simplification of (2.27) into six separate *Brouwer regimes*. Each Brouwer regime is characterized by a single dominant positive defect and a single dominant negative defect. While finding the right order of the different Brouwer regimes as a function of $p(O_2)$ may pose a bit of a challenge, one can derive all defect concentrations as a function of $p(O_2)$ by hand and plot these in a so-called "Brouwer diagram" (sometimes also

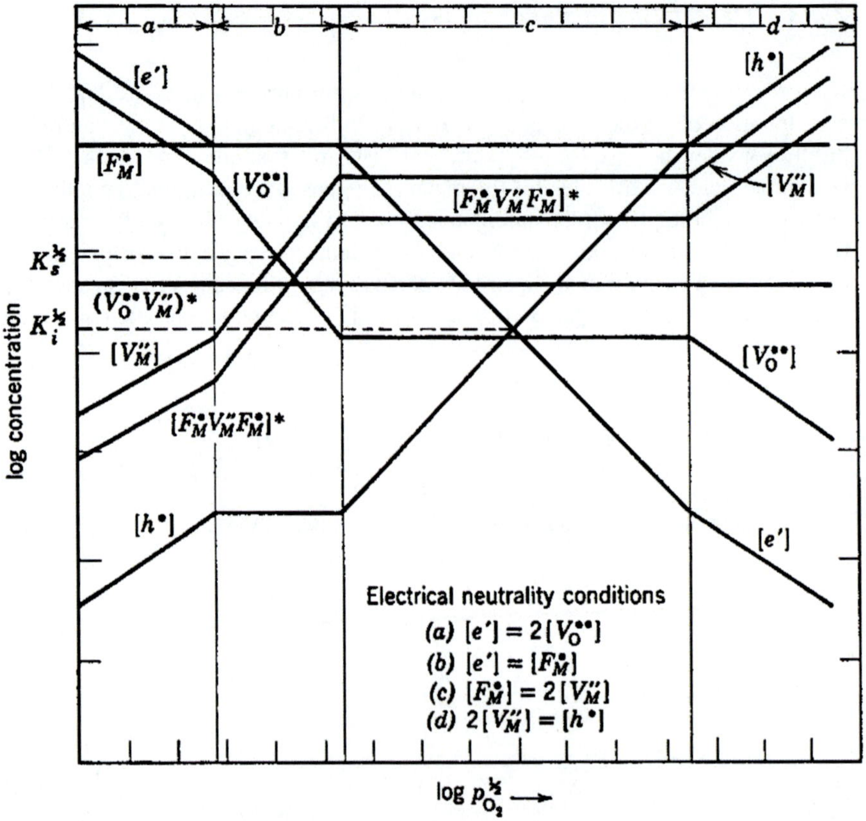

Fig. 2.10 Brouwer diagram for a metal oxide MO doped with a donor species (FO_2, with F being a metal ion) in which Schottky disorder dominates. Also indicated are defect associates (marked by an *asterisk*), clusters of defects that are held together by electrostatic forces. Such clusters are known to form at high defect concentrations. From [27], with permission

called a "Kröger–Vink diagram"). An example of such a diagram is shown in Fig. 2.10. Such diagrams can be very helpful in the study of metal oxide photoelectrodes since they immediately reveal which defects dominate in a certain $p(O_2)$ regime.

2.4.4 *Doping of Complex Metal Oxides*

Since only very few binary oxides show promise as photoelectrodes for water splitting, most notably WO_3 and Fe_2O_3, PEC research activities have now partly shifted toward ternary and more complex metal oxides (see, e.g., Chaps. 5 and 6). One of the best known examples is $BiVO_4$, which is an n-type semiconductor with a bandgap of ~2.4 eV [35, 36]. Recent results from the author's laboratory show that electron transport limits the performance of undoped $BiVO_4$ photoanodes [32]. To improve this, one can dope $BiVO_4$ with tungsten in order to increase the concentration of free electrons. Tungsten has a valence of 6+ and a crystal ionic radius of 42 pm. It will therefore be a substitute for tetrahedrally coordinated V^{5+}, which has a slightly smaller ionic radius (36 pm).[6] When W^{6+} substitutes for V^{5+}, one might expect the positive charge to be compensated by free electrons, thereby increasing the conductivity. However, there is a minor complication: the compensating negative charge is already provided in the form of Bi vacancies. Since tungsten only occupies sites on the vanadium sublattice, the formation of Bi vacancies is necessary in order to preserve lattice site stoichiometry. This leads to the following electronic compensation reaction:

$$WO_3 + 1/2O_2(g) + 2e^{/} \xrightarrow{BiVO_4} V_{Bi}^{///} + W_V^{\bullet} + 4O_O^{x}. \qquad (2.28a)$$

According to (2.28a), W-doping actually *decreases* the conductivity of n-type $BiVO_4$, which is a rather unexpected (and undesired) result. It is also possible to write an alternative reaction in which lattice site stoichiometry is preserved by the segregation of vanadium out of the $BiVO_4$ in the form of V_2O_5:

$$2BiVO_4 + 2WO_3 \xrightarrow{2BiVO_4} 2Bi_{Bi}^{x} + 2W_V^{\bullet} + V_2O_5 + O_O^{x} + 1/2O_2(g) + 2e^{/}. \qquad (2.28b)$$

This avoids the need for highly charged cation vacancies, and the W dopant is now indeed charge-compensated by the desired free electrons. Which reaction actually occurs – (2.28a) or (2.28b) – depends on the energy required to form a triply charged Bi vacancy vs. the energy required to form a second phase.

It should be realized that reactions (2.28a) and (2.28b) are both unfavorable from an energetic point of view. In one case a highly charged defect has to be formed, while in the other case a second phase has to segregate out. This leads to an important general observation for ternary – and more complex – metal oxides: doping these materials by simply dissolving a binary oxide is difficult to achieve in practice. An elegant solution for this is to co-dope the $BiVO_4$ with equal amounts of W and Bi:

$$2WO_3 + Bi_2O_3 \xrightarrow{2BiVO_4} 2Bi_{Bi}^{x} + 2W_V^{\bullet} + 8O_O^{x} + {}^1\!/\!_2O_2(g) + 2e^{/}. \qquad (2.29)$$

Co-doping can be easily achieved by having a small excess of Bi over V during the synthesis of the material, so that [Bi] + [W] = [V]. This is a well known and powerful trick to increase the solubility of binary oxides in ternary oxides in order to incorporate donor- or acceptor-type dopants.

---

[6] Note that the octahedrally coordinated Bi sites (103 pm) are clearly too large for W, and would cause the W^{6+} to "rattle," which is energetically very unfavorable.

2.5 Space Charges and Band Bending

One of the key features of a semiconductor is the presence of a built-in electric field, or *space charge*. In devices for solar energy conversion, this field is responsible for the efficient separation of the photogenerated electrons and holes, thereby preventing recombination. The field is present near the surface or interface of a semiconductor, and is formed by the transfer of charges from the semiconductor bulk across its interface. The reason for charge transfer and the field induced by the remaining charges inside the semiconductor is described below.

2.5.1 Origin of the Space Charge Layer

In most semiconductor textbooks, the formation of a space charge is explained by a semiconductor being brought into contact with a metal or another semiconductor [8, 37]. When both materials have different Fermi levels (i.e., electrochemical potentials), charge is transferred between them until an equilibrium is established. In metal oxide semiconductors, a built-in electric field can even be formed at the surface without making contact with another solid. This is illustrated in the schematic diagram of Fig. 2.11. When a metal oxide is exposed to air, water molecules from the air can dissociatively adsorb at its surface, resulting in -OH surface termination. Since the symmetry of the bulk lattice is broken at the surface, these -OH groups form electronic surface states within the bandgap of the semiconductor. The energy levels of these surface states are below the conduction band minimum, and free electrons from the bulk will occupy these levels. The ionized donors from which the free electrons originated will of course stay behind in the bulk, where they form a positive space charge. An electric field now forms, and the charge transfer from bulk to surface will continue until the potential barrier becomes too large for bulk electrons to cross. At this point, a dynamic equilibrium establishes at which no net electron transport takes place. The Fermi level at the surface is then located somewhere halfway the DOS of the surface state.

2.5.2 Potential Distribution in The Depletion Layer

The potential distribution and width of space charge depend on the amount of charges transferred to the surface and the density of shallow donors in the material, N_D. Quantitative expressions for these relationships are indispensible in PEC research, and to understand where they come from, we derive them below. The geometry of the problem and the relevant parameters are schematically illustrated in Fig. 2.12. The starting point for the derivation is Poisson's law, which relates the potential to the net amount of charge:

$$\frac{d^2\phi}{dx^2} = \frac{-d\xi}{dx} = \frac{-\rho(x)}{\varepsilon_0\varepsilon_r}. \tag{2.30}$$

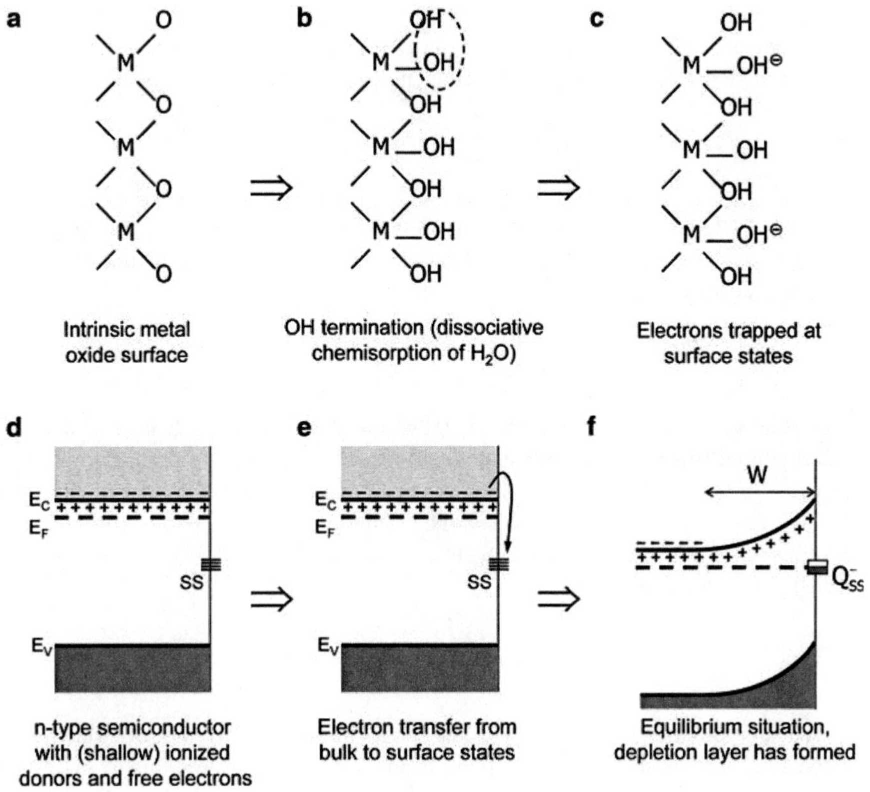

Fig. 2.11 Simplified illustration of the formation of a space charge region (SCR) at a metal oxide semiconductor surface when exposed to (humid) air

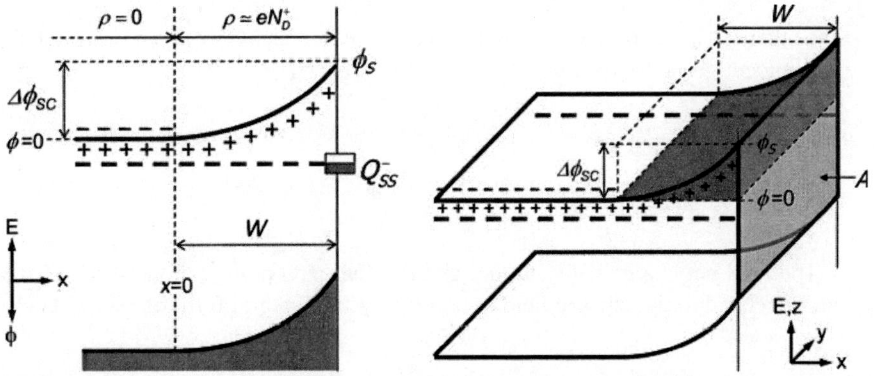

Fig. 2.12 Illustration of band bending at the surface of a n-type semiconductor in depletion. The "+" markers just below the conduction band represent the ionized donor species, whereas the "–" markers just above the conduction band level represent the free electrons. The potential in the bulk of the semiconductor is chosen as zero

Here, ξ was chosen as a symbol for the electric field to avoid confusion with symbol E for energy. The total charge density in the space charge is given by the concentration of free electrons, n, plus the concentration of ionized donors, N_D^+:

$$\rho(x) = \mathrm{e}\left(N_D^+ - n\right). \tag{2.31}$$

The concentration of free electrons is given by the distance between the conduction band minimum and the Fermi level, cf. (2.9). Inside the SCR, this distance is modified by the band bending $\phi(x)$, which leads to the following expression for n:

$$n(x) = N_C\,\mathrm{e}^{-(E_C - E_F - e\phi(x))/kT} = n_b\,\mathrm{e}^{e\phi(x)/kT}. \tag{2.32}$$

Note that the energy of an electron is related to its potential by $E = -e\phi$ and that the potential in the bulk of the semiconductor is conveniently chosen as zero (see Fig. 2.12). Since the bulk is electrically neutral, its electron concentration, n_b, must be equal to the number of ionized donors. The latter are assumed to be fully ionized, so that $N_D^+ = N_D = n_b$. Combining this with (2.31) and (2.32), the total charge density at point x can be written as

$$\rho(x) = eN_D\left(1 - \mathrm{e}^{e\phi(x)/kT}\right). \tag{2.33}$$

The combination of (2.30) and (2.33) cannot be readily solved due to the implicit dependence on $\phi(x)$. Fortunately, there is an elegant way around this by taking the derivative of the square of the electric field [38]. Using this with (2.30) yields:

$$\frac{\mathrm{d}(\xi^2)}{\mathrm{d}x} = 2\xi\frac{\mathrm{d}\xi}{\mathrm{d}x} = 2\xi\frac{\rho(x)}{\varepsilon_0\varepsilon_r} = -2\frac{\rho(x)}{\varepsilon_0\varepsilon_r}\frac{\mathrm{d}\phi}{\mathrm{d}x}. \tag{2.34}$$

Since there is a one-to-one correspondence between ϕ and x, we can now switch the independent variable from x to ϕ and rewrite (2.34) as

$$\xi^2 = \int_0^{\phi(x)} -2\frac{\rho(\phi)}{\varepsilon_0\varepsilon_r}\,\mathrm{d}\phi. \tag{2.35}$$

The total net charge between the (neutral) bulk and point x in the space charge can be related to the electric field at point x via the integral form of Gauss' law:

$$\xi = \frac{Q}{\varepsilon_0\varepsilon_r A}. \tag{2.36}$$

Here, A is the surface area of semiconductor. Combining (2.35) and (2.36) gives:

$$Q = \sqrt{-2\varepsilon_0\varepsilon_r A^2 \int_0^{\phi(x)} \rho(\phi)\,\mathrm{d}\phi}. \tag{2.37}$$

This equation provides a convenient starting point for further analysis of the space charge. It is generally applicable and can also deal with nonhomogeneous doping profiles and deep donors/acceptors. Here, we limit ourselves to homogeneously distributed shallow donors, for which the charge density is given by (2.33).

The total amount of charge in the SCR can now be obtained by combining (2.33) and (2.37) and integrating between $\phi = 0$ and $\phi = \phi_S = -\phi_{SC}$:

$$\begin{aligned} Q_{SC} &= \sqrt{-2\varepsilon_0\varepsilon_r A^2 \int_0^{-\phi_{sc}} eN_D(1 - e^{e\phi/kT})\,\mathrm{d}\phi} \\ &= \sqrt{2\varepsilon_0\varepsilon_r eN_D A^2 \left(\phi_{SC} + \frac{kT}{e}\,e^{-e\phi_{SC}/kT} - \frac{kT}{e}\right)}. \end{aligned} \tag{2.38}$$

Under normal PEC operating conditions, the potential drop across the space charge is at least 0.1 V, and the previous result can be simplified to

$$Q_{SC} = \sqrt{2\varepsilon_0\varepsilon_r eN_D A^2 \left(\phi_{SC} - \frac{kT}{e}\right)}. \tag{2.39}$$

The total amount of charge is related to the depletion layer thickness via

$$Q_{SC} = eN_D AW \tag{2.40}$$

which leads to the following widely used expression for the space charge width:

$$W = \sqrt{\frac{2\varepsilon_0\varepsilon_r}{eN_D}\left(\phi_{SC} - \frac{kT}{e}\right)}. \tag{2.41}$$

Although this important result was derived for n-type semiconductors, it is also valid for p-type materials if N_D is replaced with the shallow acceptor density, N_A.

To get an idea of typical numbers, Fig. 2.13 shows the calculated depletion layer width for α-Fe_2O_3 as a function of ϕ_{SC} for various dopant densities. Typical values range between 5 and 500 nm. The total amount of charge in the depletion layer has to be compensated by a sheet of oppositely charged species (e.g., trapped electrons) at the surface of the material. As can be seen in the right-hand part of Fig. 2.13, less

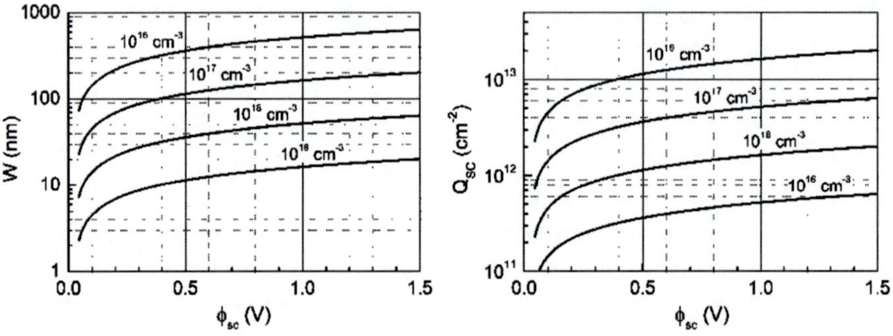

Fig. 2.13 *Left*: Depletion layer width as a function of potential drop across the space charge (ϕ_{SC}) and (shallow) dopant density. *Right*: Corresponding amount of adsorbed surface charges needed to compensate the charges in the depletion layer. The data are calculated for α-Fe_2O_3 assuming a static dielectric constant of 25 [39, 40]

than 1% of a monolayer[7] of adsorbed surface charges is already more than sufficient to induce a significant space charge layer in the material.

The actual values of the dopant density and the depletion layer width can be determined by impedance measurements. Under certain conditions one can extract the differential capacitance of the space charge, C_{SC}, from the total impedance of the system. C_{SC} can be derived by differentiating (2.39) with respect to ϕ_{SC}:

$$\frac{1}{C_{SC}^2} = \left(\frac{dQ_{SC}}{d\phi_{SC}}\right)^{-2} = \frac{2}{\varepsilon_0 \varepsilon_r e N_D A^2}\left(\phi_{SC} - \frac{kT}{e}\right). \tag{2.42}$$

This is the so-called "Mott–Schottky" equation. Typical values for C_{SC} are 10–1,000 nF/cm^2. By plotting $1/C_{SC}{}^2$ as a function of the applied potential, the donor density of the semiconductor can be determined.

2.5.3 Deep Depletion, Inversion, and Accumulation Layers

In addition to the depletion layer, other types of space charges can also form in a semiconductor. These are illustrated in Fig. 2.14. If the number of adsorbed negative (positive) surface charges increases beyond a certain number for an n-type (p-type) semiconductor, the Fermi level crosses the middle of the bandgap and the surface region becomes p-type (n-type). This is called an inversion layer. Formation of an inversion layer is not always possible; if the dominant charge carriers in the inversion layer are annihilated faster than they are (thermally)

[7] 1 monolayer (ML) corresponds to ~10^{15} atoms/cm^2.

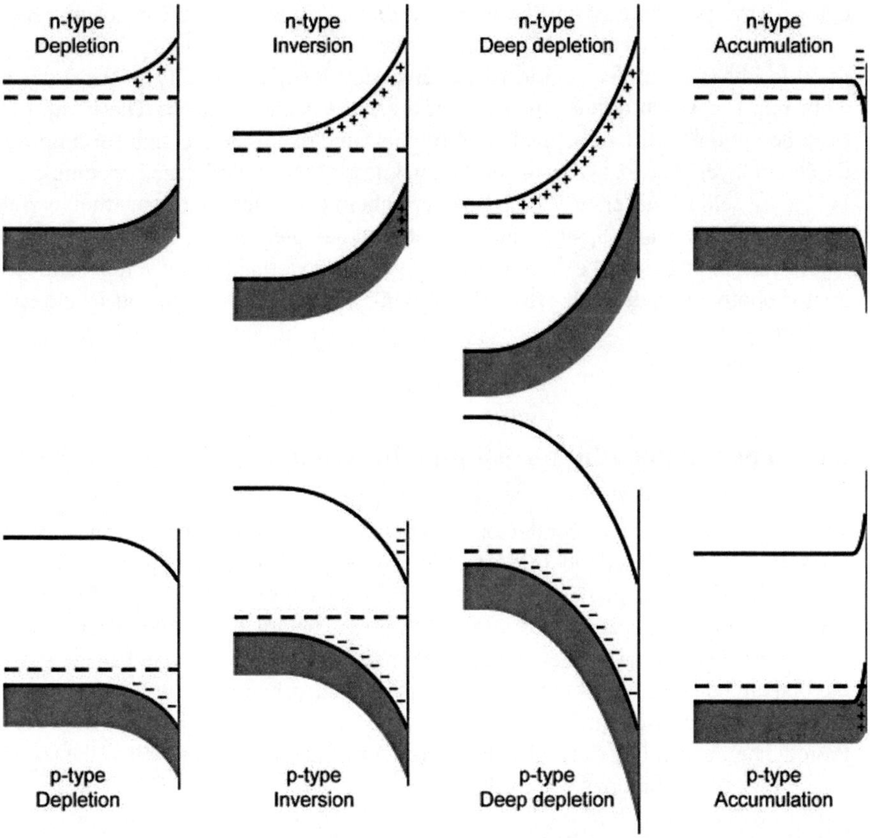

Fig. 2.14 Different types of space charges in n- and p-type semiconductors. A normal depletion layer contains only ionized donors or acceptors. An inversion layer is formed when the Fermi level crosses the midgap energy, and the minority carriers outnumber the majority carriers in a thin layer at the surface. When these minority carriers are consumed faster than they are generated, a deep depletion layer forms; under these conditions the surface is not in thermal equilibrium and the Fermi level is not well defined in this region. In an accumulation layer, the adsorbed surface charges are compensated by majority charge carriers that accumulate at the surface

generated, no free charge carriers are present and the surface remains insulating. A deep depletion layer is then formed. Deep depletion layers are fairly common in photoelectrode materials with a relatively large bandgap (>2 eV) because generation of minority carriers is difficult in these materials. They can also be formed in the presence of surface-adsorbed species that consume the minority carriers through fast oxidation or reduction reactions. An accumulation layer can form when an excess of positive (negative) charges is adsorbed at the surface of an n-type (p-type) semiconductor. To compensate these surface charges, free majority carriers will accumulate near the surface, forming the accumulation layer.

Whereas the charge density in a depletion layer is limited by the concentration of ionized dopant ions (usually 10^{16}–10^{18} cm^{-3}), the concentration of electronic charge carriers is limited by the effective density of states and can reach values up to ~10^{20} cm^{-3}. Due to this high density, the widths of the accumulation layer and the thin inverted surface region for an inversion layer are usually in the order of 1–15 nm, i.e., much thinner than that of a typical depletion layer. The width of a deep depletion layer can be evaluated by the same expression as that for a normal depletion layer. The expressions for the thickness of the inversion and accumulation layers are seldom necessary in photoelectrochemical water splitting – they can be found in the literature [8, 41]. Finally, it should be noted that all four space charge layers can, in principle, be formed by applying an external potential to a semiconductor photoelectrode. The effect of applying a potential is discussed in the next section.

2.6 The Semiconductor–Liquid Junction

When a metal oxide semiconductor is exposed to humid air, hydroxylation of the surface occurs as described in Sect. 2.5.1. A slightly more complicated situation arises when the semiconductor is immersed in an aqueous solution. Depending on the pH of the solution, specific adsorption of protons and/or hydroxide species affect the charge distribution at the semiconductor/electrolyte interface, and thereby also the potential distribution. Analysis of the potential distribution allows us to construct an energy diagram of the complete PEC cell. This diagram is indispensible for a thorough understanding of the thermodynamics of PEC devices, as well as the charge transfer processes that make it work.

2.6.1 Surface Hydroxylation

When a semiconductor is immersed in an aqueous solution, H^+ and OH^- ions in the solution will continuously adsorb and desorb from the surface. A dynamic equilibrium will be established, which can be described by the following protonation and deprotonation reactions:

$$M - OH \overset{k_a}{\rightleftharpoons} MO^- + H^+_{aq} \tag{2.43}$$

$$M - OH + H^+_{aq} \overset{k_b}{\rightleftharpoons} M - OH_2^+. \tag{2.44}$$

The equilibrium of these reactions depends on the pH of the solution and the Brønsted acidity of the surface. Depending on these conditions, the net total charge adsorbed at the surface will be positive, zero, or negative. The pH at which the net adsorbed charge is zero is called the point of zero charge (PZC) of the semiconductor.

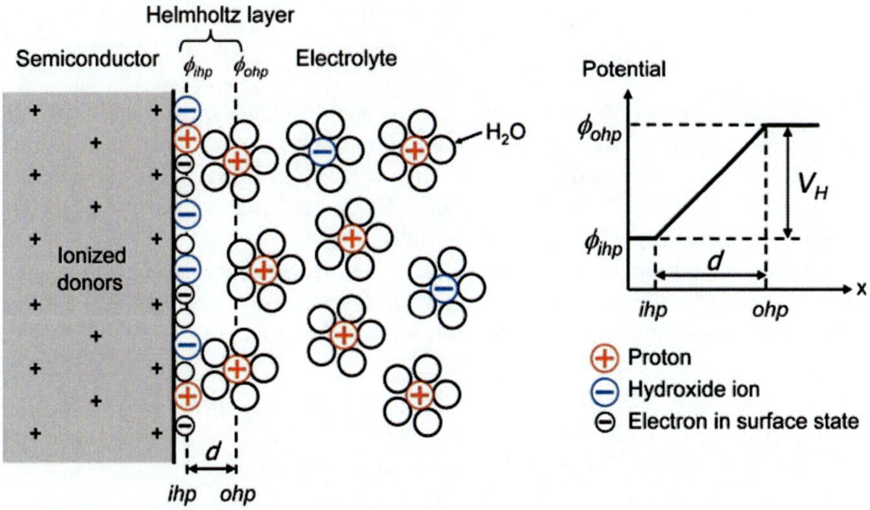

Fig. 2.15 Schematic model of the semiconductor/electrolyte interface and the Helmholtz layer. The inner Helmholtz plane (*ihp*) consists of H^+ and OH^- ions that are specifically adsorbed at the semiconductor surface. The outer Helmholtz plane (*ohp*) marks the distance of closest approach for ions still in the solution. The distance *d* is only a few Ångströms due to the solvation sheet of water molecules surrounding each ion

Note that the reactions (2.43) and (2.44) are only appropriate for hydroxylated semiconductor surfaces that are amphoteric,[8] which is indeed true for most metal oxide photoelectrodes.

2.6.2 *The Semiconductor/Electrolyte Interface*

The structure of the semiconductor/electrolyte interface is illustrated in Fig. 2.15. The charges at the surface consists of electrons (or holes) trapped in surface states, plus the specifically adsorbed ions described in reactions (2.43) and (2.44). The counter charges are provided by ionized donors or acceptors in the solid, plus an accumulation of oppositely charged ions in the solution. Due to the large dipole moment of water, all ions in the solution are surrounded by a solvation cloud of water molecules. This cloud prevents them from approaching the surface closer than a few Ångströms (see Fig. 2.15). The region between the specifically adsorbed ions and the closest ions in the solution is called the *Helmholtz layer*. The potential drop across this layer is given by

[8] "Amphoteric" means that the semiconductor surface can either donate or accept a proton, i.e., it can act both as a Brønsted acid and as a Brønsted base.

$$V_{\mathrm{H}} = \xi d = \frac{Q_{\mathrm{S}} d}{\varepsilon_0 \varepsilon_{\mathrm{r}}}. \tag{2.45}$$

The width of the Helmholtz layer is ~2–5 Å, and on such a small length scale the dielectric constant of water is estimated to be ~6, i.e., much smaller than its bulk value of 80 [42]. For a surface charge of $Q_{\mathrm{S}} = 10^{13}$ cm^{-2} (Fig. 2.13), the potential drop across the Helmholtz layer is typically in the order of ~0.1–0.5 V. The Helmholtz capacitance is 10–20 μF/cm^2 [42].

During adsorption and desorption from the semiconductor surface, the ions gain or lose energy while crossing the Helmholtz layer due to the Helmholtz voltage. This leads to the following equilibrium constants for reactions (2.43) and (2.44):

$$\frac{[\mathrm{M-O^-}][\mathrm{H^+_{aq}}]}{[\mathrm{M-OH}]} = k_{\mathrm{a}} = \exp\left(\frac{-(\Delta G_{\mathrm{a}} - eV_{\mathrm{H}})}{kT}\right) \tag{2.46}$$

$$\frac{[\mathrm{M-OH_2^+}]}{[\mathrm{M-OH}][\mathrm{H^+_{aq}}]} = k_{\mathrm{b}} = \exp\left(\frac{-(\Delta G_{\mathrm{b}} + eV_{\mathrm{H}})}{kT}\right). \tag{2.47}$$

Making use of the fact that $V_{\mathrm{H}} = 0$ and $[\mathrm{M{-}O^-}] = [\mathrm{M{-}OH_2^+}]$ at pH = PZC, the Gibbs free energies can be eliminated to yield the following expression for V_{H}:

$$V_{\mathrm{H}} = \frac{2.3kT}{e} \log\left(\frac{[\mathrm{M-O^-}]}{[\mathrm{M-OH_2^+}]}\right)^{1/2} + \frac{2.3kT}{e}(\mathrm{PZC} - \mathrm{pH}). \tag{2.48}$$

As seen above, the excess charge at the surface is usually very small, in the order of 1% of a monolayer or less. Hence, $[\mathrm{M{-}O^-}] \approx [\mathrm{M{-}OH_2^+}]$ and (2.48) can be simplified to

$$V_{\mathrm{H}} = \frac{2.3kT}{e}(\mathrm{PZC} - \mathrm{pH}). \tag{2.49}$$

This important result shows that the Helmholtz potential changes with -59 mV ($2.3kT/e$) per pH unit at 25°C.

2.6.3 *The Band Diagram*

Now we understand the structure of the semiconductor/electrolyte interface, we can draw a detailed band diagram for a complete PEC cell. An example is shown in Fig. 2.16 for a cell composed of a n-type photoanode and a metal counter electrode. As usual, the *y*-axis represent the energy of an electron at a certain point *x* in the cell. The energy of an electron in vacuum at infinite distance is chosen as a reference. It is important to note that the vacuum level bends in the presence of an

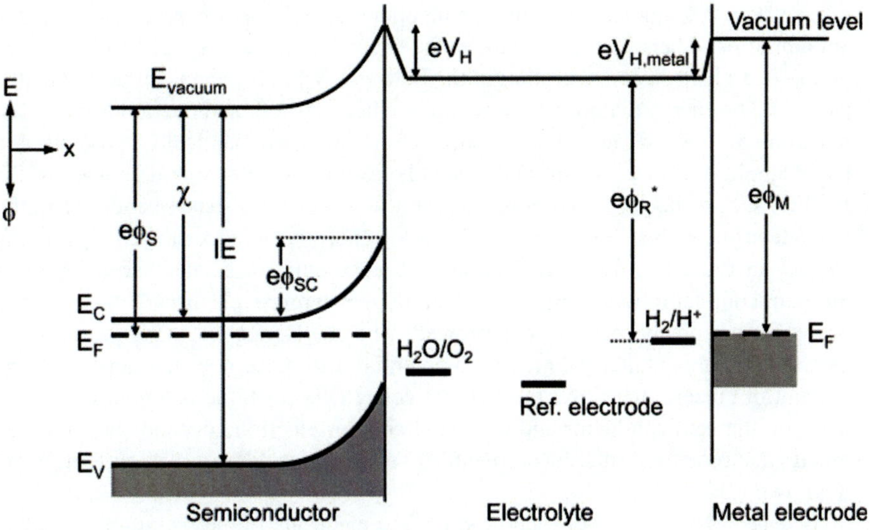

Fig. 2.16 Energy diagram for a PEC cell based on a n-type semiconductor and a metal counter electrode. The vacuum energy level is taken as a reference; this is the energy of an electron in vacuum at infinite distance. The electron affinity (χ) and ionization energy (IE) are materials constants, whereas the semiconductor work function (ϕ_S) also depends on the distance to the surface. Note that a Helmholtz layer is also present at the metal/electrolyte interface

electric field, i.e., it follows the potential gradients that are present in the PEC cell. This may seem a bit odd for a "reference" energy, but it is a consequence of the convention that the electrochemical potential (or Fermi energy) of the electrons is drawn as a constant value when the system is in equilibrium.

We now take a closer look at the electrolyte phase in Fig. 2.16. The energy of a redox couple in an electrolyte with respect to the vacuum level ($e\phi_R$*) is not accurately known. Theoretical and experimental estimates place the H_2/H^+ standard redox energy between 4.3 and 4.85 eV below E_{vacuum} [43]. Usually a value of 4.5 eV is taken, but it should be realized that this value has an uncertainty of a few tenths of an eV; Bockris suggested a value of (4.6 ± 0.2) eV [42]. The value recommended by IUPAC is (4.44 ± 0.02) eV [44]. The vacuum level has been drawn as a horizontal line in the electrolyte, since no electric field is assumed to be present in this phase. This is a realistic assumption for most water splitting applications because highly concentrated electrolytes are used in order to avoid Ohmic voltage losses. For very low electrolyte concentrations (<0.1 M), there may be insufficient ions available at the outer Helmholtz plane to compensate all the adsorbed charges at the semiconductor surface. These charges are then compensated in a region that extends much beyond the outer Helmholtz plane, the so-called *Gouy layer*. The potential distribution in this layer resembles that of a SCR in a semiconductor and is described by the Gouy–Chapman theory. Since the Gouy layer is rarely relevant for solar water splitting applications, we refer the reader to the literature for a more detailed description [42].

The key parameters for the semiconductor are the band positions, the amount of band bending, and the difference between E_C and E_F. The latter depends on the free electron concentration in the bulk ($= N_D^+ - N_A^-$) and can be calculated by (2.9). The energy diagram directly shows whether a certain reduction or oxidation reaction at the semiconductor surface is thermodynamically possible. In the example of Fig. 2.16, photogenerated holes will be able to oxidize water if the H_2O/O_2 redox energy is positioned above the top of the valence band. Similarly, only electrons with an energy above the redox energy can reduce the corresponding species in the electrolyte. By measuring the potential (i.e., the Fermi level) of the semiconductor with respect to that of a reference electrode, its reduction or oxidation power can be determined. The advantage of using a reference electrode for this is that the measured potential difference does not depend on the amount of current that flows through the cell. In contrast, the potential difference between the semiconductor and the metal counter electrode depends on V_H at the metal electrode (i.e., the overpotential), which depends on the current in an unknown way.

Because of the high concentration of free electrons in the metal counter electrode, the SCR inside the metal is extremely thin (~1 Å) and can therefore be ignored. The structure of the Helmholtz layer at the metal/electrolyte interface is similar to that described for the semiconductor, and the capacitance is also in the order of 10–20 μF/cm^2. As already hinted in the previous paragraph, the potential drop across the Helmholtz layer depends on the kinetics of electron transfer across the interface. This is because any overpotential applied to a metal electrode must fall across the Helmholtz layer. This is quite different from the case of an external potential applied to a semiconductor electrode, which is discussed in the next section.

2.6.4 Applying A Bias Potential

The operation of a PEC cell can be influenced by applying an external bias potential to the semiconductor. When the bias is applied with respect to a reference electrode, the potential difference will be distributed over the space charge layer and the Helmholtz layer. These layers act as two capacitances in series [41]:

$$\frac{1}{C_{\text{tot}}} = \frac{1}{C_{SC}} + \frac{1}{C_H}. \tag{2.50}$$

Both layers have the same charge Q associated with them (this is the charge that is located at the inner Helmholtz plane), and since $C = Q/V$ the potential distribution is given by

$$\frac{\Delta V_{SC}}{\Delta V_H} = \frac{C_H}{C_{SC}}. \tag{2.51}$$

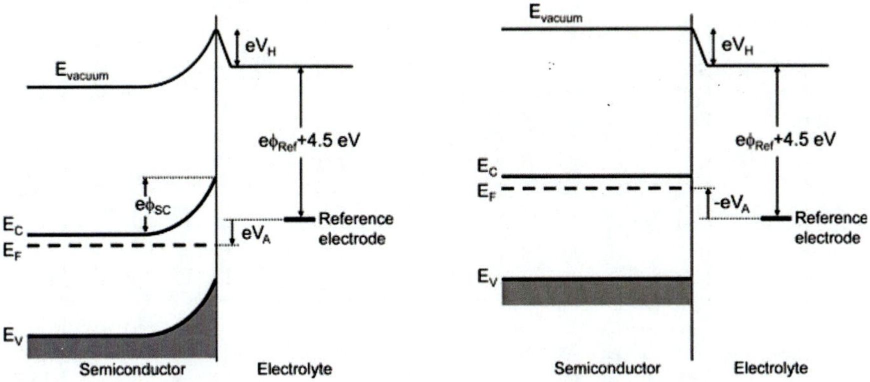

Fig. 2.17 Effect of applying a bias voltage (V_A) to an n-type semiconductor electrode. Any change in applied potential falls across the space charge layer, whereas V_H remains constant. In the picture on the left, a positive potential is applied to the semiconductor. When a sufficiently negative bias is applied, the band bending can be reduced to zero (*right*)

Because $C_H \gg C_{SC}$ (see above), any change in applied bias will fall across the depletion layer of the semiconductor.[9] This important conclusion is illustrated in Fig. 2.17. As shown in this figure, applying a positive bias to an n-type semiconductor results in an increase of the depletion layer. For a p-type semiconductor, a negative bias is required to increase the depletion layer. The ability to change the band bending in a semiconductor electrode is immensely useful in the characterization of these materials, as we see in later chapters.

2.6.5 pH Dependence of The Band Edges

While V_H is not affected by the applied bias potential, it does depend on the pH of the solution as described by (2.49). This means that the positions of the band edges shift with -59 mV per pH unit with respect to the redox potentials in the electrolyte. At first sight, this seems to be a very useful property: if reduction of a certain species is not possible because the conduction band is too low in energy, one might consider increasing the pH toward more alkaline values to make V_H more negative and raise E_C with respect to E_{redox} (a consequence is of course that this also increases E_V, so it reduces the oxidation power of the semiconductor). Unfortunately, this strategy does not help for water splitting because the reduction and oxidation potentials of water also depend on the pH via the Nernst equation. For reactions (2.1) and (2.2) in Sect. 2.2, the Nernst equations read:

[9] Note that in highly doped semiconductors ($>10^{19}$ cm^{-3}) and metals, C_{SC} can exceed C_H so that any change in the applied potential will fall across the Helmholtz layer instead of the depletion layer.

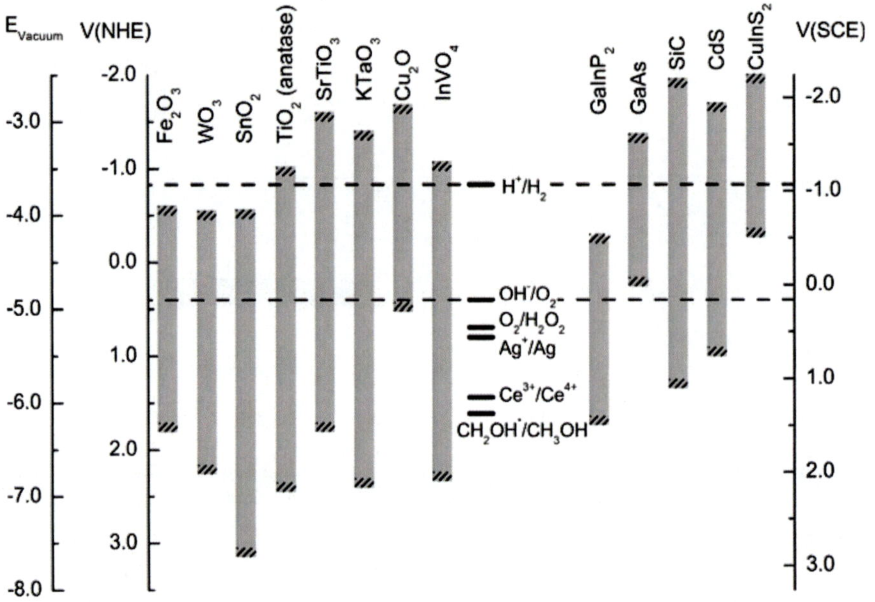

Fig. 2.18 Band edge positions for selected semiconductors at pH 14, together with some important redox potentials. It should be realized that the uncertainty in the band edge positions can amount to a few tenths of an eV for most semiconductors shown here

$$E_{\mathrm{red}} = E^0_{\mathrm{red}} - \frac{RT}{4F}\ln\left(\frac{p^2_{\mathrm{H_2}}}{[\mathrm{H}^+]^4}\right) = E^0_{\mathrm{red}} - \frac{2.3RT}{F}\left(\log\left(p^{1/2}_{\mathrm{H_2}}\right) + \mathrm{pH}\right) \tag{2.52}$$

$$E_{\mathrm{ox}} = E^0_{\mathrm{ox}} + \frac{RT}{4F}\ln\left(p_{\mathrm{O_2}}[\mathrm{H}^+]^4\right) = E^0_{\mathrm{ox}} + \frac{2.3RT}{F}(\log(p_{\mathrm{O_2}}) - \mathrm{pH}). \tag{2.53}$$

Hence, the reduction and oxidation potentials depend in the same way on the pH as the band positions, i.e., by -59 mV per pH unit. This means that the band positions of most metal oxides are fixed with respect to the water redox potentials.

Knowledge of the band edge positions at the surface is very useful, since they determine the maximum reduction and oxidation potentials of the photogenerated electrons and holes in a semiconductor. Figure 2.18 shows the band edge energies for several important semiconductors, together with the redox potentials for water splitting. The diagram is valid for a pH of 14. As outlined above, the band positions of most metal oxides show the same pH dependence as the water reduction and oxidation potentials, so relative positions remain the same at other pH values. This is, however, not the case for semiconductors and redox couples that do *not* show a -59 mV/pH dependence (MoS_2, Ce^{3+}/Ce^{4+}, Ag^+/Ag, etc.).

It should be noted that the actual reduction power of the electrons in a photoanode is always less than that suggested by the conduction band edge.

The same is true for the oxidative power of holes generated in a photocathode. This is because the photogenerated carriers lose some of their energy while traversing the SCR. Some more energy is lost when they leave the majority carrier band of the semiconductor and enter the metal counter electrode at its Fermi level. For an n-type semiconductor, the total energy loss is given by $E_{loss} = e\phi_{SC} + (E_C - E_F)$, and typically amounts to 0.1–0.5 eV.

2.6.6 The Flatband Potential

The position of the band edges with respect to the redox potentials in the electrolyte is conveniently expressed by the so-called "flatband potential", ϕ_{FB}. As the word suggests, this is the potential that needs to be applied to the semiconductor to reduce the band bending to zero. The flatband situation is illustrated in the right-hand panel of Fig. 2.17. It is important to realize that the flatband potential denotes the position of the *Fermi level* of the semiconductor with respect to the potential of the reference electrode. This means that ϕ_{FB} is slightly below the conduction band edges shown in Fig. 2.18, and that it *accurately reflects the thermodynamic ability of an n-type semiconductor to reduce water to hydrogen.*

Several techniques can be used to determine the flatband potential of a semiconductor. The most straightforward method is to measure the photocurrent onset potential, ϕ_{onset}. At potentials positive of ϕ_{FB} a depletion layer forms that enables the separation of photogenerated electrons and holes, so one would expect a photocurrent. However, the actual potential that needs to be applied before a photocurrent is observed is often several tenths of a volt more positive than ϕ_{FB}. This can be due to recombination in the space charge layer [45], hole trapping at surface defects [46], or hole accumulation at the surface due to poor charge transfer kinetics [43]. A more reliable method for determining ϕ_{FB} is electrolyte electroreflectance (EER), with which changes in the surface free electron concentration can be accurately detected [47]. The most often used method, however, is Mott–Schottky analysis. Here, the $1/C_{SC}^2$ is plotted as a function of the applied potential ϕ_A, and the value of the flatband potential is given by the intercept on the potential axis. This can be seen from a more commonly used version of equation, (2.42), in which ϕ_{SC} is replaced by $\phi_A - \phi_{FB}$:

$$\frac{1}{C_{SC}^2} = \frac{2}{\varepsilon_0 \varepsilon_r e N_D A^2}\left(\phi_A - \phi_{FB} - \frac{kT}{e}\right). \quad (2.54)$$

Any change of the potential across the Helmholtz layer is directly reflected in the flatband potential, so it is expected to shift by −59 mV per pH unit. This has indeed been experimentally observed for most metal oxides (e.g., TiO_2, ZnO, Fe_2O_3, SnO_2, WO_3) and even some nonoxide semiconductors (Ge, GaAs, $GaAs_{1-x}P$) [41, 48].

2.6.7 Surface and Interface States

Illumination of a semiconductor can lead to a change in the flatband potential, even at constant pH. This can be explained by the presence of surface states that trap photogenerated minority carriers [49]. The change in charge density at the interface (more specifically: the inner Helmholtz plane) results in a change in the Helmholtz potential and, therefore, the flatband potential. This is usually referred to as "unpinning of the band edges." The change in flatband potential can be used to determine the density of surface states involved:

$$\Delta Q_{\rm SS} = C_{\rm H}\left(\phi_{\rm FB}^{\rm light} - \phi_{\rm FB}^{\rm dark}\right). \tag{2.55}$$

The presence of surface states can also affect the band bending in the dark. If a high density of partially filled mono-energetic surface states is present, the Fermi level will be located somewhere halfway the surface state DOS. Any change in pH or applied potential will then be accommodated by a change in the occupation of this state and a concomitant change in the potential across the Helmholtz layer. If the surface state DOS is sufficiently high, the Fermi level will show only a very small shift, and the amount of band bending in the semiconductor remains constant. This is called "Fermi level pinning". It is fairly common in covalent semiconductors, where it can be recognized by a different amount of band bending upon forming a metal–semiconductor Schottky contact that would be expected from the work function difference between both phases. It can also be observed as a flat region in a Mott–Schottky plot because the capacitance of the depletion layer will remain constant, while the Fermi level crosses the surface state energy. The physical origin of surface or interface states can have structural defects (dangling bonds, vacancies, etc.) or chemisorbed species from an electrolyte solution. In metal oxide semiconductors, the surface state DOS is generally small and Fermi level pinning is rarely observed. However, it is important in tandem devices or photocatalysts based on a junction between covalent and metal oxide semiconductors, e.g., n-Si/n-TiO_2 [50].

The chemical nature of surface states depends strongly on the bonding character of the material. Covalent materials have dangling bonds that can act as electron donors or acceptors – these are called *Shockley surface states*. Moreover, overlap between neighboring bonds can result in the formation of coupled states that have bonding (donor) and antibonding (acceptor) character. In ionic materials, so-called *Tamm states* are formed by coordinatively unsaturated ions at the surface. For example, lattice oxygen at the surface (O_L^{2-}) generally forms a deep donor state just above the valence band, while lattice metal ions (M_L^{x+}) can form deep acceptor states just below the conduction band [41]. In addition to these intrinsic surface states, extrinsic surface states can be formed by the adsorption of ions from the solution.

2.6.8 *Interfacial Electron Transfer*

In the discussion so far, electron transfer from the semiconductor to the electrolyte is assumed to be possible if the energy of the electron in the solid exceeds the reduction energy of the species in the solution (and vice versa). Unfortunately, this simple picture does not hold for large energy differences. Moreover, it does not give any information on the kinetics of electron transfer. A more detailed theory of electron transfer was developed in the late 1950s by Marcus [51, 52], who won the 1992 Nobel Prize in Chemistry for his contributions in this field. His theory is mainly based on classical physics, and is applicable to both homogeneous systems and reactions at electrodes. Quantum-chemical extensions of the theory were later made by the Russian scientists Levich, Dogonadze, and Kuznetsov [43]. The most useful contribution to electrochemistry was made in the early 1960s by Gerischer [53]. Whereas Marcus derived his theory using configurational coordinate diagrams of the Gibbs free energy of the species involved, Gerischer approached the problem using electronic energy levels. We follow his approach in the description below.

A common element in all versions of electron transfer theory is the concept of fluctuating energy levels in the electrolyte. In polar solvents, such as water, every ion in the electrolyte is surrounded by a cloud of oriented solvent molecules (cf. Fig. 2.15). In addition to these nearest-neighbor solvent molecules in the so-called *inner sphere*, the ion also has Coulomb interactions with polar solvent molecules and other ions that are farther away in the *outer sphere*. Due to the thermal motions of the solution, the energy level of the central ion (as seen from the electrode) will fluctuate. Moreover, after reduction or oxidation of the ion, the surrounding solvent molecules and ions will rearrange themselves to accommodate the different charge on the ion. The electrostatic energy required for this reorganization is given by 2λ, where λ is called the *reorganization energy*. It can be shown that the chance for a reduced or oxidized species to have a certain energy E is given by [53]:

$$W_{\mathrm{ox}}(E) = \frac{1}{\sqrt{4\pi\lambda kT}}\,\mathrm{Exp}\left(\frac{-\left(E - E^0_{\mathrm{redox}} + \lambda\right)^2}{4kT\lambda}\right) \tag{2.56}$$

$$W_{\mathrm{red}}(E) = \frac{1}{\sqrt{4\pi\lambda kT}}\,\mathrm{Exp}\left(\frac{-\left(E - E^0_{\mathrm{redox}} - \lambda\right)^2}{4kT\lambda}\right). \tag{2.57}$$

Here, E^0_{redox} is the standard redox potential for the species involved. The distribution of these fluctuating energy levels is shown in Fig. 2.19. It should be emphasized that this is a *probability distribution* – it should not be confused with the DOS, which is a fundamentally different concept (only one level exists at any one time for every ion, so optical transitions between the redox levels are not possible). The reorganization energy can be quite large compared to the bandgap of the semiconductor,

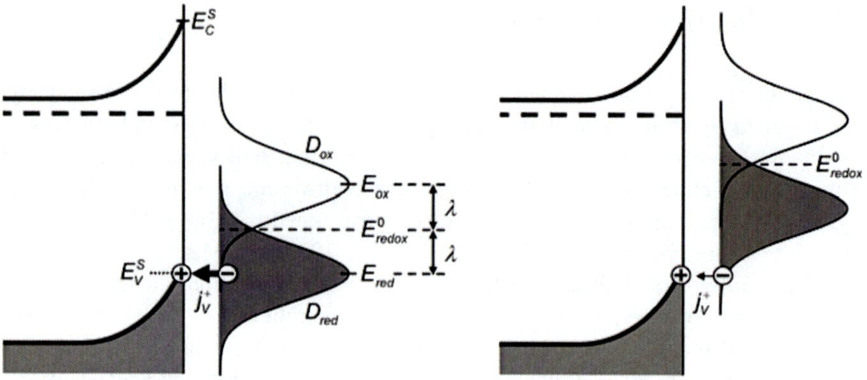

Fig. 2.19 Energy level model for electron transfer from a redox species in the electrolyte to a photogenerated hole in the semiconductor valence band. The density of states D_{red} depend on the concentration of the reductant species according to $D_{\text{red}} = c_{\text{red}}W_{\text{red}}(E)$. A large overlap of the redox DOS with the energy of the hole (*left*) gives higher currents than a small overlap (*right*)

with values ranging from 0.3 eV to more than 1 eV. At room temperature, the width of each distribution curve (at half height) is given by $\Delta E \simeq 0.53\sqrt{\lambda}$ (in eV).

For semiconductors, charge transfer takes place via either the bottom of the conduction band or the top of the valence band. In certain cases, charge transfer via surface states can also occur. Although Fig. 2.19 is drawn for the case of electron transfer to the valence band, which will only be significant if photogenerated holes are present, similar pictures can be made for other pathways, e.g., $D_{\text{red}} \rightarrow E_{\text{C}}$, $E_{\text{C}} \rightarrow D_{\text{ox}}$, or $E_{\text{V}} \rightarrow D_{\text{ox}}$. An important conclusion from Fig. 2.19 is that the probability of electron transfer actually *decreases* if E_{red} is too far above E_{V}^{S}. This is markedly different from the behavior of metal electrodes, which show a continuous increase in current with applied potential.

The actual charge transfer process takes place via tunneling. Since tunneling is an iso-energetic process, it requires the fluctuating energy level of the species in the electrolyte to be, at a certain moment in time, equal to the energy of the electron or hole in the semiconductor. During electron transfer, the energy level of the ion is assumed to remain constant. This assumption (the so-called *Frank–Condon principle*) is justified, since electron transfer processes occur on a much faster time scale than ionic reorganization. The transfer rate is then proportional to the concentration of the relevant carrier in the semiconductor, the DOS of the electrolyte species ($D_{\text{ox,red}} = c_{\text{ox,red}}W_{\text{ox,red}}(E)$), and a quantum-mechanical tunneling coefficient. With this approach, the anodic (+) and cathodic (−) valence and conduction band currents can then be written as [53]:

$$j_{\text{V}}^{+} = ek_{v}^{+}p_{\text{S}}c_{\text{red}}W_{\text{red}}(E_{\text{V}}^{S}) \quad j_{\text{V}}^{-} = ek_{v}^{-}N_{\text{V}}c_{\text{ox}}W_{\text{ox}}(E_{\text{V}}^{S}) \tag{2.58}$$

$$j_{\text{C}}^{+} = ek_{c}^{+}N_{\text{C}}c_{\text{red}}W_{\text{red}}(E_{\text{C}}^{S}) \quad j_{\text{C}}^{-} = ek_{c}^{-}n_{\text{S}}c_{\text{ox}}W_{\text{ox}}(E_{\text{C}}^{S}). \tag{2.59}$$

Here, N_C and N_V are the effective density of states in the semiconductor, and n_S and p_S are the surface concentrations of free electrons and holes in the conduction and valence bands, respectively. Further analysis leads to the following expressions for the overall conduction and valence band currents as a function of the overpotential η in the presence of a redox species:

$$j_V = j_V^0\left(\exp\left(\frac{-e\eta}{kT}\right) - 1\right) \quad \text{and} \quad j_C = -j_C^0\left(\exp\left(\frac{-e\eta}{kT}\right) - 1\right). \tag{2.60}$$

It is instructive to compare this with the Butler–Volmer equation for the current at a *metal* electrode:

$$j = j_0\left(\exp\left(\frac{(1-\alpha)e\eta}{kT}\right) - \exp\left(\frac{-\alpha e\eta}{kT}\right)\right). \tag{2.61}$$

Here, α is the so-called transfer coefficient, which is ~0.5 for metal electrodes. A Tafel plot of $\log(j/j_0)$ vs. η gives a slope of ~60 mV per current decade for a semiconductor, whereas a slope of ~120 mV/decade is found for a metal. The difference is caused by the fact that the applied overpotential falls across the Helmholtz layer for a metal, and across the SCR for a semiconductor. An even more crucial difference is not immediately apparent from (2.60) and (2.61): the exchange current density for a semiconductor is much smaller than that for a metal. This is a direct consequence of the high density of states of a metal near its Fermi level, which can be 3–4 orders of magnitude higher than N_C and N_V in a semiconductor. In fact, the current flowing at a semiconductor photoelectrode in the dark is usually negligible (often <10 nA/cm^2). This dark current is mainly carried by the *majority* carriers, i.e., electrons in the conduction band for an n-type semiconductor, and holes in the valence band for a p-type material. Under illumination, the current increases and is dominated by the transfer of *minority* carriers across the semiconductor/electrolyte interface. This is discussed in the next section.

2.7 The Photoelectrochemical Cell Under Operating Conditions

Most of the discussions above concern the properties of the semiconductor and semiconductor/electrolyte interface under equilibrium conditions, i.e., in the dark. We now explore how the system behaves under illumination.

2.7.1 The Quasi-Fermi Level

Figure 2.20 shows the energy diagram of a PEC cell in the dark and under illumination. In this example, the H_2O/O_2 redox couple is assumed to be the most

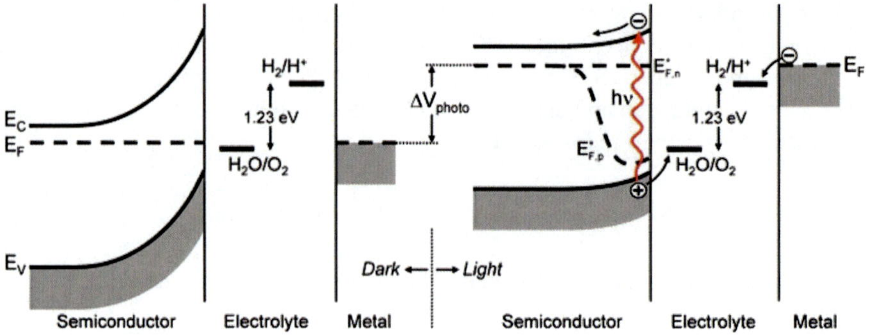

Fig. 2.20 Band diagram for a PEC cell based on an n-type semiconducting photoanode that is electrically connected to a metal counter electrode; in equilibrium in the dark (*left*) and under illumination (*right*). Illumination raises the Fermi level and decreases the band bending. Near the semiconductor/electrolyte interface, the Fermi level splits into quasi-Fermi levels for the electrons and holes

active species and therefore dominates the electrochemical potential of the solution in the dark. The Fermi levels of the semiconductor and metal, which are electrically connected, adjust to a value close to E_{ox}. Upon illumination, electron–hole pairs are created and the Fermi level increases with ΔV_{photo}, the internal photovoltage. Since the system is no longer in equilibrium, particularly in the SCR where the electrons and holes are generated, the use of a single Fermi level is no longer appropriate. Instead, the concept of *quasi-Fermi levels* is more useful. The quasi-Fermi levels are a direct measure of the concentration of electrons and holes at a certain point x in the semiconductor, and are defined as:

$$n = n_0 + \Delta n = N_C\, e^{-\left(E_C - E^*_{F,n}\right)/kT} \tag{2.62}$$

$$p = p_0 + \Delta p = N_V\, e^{-\left(E^*_{F,p} - E_V\right)/kT}. \tag{2.63}$$

Here, n_0 and p_0 are the equilibrium carrier concentrations in the dark, and Δn and Δp are the additional carriers created by illumination. For an n-type semiconductor, $n = n_0 + \Delta n \approx n_0$ and $p = p_0 + \Delta p \approx \Delta p$ so that $E^*_{F,n}$ remains horizontal whereas $E^*_{F,p}$ departs from the bulk Fermi level in the active region.

The quasi-Fermi level is often interpreted as a thermodynamic driving force. Whether or not this is appropriate is a matter of some debate. While it may provide useful insights, the concept has been derived from kinetic arguments, and can at best provide a quasi-thermodynamic description. Whereas true equilibrium thermodynamics are universally valid, the predictive value of the quasi-Fermi level as a thermodynamic driving force for, e.g., chemical reactions may depend on the reaction mechanism. One particular aspect to be noted in this respect is that the quasi-Fermi level definitions in (2.62) and (2.63) only consider electrons and holes in the conduction and valence bands. They do not account for any changes in the occupation of surface or bulk defect states, which limits their predictive value for reactions in which these states are involved.

2.7.2 Photocurrent–Voltage Characteristics

The theoretical description of semiconductor photocurrent–voltage characteristics have received considerable attention in the literature. One of the most often used models was reported by Gärtner [45], who derived the following expression for the photocurrent in a semiconductor under reverse bias:

$$j_G = j_0 + e\Phi\left(1 - \frac{\mathrm{Exp}(-\alpha W)}{1 + \alpha L_\mathrm{p}}\right). \tag{2.64}$$

Here, Φ is the incident light flux, α is the absorption coefficient (assuming monochromatic illumination), W is the depletion layer width, L_p is the hole diffusion length, and j_0 is the saturation current density. The model assumes that there is no recombination in the SCR and at the interface.

Gärtner's model was improved by Reichman [45], who used more appropriate boundary conditions for his derivation of the total valence band photocurrent in an n-type semiconductor:

$$j_\mathrm{V} = \frac{j_\mathrm{G} - j_0\,\mathrm{Exp}\left(\frac{-e\eta}{kT}\right)}{1 + \frac{j_0}{j_\mathrm{V}^0}\,\mathrm{Exp}\left(\frac{-e\eta}{kT}\right)}. \tag{2.65}$$

Here, j_V^0 is the hole transfer rate at the interface, and j_0 is the saturation current density, i.e., the hole current in the valence band at $x = W$ when $\Phi = 0$. η is the overpotential, which is defined as the difference between the applied potential and the open-circuit potential under illumination. Reichman's model includes the possibility of recombination in the SCR, which he showed could become important when the photovoltage becomes large enough for the bands to flatten. Another useful aspect of (2.65) is that it can be used to model the effect of the slow hole transfer kinetics that are often observed for metal oxides such as α-Fe_2O_3 [54]. The value of the saturation current density, $j_0 = (eL_\mathrm{p}N_\mathrm{C}N_\mathrm{V}/\tau N_\mathrm{D})\mathrm{Exp}\,(-E_\mathrm{g}/kT)$, plays an important role in Reichman's model, especially for narrow-bandgap semiconductors with large hole diffusion lengths and low donor densities. If the bandgap exceeds ~1.8 eV, which is the case for almost all metal oxides, j_0 becomes negligible and (2.64) and (2.65) give almost indistinguishable results.

At about the same time as Reichman, Jarrett also published a model that explicitly included recombination in the SCR [55]. For a given impurity concentration, the numerical solutions for the photoconversion efficiency he obtained are completely defined by three parameters: the optical absorption coefficient, the minority carrier lifetime in the depletion layer, τ_R, and the charge transfer rate at the surface. In Jarrett's model the amount of recombination in the space charge is expressed as (τ_t/τ_R), where τ_t is the time needed for the minority carriers to traverse the SCR. The usefulness of this approach is demonstrated in Chap. 9, where Jarrett's model is used to establish design criteria for photoelectrodes based on low-mobility (i.e., small polaron) semiconducting metal oxides.

It should be emphasized that the previous models only describe the current due to *minority carriers*. Under near-flatband conditions, the majority carriers start to contribute to the overall current. This can be recognized by an increase in the dark current, and under those conditions (2.64) and (2.65) are no longer useful. Another implicit assumption is the absence of mass transport limitations in the electrolyte, but this is rarely an issue in photoelectrochemistry due to the high electrolyte concentrations and modest efficiencies reported thus far. Finally, it should be realized that the models described above cannot account for recombination at the semiconductor/electrolyte interface. Wilson published an extension of Gärtner's original model that included interfacial recombination effects [56]. Although this somewhat complicated model is not often used because it does not account for the dark current and recombination in the space charge, it is of value for metal oxides that show extensive surface recombination.

2.7.3 Reaction Mechanisms

One of the main bottlenecks for water splitting is the oxidation half-reaction, which is a complicated process involving four separate electron transfer steps. Little is known about the exact molecular reaction mechanism of water oxidation on metal oxide surfaces. The only exception is TiO_2, which has been extensively investigated. Early work suggested that surface Ti–OH^- groups are photo-oxidized to Ti-$OH^{\bullet}$ radicals, which further react to adsorbed H_2O_2 species and finally to O_2 [57]. Nakato et al. proposed an alternative mechanism based on a nucleophilic attack of a H_2O molecule on a surface-trapped hole on a lattice oxygen site. The proposed reaction mechanism is shown in Fig. 2.21 [58, 59]. This mechanism served as a basis for recent theoretical DFT studies on the water oxidation reaction at various metal oxide surfaces [60–62]. For TiO_2, the reaction of water molecules with surface vacancies to form adsorbed hydroxyl groups was proposed to be the rate-determining step [61]. In contrast, the rate-limiting step for WO_3 appears to be the transfer of a proton from the surface adsorbed hydroxyl group to the electrolyte [62]. Although there remains some uncertainty about the actual molecular reaction mechanisms, these theoretical studies can provide valuable pointers for rational photoanode design [60].

Since the overpotential for water oxidation can easily exceed 0.6 V for a metal oxide photoanode, co-catalysts are often used to enhance the reaction kinetics. IrO_2 and RuO_2 are well-known oxidation catalysts, but they are either prohibitively expensive or prone to photocorrosion [63]. Recently, low-cost and stable

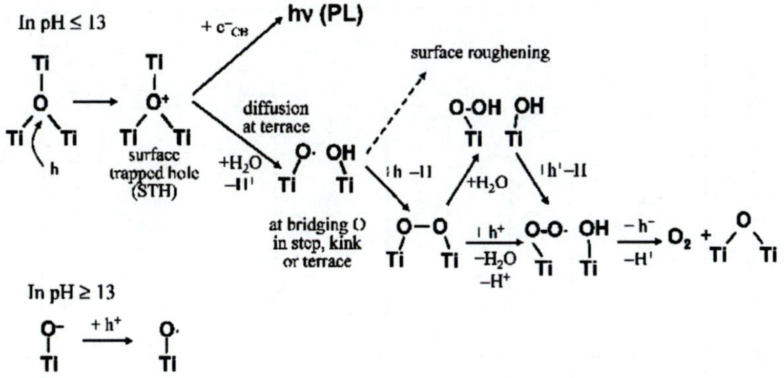

Fig. 2.21 Reaction mechanism for water oxidation on a rutile TiO_2 surface proposed by Nakato et al. Reprinted with permission from [59]. Copyright 2007 American Chemical Society

alternatives based on, e.g., cobalt have been found [64, 65]. Such catalysts have indeed been reported to enhance the efficiency of hematite photoanodes [66].

The reduction of water is a much simpler two-electron process. Many n-type semiconductors can evolve hydrogen at small overpotentials and without a co-catalyst. In contrast, p-type photoelectrodes generally require a large overpotential for the onset of cathodic photocurrents, and noble metal co-catalysts are often required [43]. Few p-type semiconductors have a conduction band that is well above the hydrogen reduction energy. Notable exceptions are SiC [67] and Cu_2O [68]. An important advantage of p-type photoelectrodes is that anodic decomposition, which is often observed for n-type photoelectrodes, does not occur. Having said this, it should be noted that the stability of many p-type oxides (Cu_2O, delafossites) in aqueous environments is notoriously poor.

2.7.4 Photocorrosion

Photocorrosion occurs when the photogenerated holes do not oxidize water, but the semiconductor itself. The general form of this anodic decomposition reaction can be written as

$$M_xO_y + \frac{2y}{x}h^+ \xrightleftharpoons{\text{electrolyte}} xM^{(2y/x)+}_{(aq)} + \frac{y}{2}O_2. \quad (2.66)$$

A well-known oxide that suffers from severe photocorrosion in aqueous environments is ZnO. For p-type oxides, cathodic photoreduction can lead to the formation of metal deposits:

$$M_xO_y + 2ye^- + 2yH^+_{(aq)} \xrightleftharpoons{\text{electrolyte}} xM_{(s)} + yH_2O. \quad (2.67)$$

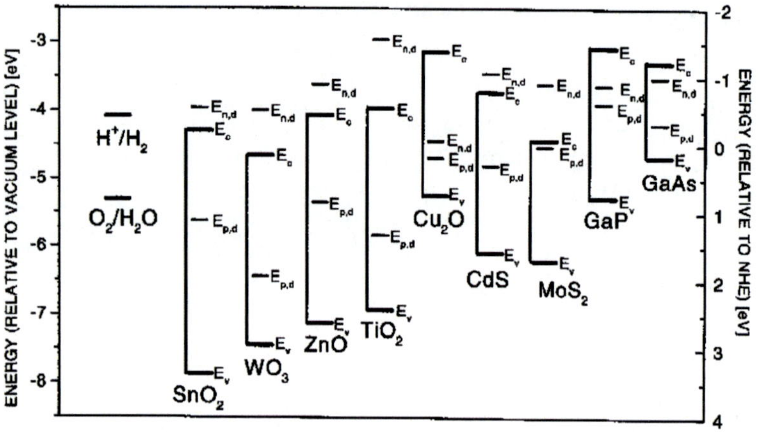

Fig. 2.22 Position of decomposition potentials of selected semiconductors. Reprinted from [70], with permission from Elsevier

In an alkaline environment, the reaction can be written as

$$M_xO_y + 2ye^- + yH_2O \xrightleftharpoons{\text{electrolyte}} xM_{(s)} + 2yOH^-_{(aq)}. \tag{2.68}$$

This has been observed for, e.g., Cu_2O. Alternatively, protons can enter the metal oxide lattice when free electrons are available at or near the surface, a process known as *intercalation*. It is worth mentioning that III–V semiconductors, such as GaAs and GaP, are prone to anodic photocorrosion, but appear to be relatively stable against cathodic photocorrosion in practice [69].

Whether or not these highly undesired reactions occur depends on their equilibrium potentials, which are in turn related to the Gibbs free energy of the reactions, cf. (2.5). If the energy level for anodic decomposition, $E_{p,decomp}$, is below the valence band maximum, no anodic decomposition can occur. Conversely, if $E_{n,decomp}$ is above the conduction band minimum, no cathodic photocorrosion can occur [41]. An overview of the anodic and cathodic decomposition potentials for various semiconductors is shown in Fig. 2.22. The actual values of the decomposition potentials depends on the concentration of the reaction products via the Nernst equation, so that they may depend on the pH of the solution [71]. A somewhat dated but useful overview of the stability of various metal oxides against photoanodic decomposition as a function of pH has been given by Scaife [72]. While some metal oxides are thermodynamically stable against cathodic photocorrosion (e.g., TiO_2), most – if not all – known metal oxides are thermodynamically unstable against anodic photocorrosion. This is also evident from the positions of $E_{p,decomp}$ in Fig. 2.22. This is even the case for n-type photoanodes that are found to be very stable, such as TiO_2 and α-Fe_2O_3; decomposition of these materials is simply inhibited by slow reaction kinetics. If a material shows signs of photocorrosion, the addition of suitable co-catalysts may alleviate the problem to some extent by favoring the water oxidation route. In practice, photocorrosion is difficult to predict and one usually has to put new materials to the test under actual operating conditions.

2.8 Energy and Quantum Conversion Efficiencies

The energy conversion efficiency is one of the key performance indicators of the PEC cell. Assuming that all the photogenerated electrons and holes are used for the water splitting reaction, the overall solar-to-hydrogen efficiency is given by

$$\eta_{\mathrm{STH}} = \frac{P_{\mathrm{electrical}}^{\mathrm{out}} - P_{\mathrm{electrical}}^{\mathrm{in}}}{P_{\mathrm{light}}} = \frac{j_{\mathrm{photo}}(V_{\mathrm{redox}} - V_{\mathrm{bias}})}{P_{\mathrm{light}}}. \quad (2.69)$$

V_{redox} is usually taken to be 1.23 V (at room temperature), based on a Gibbs free energy change for water splitting of 237 kJ/mol. Alternatively, the enthalpy change (286 kJ/mol) is sometimes used, which corresponds to a redox potential of 1.48 V. The latter value is appropriate when the hydrogen produced will be used in a combustion process, whereas the former value is more appropriate when the hydrogen is converted into electricity in a fuel cell [69]. When the end-use of the produced H_2 is not clear, it is probably best to use the Gibbs free energy value since it yields more conservative efficiency numbers. Assuming AM1.5G incident sunlight ($P_{\mathrm{light}} = 1{,}000$ W/m^2) and no applied bias, the efficiency can be directly related to the photocurrent by $\eta_{\mathrm{STH}} = 1.23 \times j_{\mathrm{photo}}$, with the efficiency in % and the photocurrent in mA/cm^2. Hence, a photocurrent of ~8 mA/cm^2 is required to reach the 10% efficiency target for economically viable PEC cells (see Chap. 8).

It should be emphasized that V_{bias} in (2.69) refers to the actual potential difference between the working and counter electrodes. Attempts to indirectly determine this potential difference from the results of a three-electrode measurement (in which the working electrode potential is reported relative to that of a reference electrode) generally leads to errors, and should be avoided (see also Chap. 3).

A more direct method to determine the solar-to-hydrogen efficiency is to measure the amount of evolved hydrogen by mass spectrometry, gas chromatography [73, 74], or by following the water displacement in an inverted burette. The STH efficiency for this approach is given by

$$\eta_{\mathrm{STH}} = \frac{\Phi_{\mathrm{H_2}} G^0_{\mathrm{f,H_2}}}{P_{\mathrm{light}}}, \quad (2.70)$$

where $\Phi_{\mathrm{H_2}}$ is rate of H_2 evolution at the illuminated area (mol/s/m^2) and $G^0_{\mathrm{f,H_2}}$ is the Gibbs free energy of formation of hydrogen (237 kJ/mol).

When reporting STH efficiencies, care should be taken to ensure that (a) the water splitting reaction is stoichiometric (H_2:O_2 = 2:1), (b) no other reactions due to, e.g., sacrificial donors/acceptors occur, and (c) the light source is well specified and closely matches the AM1.5G spectrum in intensity and spectral distribution [75]. For more detailed information on STH efficiency measurements, the reader is referred to some useful guidelines reported by a US Department of Energy (DOE) working group on photoelectrochemical hydrogen production [76] and to a careful study on the influence of the type of light source used [77].

When trying to identify performance-limiting factors in PEC photoelectrodes, the quantum efficiency as a function of wavelength is a particularly useful parameter. The definition of the *external* quantum efficiency is the fraction of the incident photons that is converted to electrons that can be measured by the outer circuit. This is the IPCE (*incident photon-to-current conversion efficiency*), which is given by

$$\mathrm{IPCE}(\lambda) = \frac{hc}{e}\left(\frac{j_{\mathrm{photo}}(\lambda)}{\lambda P(\lambda)}\right). \tag{2.71}$$

IPCE values in excess of 80% for the photo-oxidation of water have been reported for, e.g., WO_3 [78] and for TiO_2 under UV illumination [79]. Another useful parameter is the APCE, or *absorbed photon-to-current conversion efficiency*. In contrast to the IPCE, the APCE also corrects for reflection losses. Often referred to as the *internal* quantum efficiency, it is related to the IPCE via

$$\mathrm{APCE}(\lambda) = \frac{\mathrm{IPCE}(\lambda)}{A(\lambda)} = \frac{\mathrm{IPCE}(\lambda)}{1 - R - T} \tag{2.72}$$

A, R, and T are the optical absorption, reflection, and transmission, respectively. The optical absorption is not to be confused with the *absorbance* or *optical density*, which is given by

$$\mathrm{OD} = -\mathrm{Log}\left(\frac{T}{1-R}\right) = \frac{\alpha L}{2.303}. \tag{2.73}$$

Here, α is the absorption coefficient and L is the film thickness. The APCE is the more useful parameter when evaluating recombination within the semiconductor, while the IPCE is more appropriate to evaluate device-level conversion efficiencies.

Of particular interest is the use of IPCE values measured under monochromatic irradiation to predict the photocurrent under actual solar irradiation:

$$J_{\mathrm{solar}} = \int (\mathrm{IPCE}(\lambda) \times \Phi(\lambda) \times e)\,\mathrm{d}\lambda. \tag{2.74}$$

Here, J_{solar} is the total solar photocurrent in A/m^2 and $\Phi(\lambda)$ is the photon flux of sunlight in photons/m^2/s. The photon flux can be calculated from tabulated solar irradiance data, $E(\lambda)$, via $\Phi(\lambda) = E(\lambda)/(hc/\lambda)$ [75]. An important underlying assumption for (2.74) is that there is a linear relationship between the monochromatic photocurrent and the light intensity. This is often the case, but there are exceptions. A super-linear increase of the photocurrent with light intensity is sometimes observed for nanostructured materials with high concentrations of relatively shallow surface or interface traps; these traps first need to be filled before the charge carriers can reach the electrolyte or the back contact [80]. Conversely, slow charge transport or charge transfer across the interface may lead to an accumulation of free charge carriers at or near the interface. This increases the amount of recombination and will result in a sublinear increase of the photocurrent with light intensity.

2.9 Photoelectrode Requirements and Cell Configurations

The most critical aspect of the design of a photoelectrochemical device for water splitting is the choice of suitable photoanode and/or photocathode materials. Several of the requirements imposed on these materials appear to be in conflict, and certain trade-offs have to be made. In some cases, these trade-offs can be avoided by adopting smart architectures and materials combinations. The remainder of this chapter gives a few general considerations and approaches; more detailed accounts of strategies to cope with these trade-offs are given in later chapters.

2.9.1 Requirements and Trade-offs

Most of the requirements for suitable water-splitting photoanode and/or photocathode materials have already been alluded to in the previous sections of this chapter. They can be summarized as follows [81]:

- Good (visible) light absorption
- High chemical stability in the dark and under illumination
- Band edge positions that straddle the water reduction and oxidation potentials
- Efficient charge transport in the semiconductor
- Low overpotentials for reduction/oxidation of water
- Low cost

The spectral region in which the semiconductor absorbs light is determined by the bandgap of the material. The minimum bandgap is determined by the energy required to split water (1.23 eV) plus the thermodynamic losses (0.3–0.4 eV [82]) and the overpotentials that are required at various points in the system to ensure sufficiently fast reaction kinetics (0.4–0.6 eV) [77, 83]. As a result, the bandgap should be at least 1.9 eV, which corresponds to an absorption onset at 650 nm. Below 400 nm the intensity of sunlight drops rapidly, imposing an upper limit of

3.1 eV on the bandgap. Hence, the optimum value of the bandgap should be somewhere between 1.9 and 3.1 eV, which is within the visible range of the solar spectrum. In a thorough analysis, Murphy et al. suggested an optimum bandgap of 2.03 eV, which would lead to a solar-to-hydrogen efficiency of 16.8% [77].

The chemical stability requirement is a severe one that limits the usefulness of many photo-active materials. Most nonoxide semiconductors either dissolve or form a thin oxide layer that prevents charge transfer across the semiconductor/electrolyte interface. Oxide semiconductors are more stable, but may be prone to anodic or cathodic decomposition, as outlined in Sect. 2.7.4. The general trend is that the stability against (photo)corrosion increases with increasing bandgap. Although this conflicts with the requirement of visible light absorption, a small bandgap and good chemical stability are not necessarily mutually exclusive (although one could argue that a small bandgap is often accompanied by a high valence band energy, which would indeed result in a less stable material).

Few semiconductors fulfill the third requirement, which states that the band edges should straddle the water reduction and oxidation potentials. As shown in Fig. 2.18, the ones that do either have a bandgap that is too large (e.g., $SrTiO_3$, SiC) or are unstable in aqueous solutions (Cu_2O, CdS). This figure also suggests that the band edges of nonoxide semiconductors tend to be better suited toward the reduction of water (high E_C), whereas those of oxide semiconductors favor water oxidation (low E_V). It should be mentioned that the third requirement is actually a bit more stringent than Fig. 2.18 suggests. As pointed out originally by Gerischer [84], and reiterated later by Weber [85], it is the *quasi-Fermi levels* that should straddle the water redox potentials. Although the interpretation of the quasi-Fermi level as a thermodynamic driving force may not be entirely appropriate under all conditions (cf. Sect. 2.7.1), using it within the context of the third requirement seems reasonable.

The fourth requirement, that of efficient charge transport, is easily fulfilled in some materials (TiO_2, WO_3), while in others it is one of the main causes of poor overall conversion efficiencies. A particularly important example of the latter is that of α-Fe_2O_3, which will be discussed in detail in Chap. 4. One can distinguish intrinsic and extrinsic charge transport factors. As outlined in Sect. 2.2, the electronic band structure of the material gives important clues with regard to the intrinsic charge carrier mobilities. Extensive overlap of metal 3d orbitals usually leads to high electron mobilities, whereas the overlap of O-2p orbitals determines the hole mobility in most metal oxides. Extrinsic factors, in particular shallow donors/acceptors and recombination centers, are perhaps even more important. For example, electron transport was found to be rate limiting in spray-deposited, undoped films of α-Fe_2O_3, whereas hole transport was found to be rate limiting after donor doping the material with silicon [86]. The susceptibility to recombination is often expressed as the carrier lifetime, τ_R, or the minority carrier diffusion length, L_D. They are related through

$$L_D \simeq \sqrt{D\tau_R}. \tag{2.75}$$

The diffusivity of the free carriers, D, is related to their mobility, μ, via the Nernst–Einstein equation:

$$D = \frac{kT\mu}{e}. \tag{2.76}$$

Values of L_D range from several nanometers in some oxides to several hundred micrometers in high-purity silicon.

For n-type semiconductors, the fifth requirement implies that hole transfer across the semiconductor/electrolyte interface should be sufficiently fast in order to compete with the anodic decomposition reaction. More generally, interfacial charge transfer should be fast enough to avoid the accumulation carriers at the surface, as this would lead to a decrease of the electric field and a concomitant increase in electron–hole recombination. To improve the kinetics of charge transfer, catalytically active surface species can be added. Examples of effective oxygen evolution catalysts are RuO_2 [87], IrO_x [88], and Co-based compounds [64], whereas Pt, Rh [89], Cr–Rh, RuO_2, or NiO_x [73, 90] are usually employed as a catalysts for hydrogen evolution.

At this time, no single material has been found that meets all these requirements. Four main trade-offs can be identified:

- Bandgap ↔ stability
- Photon collection ↔ charge transport
- Recombination ↔ catalysis
- Performance ↔ cost

The bandgap vs. stability issue has been discussed above. The second trade-off is especially important for semiconductors with a small absorption coefficient, such as metal oxides with an indirect bandgap. In these materials most of the electron–hole pairs are generated far away from the surface, and the carriers recombine before reaching the interface (cf. Fig. 2.23). The third trade-off reflects the tendency of catalytically active surface sites to also act as efficient recombination centers for electronic charge carriers. Moreover, certain co-catalysts for hydrogen or oxygen evolution may also enhance the back-reaction between H_2 and O_2. A well-known example is Pt. This may be a problem for cell designs in which the cross-over of gases generated at the (photo)anode and (photo)cathode cannot be avoided. Interestingly, NiO is an efficient hydrogen evolution catalyst at which the $H_2 + O_2$ back-reaction does not seem to occur [73]. The final trade-off reflects the main reason for the interest in low-cost metal oxides, since the economic viability of state-of-the-art multijunction devices based on, e.g., high quality III–V materials grown by MO-CVD or MBE is doubtful [91].

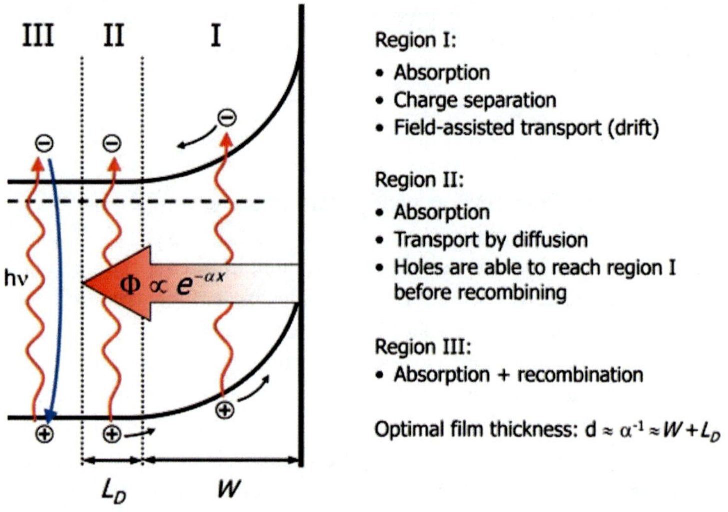

Fig. 2.23 Absorption regions in a semiconducting photoanode. The intensity of the light (Φ) decreases exponentially with distance from the surface (Lambert-Beer law). All charge carriers generated within a distance $L_D + W$ from the surface contribute to the photocurrent (regions I and II), whereas all carriers generated in region III recombine

2.9.2 Nanostructured Electrodes

Nanostructured electrode morphologies can be used to address some of the intrinsic material's limitations and trade-offs mentioned above. The most obvious advantage of a nanostructured morphology is the increase in specific surface area. The concomitant increase in the number of surface sites greatly enhances the overall charge transfer kinetics at the semiconductor/electrolyte interface.

A second advantage is the shorter diffusion path lengths for the photogenerated charge carriers. This is illustrated by the nanowire array photoelectrode shown in Fig. 2.24 (left), where the minority carriers only have to travel half the diameter of the nanowire in order to reach the semiconductor/electrolyte interface [92]. If transport of majority carriers is also an issue, one could employ an array of a conducting wires, which are coated with a thin film of the photoactive semiconductor (Fig. 2.24, right). Here, both minority and majority carriers benefit from the short distance ($\leq d$) that they have to travel before reaching the electrolyte and conducting core, respectively. This is sometimes referred to as a "guest–host" structure, in which the light absorbing nanosized guest material is attached to a nanostructured host scaffold [93]. The advantages are not limited to the highly regular nanowire array architecture shown in Fig. 2.24; nanostructures based on dendritic or cauliflower-type morphologies or randomly packed spheres offer similar advantages [78, 93, 94]. Such structures are discussed in more detail in Chaps. 4 and 9.

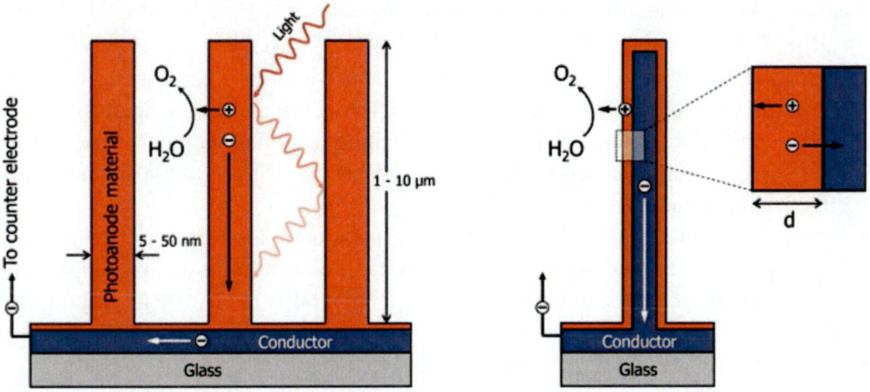

Fig. 2.24 Nanowire array photoanode (*left*), and an array based on highly conducting nanowires coated with a thin photoanode film (*right*)

A third possible advantage of using materials with nanoscale dimensions is the occurrence of quantum size effects. As first described by Brus in the early 1980s, spatial confinement of charge carriers to a volume that is less than their De Broglie wavelength results in a widening of the bandgap [9]. An approximate expression for the bandgap enlargement as a function of particle radius (R) is given by [95]:

$$E_g \simeq E_g^0 + \frac{h^2\pi^2}{2R^2}\left(\frac{1}{m_e^*} + \frac{1}{m_h^*}\right) - \frac{1.8e^2}{\varepsilon_0\varepsilon_r R}. \tag{2.77}$$

Here, E_g^0 is the bulk bandgap and m_e^* and m_h^* are the effective masses for the electrons and holes, respectively. The first term describes the quantum confinement effect, while the second (smaller) term describes the attractive Coulomb interaction between the electrons and holes. The negative sign of the second term shows that it in fact acts against the bandgap enlargement. While the resulting blue shift of the absorption spectrum is usually not desirable for PEC materials, the widening of the gap shifts the conduction band edge positively with respect to the hydrogen reduction energy. This would decrease the required bias potential and increase the solar-to-hydrogen conversion efficiency (2.69) of materials such as α-Fe_2O_3 and WO_3, for which E_C is located below $E^0(H_2/H^+)$ – see Fig. 2.18. Although one recent report claims a 0.3–0.6 eV upward shift of the conduction band of α-Fe_2O_3 nanowires [96], no direct photoelectrochemical evidence for such a shift in α-Fe_2O_3 has been reported so far.

It should be noted that quantum size effects are difficult to induce in most metal oxides. This is due to their relatively large electron and hole effective masses (cf. Table 2.1), which would require exceedingly small particle sizes (<2–3 nm) for any confinement effect to occur. ZnO is a well-known exception that shows appreciable bandgap enlargement at particle sizes below 6 nm [11, 97]. The bandgap enlargement is generally asymmetric, with $\Delta E_C = h^2\pi^2/(2R^2m_e^*)$ and $\Delta E_V = h^2\pi^2/(2R^2m_h^*)$. This asymmetry is less evident in ZnO ($m_e^* = 0.24m_e$, $m_h^* = 0.24m_e$) than in anatase TiO_2 ($m_h^* = 0.8m_e$, $m_e^* = 10m_e$), for which any bandgap enlargement appears to be solely due to a shift in E_V [11]. Clearly, if quantum confinement is to be used to shift the conduction band edge to higher energies without increasing the bandgap too much, one would prefer to have a semiconductor with a small electron effective mass and a large hole effective mass.

An important aspect of nanosized semiconductors is the absence of band bending. Since the number of ionized donors or acceptors in nanosized particles is very small, such particles cannot sustain a large built-in electric field (cf. (2.41) and Fig. 2.13, left). The maximum attainable potential drop within a spherical semiconductor particle with radius R and a donor (or acceptor) density N_D is given by [98]:

$$\phi_{SC} = \frac{-eN_D R^2}{6\varepsilon_0\varepsilon_r}. \tag{2.78}$$

From this expression, the band bending in a 10 nm α-Fe_2O_3 particle ($R = 5$ nm, $\varepsilon_r = 25$) with a relatively high donor density of 10^{18} cm^{-3} is limited to ~3 mV. This clearly shows that the electric field in the space charge of a nanoparticle cannot play a significant role in the separation of photogenerated electrons and holes. Fortunately, most carriers are nevertheless able to reach the interface since the minority carrier diffusion length is – with few exceptions – much larger than the radius of the nanoparticle. As a result, the charge transfer kinetics at the semiconductor/electrolyte interface usually dominate the overall charge separation and recombination kinetics in nanoparticles.

2.9.3 PEC Device Configurations

When the conduction and valence bands do not straddle the water reduction and oxidation potentials, or if the available overpotentials for water splitting are insufficient, a combination of two or more semiconductors may offer a solution. In such a device, one semiconductor acts as a regular photoanode or photocathode. The other semiconductor(s) function(s) as either a complementary photoelectrode, or as an integrated photovoltaic device that provides a bias voltage. Figure 2.25 shows some examples of possible configurations.

Configuration (A) is the one we have used throughout this chapter to illustrate the main principles of photoelectrochemical water splitting. Strontium titanate, $SrTiO_3$, is one of very few materials that have been observed to split water in this configuration [99]. However, the efficiency is less than 1% due to its large bandgap (3.2 eV). Cuprous oxide, Cu_2O, and p-type SiC are candidates for configuration (B). While Cu_2O has been found to suffer from photoreduction to metallic Cu [100, 101],

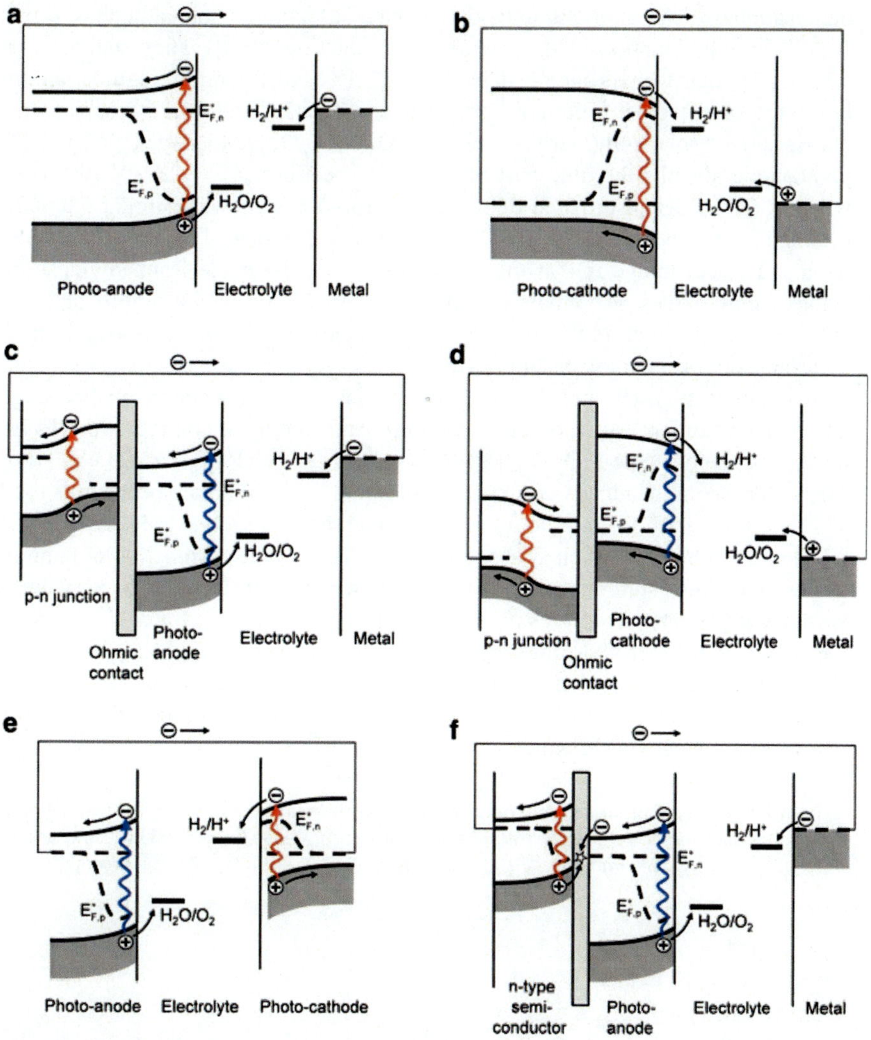

Fig. 2.25 Examples of possible PEC configurations under illumination. *Top row*: Standard single-semiconductor devices based on a photoanode (**a**) or photocathode (**b**) with a metal counter electrode. *Middle row*: Monolithic devices based on a photoanode (**c**) or photocathode (**d**) biased with an integrated p–n junction. *Bottom row*: p–n junction photoelectrochemical device (**e**), and an n–n heterojunction PEC device based on a photoanode deposited on top of a second n-type semiconductor that "boosts" the energy of the electrons (**f**)

other reports indicate that it may be stable under the right conditions [68, 102]. The 4H polytype of silicon carbide is one of the few chemically stable photocathode materials able to split water, but its efficiency is also low ($E_g = 3.26$ eV) [67].

In configurations (C) and (D) a bias voltage is applied to the photoanode and photocathode, respectively, through an integrated p–n junction. Turner et al. have demonstrated such a monolithic tandem device based on a photocathode of p-type $GaInP_2$ in combination with a GaAs p–n junction [103]. They achieved an impressive solar-to-hydrogen efficiency of 12.4%, but the rapid degradation and high cost of the device limit its practical use. Grätzel and Augustynski used a dye-sensitized solar cell (DSSC) to bias a WO_3 or Fe_2O_3 photoanode [91, 104]. The photoanode absorbs the blue part of the solar spectrum ($E_g = 2.6$ eV for WO_3), whereas the DSSC absorbs red light. This approach is somewhat similar to configuration (C), except for the fact that the p–n junction is replaced by a separate DSSC device. The advantage is that the entire system can be made from low-cost and stable metal oxides. A remaining challenge is to address the significant light scattering of the individual (nanostructured) components, which severely limits the efficiency of the total system [105].

In configurations (E) and (F) an additional bias potential is generated by a single additional absorber instead of a p–n junction. An example of configuration (E) is a device based on n-type TiO_2 ($E_g = 3.2$ eV) and p-type GaP ($E_g = 2.25$ eV) [106]. The difference in bandgaps ensures that a large part of the solar spectrum is used. This is not the case for p–n homojunctions, and devices such as n-Fe_2O_3/p-Fe_2O_3 [107] will not be very efficient because of this. The n–n heterojunction of configuration (F) was first studied many years ago, and was not considered to be very useful since the photogenerated holes in the buried n-type semiconductor could not reach the photoanode/electrolyte interface [108]. More recently, however, Yang et al. showed that n-type Si underneath n-type TiO_2 "boosts" the energy of TiO_2 conduction band electrons [50]. This increases the available overpotential for the reduction of water and improves the kinetics.

Acknowledgments The author gratefully acknowledges Fatwa F. Abdi for critical reading of the manuscript, and the NWO-ACTS Sustainable Hydrogen program (project 053.61.009) and the European Commission's Framework 7 program (NanoPEC, Project 227179) for support.

References

1. Chelikowsky, J.R., Cohen, M.L.: Electronic-structure of silicon. Phys. Rev. B **10**, 5095–5107 (1974)
2. Fujishima, A., Honda, K.: Electrochemical photolysis of water at a semiconductor electrode. Nature **238**, 37–38 (1972)
3. Osterloh, F.E.: Inorganic materials as catalysts for photochemical splitting of water. Chem. Mater. **20**, 35–54 (2008)
4. Stoyanov, E., Langenhorst, F., Steinle-Neumann, G.: The effect of valence state and site geometry on Ti L-3, L-2 and O-K electron energy-loss spectra of Ti_xO_y phases. Am. Miner. **92**, 577–586 (2007)
5. Fischer, D.W.: X-ray band spectra and molecular-orbital structure of rutile TiO_2. Phys. Rev. B **5**, 4219 (1972)
6. Hoffmann, R.: Solids and Surfaces – A Chemist's View of Bonding in Extended Structures. Wiley-VCH, Weinheim (1988)

7. Tauc, J., Grigorov, R., Vancu, A.: Optical properties and electronic structure of amorphous germanium. Phys. Stat. Sol. **15**, 627 (1966)
8. Sze, S.M.: Physics of Semiconductor Devices. Wiley, New York (1981)
9. Brus, L.E.: Electron electron and electron–hole interactions in small semiconductor crystallites – the size dependence of the lowest excited electronic state. J. Chem. Phys. **80**, 4403–4409 (1984)
10. Deskins, N.A., Dupuis, M.: Intrinsic hole migration rates in TiO_2 from density functional theory. J. Phys. Chem. C **113**, 346–358 (2009)
11. Enright, B., Fitzmaurice, D.: Spectroscopic determination of electron and hole effective masses in a nanocrystalline semiconductor film. J. Phys. Chem. **100**, 1027–1035 (1996)
12. Berak, J.M., Sienko, M.J.: Effect of oxygen-deficiency on electrical transport properties of tungsten trioxide crystals. J. Solid State Chem. **2**, 109–133 (1970)
13. Bosman, A.J., Vandaal, H.J.: Small-polaron versus band conduction in some transition-metal oxides. Adv. Phys. **19**, 1 (1970)
14. Matsuzaki, K., Nomura, K., Yanagi, H., Kamiya, T., Hirano, M., Hosono, H.: Epitaxial growth of high mobility Cu_2O thin films and application to p-channel thin film transistor. Appl. Phys. Lett. **93**, 202107 (2008)
15. Kawazoe, H., Yasukawa, M., Hyodo, H., Kurita, M., Yanagi, H., Hosono, H.: P-type electrical conduction in transparent thin films of $CuAlO_2$. Nature **389**, 939–942 (1997)
16. Kudo, A., Yanagi, H., Hosono, H., Kawazoe, H.: $SrCu_2O_2$: a p-type conductive oxide with wide band gap. Appl. Phys. Lett. **73**, 220–222 (1998)
17. Bessekhouad, Y., Gabes, Y., Bouguelia, A., Trari, M.: The physical and photo electrochemical characterization of the crednerite $CuMnO_2$. J. Mater. Sci. **42**, 6469–6476 (2007)
18. Pierret, R.F.: Advanced semiconductor fundamentals. Addison-Wesley, Reading (1989)
19. Asahi, R., Morikawa, T., Ohwaki, T., Aoki, K., Taga, Y.: Photocatalysts sensitive to visible light – response. Science **295**, 627–627 (2002)
20. Maruska, H.P., Ghosh, A.K.: Transition-metal dopants for extending the response of titanate photoelectrolysis anodes. Solar Energy Mater. **1**, 237–247 (1979)
21. Salvador, P.: Analysis of the physical properties of TiO_2-Be electrodes in the photoassisted oxidation of water. Solar Energy Mater. **6**, 241–250 (1982)
22. Henrich, V.E., Cox, P.A.: The surface science of metal oxides. Cambridge University Press, Cambridge (1994)
23. Kato, H., Kudo, A.: Visible-light-response and photocatalytic activities of TiO_2 and $SrTiO_3$ photocatalysts codoped with antimony and chromium. J. Phys. Chem. B **106**, 5029–5034 (2002)
24. Howe, A.T., Hawkins II, R.T., Fleisch, T.H.: Photoelectrochemical cells of the electrolyte–metal–insulator–semiconductor (EMIS) configuration. I. Metal thickness and coverage effects in the Pt/silicon oxide/n-Si system. J. Electrochem. Soc. **133**, 1369–1375 (1986)
25. Piazza, F., Pavesi, L., Henini, M., Johnston, D.: Effect of as overpressure on Si-doped (111) a gaas grown by molecular-beam epitaxy – a photoluminescence study. Semicond. Sci. Technol. **7**, 1504–1507 (1992)
26. Kröger, F.A.: The chemistry of imperfect crystals. North-Holland Publishing Co., Amsterdam (1964)
27. Chiang, Y.-M., Birnie III, D., Kingery, D.W.: Physical Ceramics. Wiley, New York (1997)
28. Smyth, D.M.: The Defect Chemistry of Metal Oxides. Oxford University Press, New York (2000)
29. Jorand Sartoretti, C., Alexander, B.D., Solarska, R., Rutkowska, W.A., Augustynski, J., Cerny, R.: Photoelectrochemical oxidation of water at transparent ferric oxide film electrodes. J. Phys. Chem. B **109**, 13685–13692 (2005)
30. Liang, Y., Enache, C.S., Van de Krol, R.: Photoelectrochemical characterization of sprayed α-Fe_2O_3 thin films: Influence of Si-doping and SnO_2 interfacial layer. Int. J. Photoenergy (2008). doi:10.1155/2008/739864

31. Kennedy, J.H., Shinar, R., Ziegler, J.P.: Alpha-Fe_2O_3 photoanodes doped with silicon. J. Electrochem. Soc. **127**, 2307–2309 (1980)
32. Liang, Y., Tsubota, T., Mooij, L.P.A., van de Krol, R.: Highly improved quantum efficiencies for thin film $BiVO_4$ photoanodes. J. Phys. Chem. C **115**, 17594–17598 (2011)
33. Das Mulmi, D., Sekiya, T., Kamiya, N., Kurita, S., Murakami, Y., Kodaira, T.: Optical and electric properties of Nb-doped anatase TiO_2 single crystal. J. Phys. Chem. Solids **65**, 1181–1185 (2004)
34. Anpo, M., Takeuchi, M.: The design and development of highly reactive titanium oxide photocatalysts operating under visible light irradiation. J. Catal. **216**, 505–516 (2003)
35. Kudo, A., Ueda, K., Kato, H., Mikami, I.: Photocatalytic O_2 evolution under visible light irradiation on $BiVO_4$ in aqueous $AgNO_3$ solution. Catal. Lett. **53**, 229–230 (1998)
36. Sayama, K., Nomura, A., Arai, T., Sugita, T., Abe, R., Yanagida, M., Oi, T., Iwasaki, Y., Abe, Y., Sugihara, H.: Photoelectrochemical decomposition of water into H_2 and O_2 on porous $BiVO_4$ thin-film electrodes under visible light and significant effect of Ag ion treatment. J. Phys. Chem. B **110**, 11352–11360 (2006)
37. Pierret, R.F.: Semiconductor Device Fundamentals. Addison Wesley, Reading (1996)
38. Nicollian, E.H., Brews, J.R.: MOS (metal oxide semiconductor) Physics and Technology. Wiley, Hoboken (2003)
39. Onari, S., Arai, T., Kudo, K.: Ir lattice-vibrations and dielectric-dispersion in alpha-Fe_2O_3. Phys. Rev. B **16**, 1717–1721 (1977)
40. Rosso, K.M., Smith, D.M.A., Dupuis, M.: An ab initio model of electron transport in hematite (alpha-Fe_2O_3) basal planes. J. Chem. Phys. **118**, 6455–6466 (2003)
41. Morrison, S.R.: Electrochemistry of semiconductor and oxidized metal electrodes. Plenum, New York (1980)
42. Bockris, J.O.M., Reddy, A.K.N., Galboa-Aldeco, M.E.: Modern Electrochemistry 2A – Fundamentals of Electrodics. Springer, New York (2001)
43. Memming, R.: Semiconductor Electrochemistry. Wiley, New York (2000)
44. McNaught, A.D., Wilkinson, A.: IUPAC Compendium of Chemical Terminology, 2nd edn. (the "Gold Book"). Blackwell Scientific Publications, Oxford (1997). http://goldbook.iupac.org/S05917.html. Accessed 8 Sep 2010
45. Reichman, J.: The current–voltage characteristics of semiconductor-electrolyte junction photo-voltaic cells. Appl. Phys. Lett. **36**, 574–577 (1980)
46. Peter, L.M., Li, J., Peat, R.: Surface recombination at semiconductor electrodes. 1. Transient and steady-state photocurrents. J. Electroanal. Chem. **165**, 29–40 (1984)
47. Gandia, J., Pujadas, M., Salvador, P.: Electrolyte electroreflectance – easy and reliable flat-band potential measurements. J. Electroanal. Chem. **244**, 69–79 (1988)
48. Radecka, M., Sobas, P., Wimbicka, M., Rekas, M.: Photoelectrochemical properties of undoped and Ti-doped WO_3. Phys. B Condens. Matter **364**, 85–92 (2005)
49. Hagfeldt, A., Björksten, U., Grätzel, M.: Photocapacitance of nanocrystalline oxide semiconductor films: band-edge movement in mesoporous TiO_2 electrodes during UV illumination. J. Phys. Chem. **100**, 8045–8048 (1996)
50. Hwang, Y.J., Boukai, A., Yang, P.D.: High density n-Si/n-TiO_2 core/shell nanowire arrays with enhanced photoactivity. Nano Lett. **9**, 410–415 (2009)
51. Marcus, R.A.: Theory of oxidation–reduction reactions involving electron transfer 1. J. Chem. Phys. **24**, 966–978 (1956)
52. Marcus, R.A.: Chemical + electrochemical electron-transfer theory. Ann. Rev. Phys. Chem. **15**, 155 (1964)
53. Gerischer, H.: Charge transfer processes at semiconductor-electrolyte interfaces in connection with problems of catalysis. Surf. Sci. **18**, 97 (1969)
54. Duret, A., Grätzel, M.: Visible light-induced water oxidation on mesoscopic alpha-Fe_2O_3 films made by ultrasonic spray pyrolysis. J. Phys. Chem. B **109**, 17184–17191 (2005)
55. Jarrett, H.S.: Photocurrent conversion efficiency in a Schottky-barrier. J. Appl. Phys. **52**, 4681–4689 (1981)

56. Wilson, R.H.: Model for current–voltage curve of photoexcited semiconductor electrodes. J. Appl. Phys. **48**, 4292–4297 (1977)
57. Salvador, P.: Kinetic approach to the photocurrent transients in water photoelectrolysis at n-TiO_2 electrodes. I. Analysis of the ratio of the instantaneous to steady-state photocurrent. J. Phys. Chem. **89**, 3863–3869 (1985)
58. Kisumi, T., Tsujiko, A., Murakoshi, K., Nakato, Y.: Crystal-face and illumination intensity dependences of the quantum efficiency of photoelectrochemical etching, in relation to those of water photooxidation, at n-TiO_2 (rutile) semiconductor electrodes. J. Electroanal. Chem. **545**, 99–107 (2003)
59. Imanishi, A., Okamura, T., Ohashi, N., Nakamura, R., Nakato, Y.: Mechanism of water photooxidation reaction at atomically flat TiO_2 (rutile) (110) and (100) surfaces: dependence on solution pH. J. Am. Chem. Soc. **129**, 11569–11578 (2007)
60. Rossmeisl, J., Qu, Z.W., Zhu, H., Kroes, G.J., Norskov, J.K.: Electrolysis of water on oxide surfaces. J. Electroanal. Chem. **607**, 83–89 (2007)
61. Valdes, A., Qu, Z.W., Kroes, G.J., Rossmeisl, J., Norskov, J.K.: Oxidation and photo-oxidation of water on TiO_2 surface. J. Phys. Chem. C **112**, 9872–9879 (2008)
62. Valdes, A., Kroes, G.J.: First principles study of the photo-oxidation of water on tungsten trioxide (WO_3). J. Chem. Phys. **130**(11), 114701 (2009)
63. Harriman, A., Pickering, I.J., Thomas, J.M., Christensen, P.A.: Metal-oxides as heterogeneous catalysts for oxygen evolution under photochemical conditions. J. Chem. Soc. Faraday Trans. I **84**, 2795–2806 (1988)
64. Kanan, M.W., Nocera, D.G.: In situ formation of an oxygen-evolving catalyst in neutral water containing phosphate and Co^{2+}. Science **321**, 1072–1075 (2008)
65. Jiao, F., Frei, H.: Nanostructured cobalt oxide clusters in mesoporous silica as efficient oxygen-evolving catalysts. Angew. Chem. Int. Ed. **48**, 1841–1844 (2009)
66. Kay, A., Cesar, I., Grätzel, M.: New benchmark for water photooxidation by nanostructured alpha-Fe_2O_3 films. J. Am. Chem. Soc. **128**, 15714–15721 (2006)
67. Van Dorp, D.H., Hijnen, N., Di Vece, M., Kelly, J.J.: SiC: a photocathode for water splitting and hydrogen storage. Angew. Chem. Int. Ed. **48**, 6085–6088 (2009)
68. De Jongh, P.E., Vanmaekelbergh, D., Kelly, J.J.: Photoelectrochemistry of electrodeposited Cu_2O. J. Electrochem. Soc. **147**, 486–489 (2000)
69. Rajeshwar, K., McConnell, R., Licht, S.: Solar hydrogen generation – toward a renewable energy future. Springer, New York (2008)
70. Bak, T., Nowotny, J., Rekas, M., Sorrell, C.C.: Photo-electrochemical hydrogen generation from water using solar energy. Materials-related aspects. Int. J. Hydrogen Energy **27**, 991–1022 (2002)
71. Park, S.M., Barber, M.E.: Thermodynamic stabilities of semiconductor electrodes. J. Electroanal. Chem. **99**, 67–75 (1979)
72. Scaife, D.E.: Oxide semiconductors in photoelectrochemical conversion of solar energy. Solar Energy **25**, 41–54 (1980)
73. Kudo, A., Miseki, Y.: Heterogeneous photocatalyst materials for water splitting. Chem. Soc. Rev. **38**, 253–278 (2009)
74. Ritterskamp, P., Kuklya, A., Wustkamp, M.A., Kerpen, K., Weidenthaler, C., Demuth, M.: A titanium disilicide derived semiconducting catalyst for water splitting under solar radiation – reversible storage of oxygen and hydrogen. Angew. Chem. Int. Ed. **46**, 7770–7774 (2007)
75. Solar Irradiance Data, ASTM-G173-03 (AM1.5, global tilt): http://rredc.nrel.gov/solar/spectra/am1.5/. Accessed 11 Aug 2010
76. Chen, Z.B., Jaramillo, T.F., Deutsch, T.G., Kleiman-Shwarsctein, A., Forman, A.J., Gaillard, N., Garland, R., Takanabe, K., Heske, C., Sunkara, M., McFarland, E.W., Domen, K., Miller, E.L., Turner, J.A., Dinh, H.N.: Accelerating materials development for photoelectrochemical hydrogen production: standards for methods, definitions, and reporting protocols. J. Mater. Res. **25**, 3–16 (2010)

77. Murphy, A.B., Barnes, P.R.F., Randeniya, L.K., Plumb, I.C., Grey, I.E., Horne, M.D., Glasscock, J.A.: Efficiency of solar water splitting using semiconductor electrodes. Int. J. Hydrogen Energy **31**, 1999–2017 (2006)
78. Santato, C., Ulmann, M., Augustynski, J.: Enhanced visible light conversion efficiency using nanocrystalline WO_3 films. Adv. Mater. **13**, 511 (2001)
79. Kavan, L., Grätzel, M.: Highly efficient semiconducting TiO_2 photoelectrodes prepared by aerosol pyrolysis. Electrochim. Acta **40**, 643–652 (1995)
80. De Jongh, P.E., Vanmaekelbergh, D.: Trap-limited electronic transport in assemblies of nanometer-size TiO_2 particles. Phys. Rev. Lett. **77**, 3427–3430 (1996)
81. Van de Krol, R., Liang, Y.Q., Schoonman, J.: Solar hydrogen production with nanostructured metal oxides. J. Mater. Chem. **18**, 2311–2320 (2008)
82. Weber, M.F., Dignam, M.J.: Splitting water with semiconducting photoelectrodes efficiency considerations. Int. J. Hydrogen Energy **11**, 225–232 (1986)
83. Bolton, J.R., Strickler, S.J., Connolly, J.S.: Limiting and realizable efficiencies of solar photolysis of water. Nature **316**, 495–500 (1985)
84. Gerischer, H.: Electrochemical behavior of semiconductors under illumination. J. Electrochem. Soc. **113**, 1174 (1966)
85. Weber, M.F., Dignam, M.J.: Efficiency of splitting water with semiconducting photoelectrodes. J. Electrochem. Soc. **131**, 1258–1265 (1984)
86. Liang, Y., Enache, C.S., Van de Krol, R.: Photoelectrochemical characterization of sprayed α-Fe_2O_3 thin films: Influence of Si-doping and SnO_2 interfacial layer. Int. J. Photoenergy (2008). doi:10.1155/2008/739864
87. Licht, S., Wang, B., Mukerji, S., Soga, T., Umeno, M., Tributsch, H.: Efficient solar water splitting, exemplified by RuO_2-catalyzed AlGaAs/Si photoelectrolysis. J. Phys. Chem. B **104**, 8920–8924 (2000)
88. Youngblood, W.J., Lee, S.H.A., Kobayashi, Y., Hernandez-Pagan, E.A., Hoertz, P.G., Moore, T.A., Moore, A.L., Gust, D., Mallouk, T.E.: Photoassisted overall water splitting in a visible light-absorbing dye-sensitized photoelectrochemical cell. J. Am. Chem. Soc. **131**, 926 (2009)
89. Maeda, K., Teramura, K., Lu, D.L., Takata, T., Saito, N., Inoue, Y., Domen, K.: Photocatalyst releasing hydrogen from water – enhancing catalytic performance holds promise for hydrogen production by water splitting in sunlight. Nature **440**, 295–295 (2006)
90. Kato, H., Asakura, K., Kudo, A.: Highly efficient water splitting into H_2 and O_2 over lanthanum-doped $NaTaO_3$ photocatalysts with high crystallinity and surface nanostructure. J. Am. Chem. Soc. **125**, 3082–3089 (2003)
91. Grätzel, M.: Photoelectrochemical cells. Nature **414**, 338–344 (2001)
92. Mor, G.K., Prakasam, H.E., Varghese, O.K., Shankar, K., Grimes, C.A.: Vertically oriented Ti-Fe-O nanotube array films: toward a useful material architecture for solar spectrum water photoelectrolysis. Nano Lett. **7**, 2356–2364 (2007)
93. Sivula, K., Le Formal, F., Grätzel, M.: WO_3–Fe_2O_3 photoanodes for water splitting: a host scaffold guest absorber approach. Chem. Mater. **21**, 2862–2867 (2009)
94. Cesar, I., Kay, A., Martinez, J.A.G., Grätzel, M.: Translucent thin film Fe_2O_3 photoanodes for efficient water splitting by sunlight: nanostructure-directing effect of Si-doping. J. Am. Chem. Soc. **128**, 4582–4583 (2006)
95. Brus, L.: Electronic wave-functions in semiconductor clusters – experiment and theory. J. Phys. Chem. **90**, 2555–2560 (1986)
96. Vayssieres, L., Sathe, C., Butorin, S.M., Shuh, D.K., Nordgren, J., Guo, J.H.: One-dimensional quantum-confinement effect in alpha-Fe_2O_3 ultrafine nanorod arrays. Adv. Mater. **17**, 2320 (2005)
97. Lin, K.F., Cheng, H.M., Hsu, H.C., Lin, L.J., Hsieh, W.F.: Band gap variation of size-controlled ZnO quantum dots synthesized by sol-gel method. Chem. Phys. Lett. **409**, 208–211 (2005)

98. Goossens, A.: Potential distribution in semiconductor particles. J. Electrochem. Soc. **143**, L131–L133 (1996)
99. Mavroides, J.G., Kafalas, J.A., Kolesar, D.F.: Photoelectrolysis of water in cells with $SrTiO_3$ anodes. Appl. Phys. Lett. **28**, 241–243 (1976)
100. Schoppel, H.R., Gerischer, H.: Cathodic reduction of Cu-I oxide electrodes as example for reduction mechanism of semiconductor crystal. Ber. Bunsenges. Phys. Chem. **75**, 1237 (1971)
101. Nagasubramanian, G., Gioda, A.S., Bard, A.J.: Semiconductor electrodes. 37. Photoelectrochemical behavior of p-type Cu_2O in acetonitrile solutions. J. Electrochem. Soc. **128**, 2158–2164 (1981)
102. Hara, M., Kondo, T., Komoda, M., Ikeda, S., Shinohara, K., Tanaka, A., Kondo, J.N., Domen, K.: Cu_2O as a photocatalyst for overall water splitting under visible light irradiation. Chem. Commun. **3**, 357–358 (1998)
103. Khaselev, O., Turner, J.A.: A monolithic photovoltaic-photoelectrochemical device for hydrogen production via water splitting. Science **280**, 425–427 (1998)
104. Grätzel, M.: Mesoscopic solar cells for electricity and hydrogen production from sunlight. Chem. Lett. **34**, 8–13 (2005)
105. Brillet, J., Cornuz, M., Le Formal, F., Yum, J.H., Grätzel, M., Sivula, K.: Examining architectures of photoanode-photovoltaic tandem cells for solar water splitting. J. Mater. Res. **25**, 17–24 (2010)
106. Nozik, A.J.: p–n photoelectrolysis cells. Appl. Phys. Lett. **29**, 150–153 (1976)
107. Turner, J.E., Hendewerk, M., Somorjai, G.A.: The photodissociation of water by doped iron-oxides – the unbiased P/N assembly. Chem. Phys. Lett. **105**, 581–585 (1984)
108. Kohl, P.A., Frank, S.N., Bard, A.J.: Semiconductor electrodes. 11. Behavior of n-type and p-type single-crystal semiconductors covered with thin normal-TiO_2 films. J. Electrochem. Soc. **124**, 225–229 (1977)

Chapter 3
Photoelectrochemical Measurements

Roel van de Krol

3.1 Introduction

Despite the many advances in computational materials science in the last decade, photoelectrochemical water splitting is still very much an experimental field. This is because many of the relevant properties of photoelectrodes are determined by factors that are either difficult to control or not yet well understood, which makes it difficult to predict a priori their influence on the actual performance. An example of this is the presence of certain defects, such as dislocations or impurities. These are often impossible to avoid, especially when low-temperature and low-cost synthesis methods are used. An important stage in the development process of a photoelectrode is therefore to systematically optimize the synthesis procedure[1] in order to achieve a maximum photoresponse.

A common metric to determine the optimal photoresponse is the photocurrent density under 1 sun illumination (1,000 W/m^2). To be able to compare the photocurrent for different samples, two criteria should be fulfilled. The main one is that the measured data reflect the properties of the sample, as opposed to artifacts or limitations caused by the measurement setup. For example, mass transport limitations should be avoided by careful design of the electrochemical cell, and Ohmic losses and pH fluctuations can be avoided by choosing a suitable electrolyte. For a meaningful comparison of performance numbers reported by different laboratories one should also fulfill the second criterion, which is that the measurements must be

[1] Examples of optimization parameters for photoelectrode synthesis are deposition temperature, substrate cleaning method, additives used to optimize the boiling point, surface tension, water content or viscosity of a solution, postdeposition heat treatments to improve the crystallinity, dipping as-prepared samples in a precursor solution to improve interparticle contacts, etc.

R. van de Krol (✉)
Faculty of Applied Sciences, Department Chemical Engineering/Materials for Energy Conversion and Storage, Delft University of Technology, P.O. Box 5045, 2600 GA Delft, The Netherlands
e-mail: r.vandekrol@tudelft.nl

R. van de Krol and M. Grätzel (eds.), *Photoelectrochemical Hydrogen Production*, Electronic Materials: Science & Technology 102, DOI 10.1007/978-1-4614-1380-6_3,

performed under the exact same conditions. While seemingly trivial, there are many performance claims in the literature that fail to meet this requirement. This can be due to lack of, e.g., proper equipment (calibrated light sources in particular), but also to missing information in the paper to which the authors compare their results. To avoid this, one should be careful to mention at least the sample area and thickness, the intensity and type of light source used, the electrolyte composition and pH, the cell geometry and counter electrode area, and the applied potential.

In addition to performance evaluations, many photoelectrochemical experiments are aimed to identify performance-limiting steps or to determine certain materials properties. Examples of the latter are donor or acceptor densities and the flatband potential of a material, which can be determined by electrochemical impedance measurements. The challenge with these measurements is that they always yield data, but that it can be difficult – and sometimes even impossible – to translate the measured data to the desired materials parameters. Carefully performed control experiments and a good basic understanding of the measurement equipment – in particular, the potentiostat and the frequency response analyzer (FRA) – are essential for obtaining meaningful results.

Much information on photoelectrochemical measurements and measurement equipment is available, but it is often scattered throughout the various papers, books, and technical notes of manufacturers. Moreover, many practical methods and tricks are passed on by communication between colleagues or between supervisors and students, and cannot be found in the literature. This chapter attempts to bring some of the basic information on photoelectrochemical measurement techniques and equipment together and aims to serve as a quick start for students and other newcomers in this exciting field.

3.2 The Photoelectrochemical Cell

One of the key components of a photoelectrochemical setup is the photoelectrochemical cell (further abbreviated as *PEC* cell) in which the photoanode or photocathode is mounted.[2] At bare minimum, a PEC cell consists of a reservoir to hold the electrolyte, the sample to be studied (the *working* electrode), a *counter* electrode to supply current, an optically transparent window that allows the sample to be illuminated, and facilities to electrically connect both electrodes to the outside world. Other components that are often present are a *reference* electrode, a quasi-reference electrode (see Sects. 3.2.4 and 3.6.6), and one or more inputs and outputs for a gas circulation system and/or gas purging of the electrolyte. In addition, one may encounter a magnetic stir bar, and a membrane that separates the anode and cathode compartments in order to avoid mixing of the evolved oxygen and hydrogen gases.

---

[2] Most photoelectrochemical cells used in PEC research are designed to study the properties of a single photoelectrode. PEC cells in which more than one electrode is illuminated at once (e.g., tandem cells) are somewhat more complicated and will not be considered here.

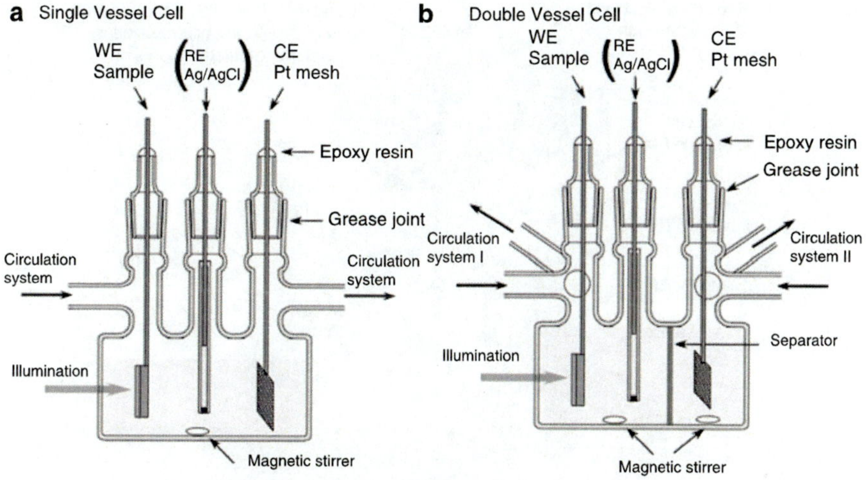

Fig. 3.1 Example of a conventional photoelectrochemical cell with a single (**a**) or double (**b**) compartment. From ref. [1], reproduced with permission

3.2.1 Cell Designs

An example of a common PEC cell design is shown in Fig. 3.1. This design is based on that of conventional electrochemical cells used for, e.g., corrosion studies. The main aspect that distinguishes the PEC cell from a standard electrochemical cell is the presence of an optically transparent window through which the sample can be illuminated. Since normal glass shows a transmission cut-off below ~350 nm, fused silica (amorphous SiO_2) is most often used as a window material in PEC cells. UV-grade fused silica is transparent in the entire range from UV to near-IR with a transmission >90% between 200 nm and 2.2 μm.[3] It shows excellent chemical stability in both acidic and alkaline aqueous solutions (except in HF) and is relatively cheap.[4] Cells such as the one shown in Fig. 3.1 are commercially available, fairly robust (despite being constructed entirely from glass), and easy to use when it comes to mounting the electrodes. There are also some disadvantages. As it is made from glass, the cell can be rather large and may require a large amount of electrolyte (>100 mL). This quickly becomes impractical when screening large numbers of photoelectrodes. Moreover, the distance between the working and counter electrodes usually exceeds 5 cm and may give rise to Ohmic voltage losses at higher current densities (see Sect. 3.2.5). A more practical issue is sample

[3] Except for a small dip around ~1400 nm, depending on the purity grade.

[4] A ∅50 mm, 3 mm thick window of UV-grade fused silica costs ~40 EUR (in 2010). Note that fused silica is often incorrectly referred to as "quartz," which is the crystalline form of SiO_2 that shows slightly better transmission than fused silica at a much higher price.

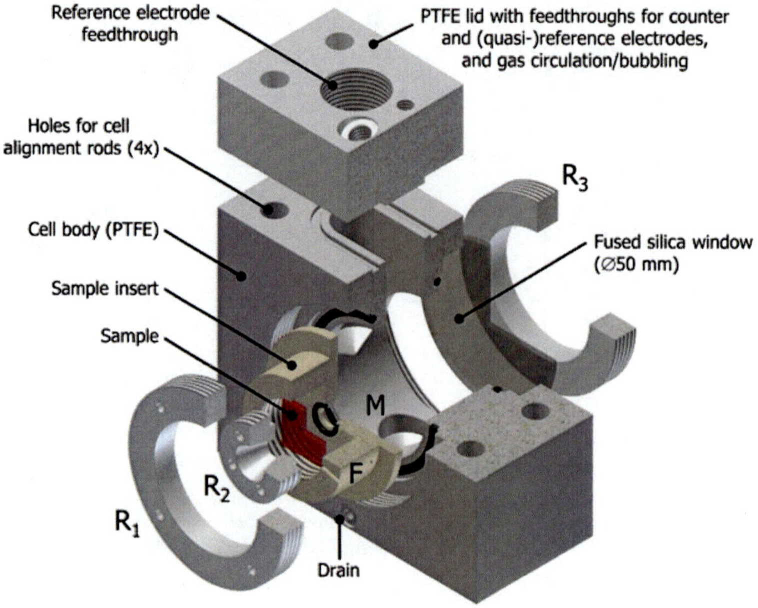

Fig. 3.2 Photoelectrochemical cell designed by the author's group at TU Delft

mounting: since the sample is fully immersed in the electrolyte, it needs to be partially masked with, e.g., epoxy resin to avoid any direct contact between the electrolyte and the conducting back-contact and/or the connecting wire.

To address these issues, many groups design their own photoelectrochemical cells that are specifically optimized for the types of samples and experiments that are carried out in their laboratories. Figure 3.2 shows an example of a cell that was developed in the author's laboratory. It has a modest volume (~30 mL), and the sample is mounted in a detachable sample insert for ease of handling. The sample insert is pressed against an O-ring seal on the cell body via ring that is threaded at the outside (R_1). A similar ring (R_3) is used to mount the fused silica window at the front of the cell. The sample surface is pressed against an O-ring with another threaded insert (R_2). Electrical contact to the sample is made with a thin copper wire that is glued to the sample with carbon or silver paste. The wire is guided to the outside via a narrow slot (F) in the sample insert. The design ensures that the sample area wetted by the electrolyte is the same as the illuminated area. The sample can be illuminated from the front or from the back. Since the center of the sample coincides with the center of the four alignment rods, switching from front- to back-illumination can be done by rotating the cell 180° without risk of losing optical alignment. At the bottom of the cell a cylindrical hole is made to accommodate a small magnetic stirbar, and the same hole is connected to a drain at the front of the cell. The drain allows the electrolyte to be refreshed without removing the cell or the cell lid from the setup. The feedthroughs in the lid of the cell are threaded on the inside, with an O-ring seal to ensure gas tightness.

A variation of this cell was designed by the LPI group at EPFL. Their "Cappuccino" cell has an even smaller electrolyte volume and is fabricated from PEEK (polyether ether ketone), a chemically resistant polymer that is easier to machine and has better mechanical properties than PTFE. Their simple but ingenious design allows two methods of sample mounting. One is by pressing the sample against an O-ring, similar to the approach shown in Fig. 3.2. In the other method, the top of the sample is clamped in a miniature steel vise. The vise is mounted at the top of the cell, so that the bottom part of the sample is fully immersed in the electrolyte. The top part of the sample should not be coated, so that the vise connects with the conducting back-contact and thereby serves as an electrical contact. This configuration is particularly useful for (meso)porous samples, which cannot be easily sealed with an O-ring. The illuminated area is defined by a circular hole at the front of the cell. The clamped sample – including the vise – can be easily lifted out of the cell and rotated 180° for back-side illumination. The Cappuccino design is particularly suited for fast screening of nanostructured samples. It is easier to handle than the design of Fig. 3.2, but with the limitations of not being gas tight and having a wetted area that is ~2–3 times larger than the illuminated area.

3.2.2 Sample and Substrate

From a practical point of view, the main characteristic of a photoanode or photocathode sample is whether the film is compact or porous. Compact films can be easily sealed by gently pressing them against an O-ring. The area that is exposed to the electrolyte can be accurately determined, which is essential for, e.g., donor density calculations using the Mott–Schottky relationship – see (2.54) and Sect. 3.6. Moreover, the illuminated area is the same as the wetted area, which minimizes dark current contributions to the total current. Porous films can be mounted by gluing them to a mask with a hole that defines the area to be wetted. The choice of glue or sealant is critical: it needs to be chemically resistant in the electrolyte to be used, and it also needs to have a fairly high viscosity to prevent the infiltration of the porous film in the area that is exposed to the electrolyte. In practice, this limits the choice to certain epoxy resins, or to certain thermoplastic polymers that soften between ~100 and 200°C. Examples of the latter are Surlyn® and Bynel® from Dupont, which are used to seal dye-sensitized solar cells based on mesoporous TiO_2. These sealants are sold as 25–50 μm sheets and are convenient to use, but somewhat costly. Alternatively, one can simply immerse the entire sample (except the top part) in the electrolyte, as discussed above for the Cappuccino cell. Care should be taken that the entire sample is coated in order to avoid direct exposure of the conducting substrate to the electrolyte. This is the most convenient method for photocurrent measurements on porous films. However, it is not suited for quantitative impedance analysis, since the meniscus at the sample/water/air interface results in a poorly defined sample area.

Table 3.1 Typical materials parameters for some common TCOs [2, 5–8]

TCO material	Highest conductivity (S/cm)	Carrier concentration (cm^{-3})	Mobility (cm^2/V^ls)	Work function (eV)	Thermal stability in air (°C)
In_2O_3:Sn (ITO)	~1 × 10^4	~10^{21}	~40	4.7	<350°C[a]
SnO_2:F (FTO)	~1 × 10^3	4 × 10^{20}	~30	4.9	<700°C
ZnO:Al (AZO)	~7 × 10^3	1.5 × 10^{21}	50–100	4.6	<400°C

[a]A thermal stability of up to 900°C in air has been reported for ITO on fused silica [8]

Photoelectrochemical measurements on thin film samples require a conducting substrate as a back-contact. Optically transparent substrates are usually preferred since this also allows the sample to be illuminated from the back. The substrate should form an Ohmic contact with the photoactive material to ensure uninhibited current flow. For n-type photoanodes, this requires conducting materials with a work function that is lower than that of the photoanode itself. In most cases, transparent conducting oxides (TCOs) are used. These materials have high carrier concentrations and mobilities, but still show good optical transparency (>80%) because their plasma frequencies are in the near infrared ($\lambda_p > 1$ μm). They are deposited as a 100–400 nm film on glass by sputtering or spray pyrolysis. Examples of TCOs are F-doped SnO_2 (FTO), Sb-doped SnO_2 (ATO), Sn-doped In_2O_3 (ITO), and Al-doped ZnO (AZO). Table 3.1 indicates several key materials properties of important TCOs. Many other TCOs exist, but FTO and ITO are most common ones that are commercially available. Examples of suppliers are Pilkington, Asahi, Libbey–Owens–Ford, and several smaller ones such as Solaronix and Dyesol. Typical sheet resistances of commercial TCOs on glass are 8–30 Ω/sq. ITO has a ~2 times higher conductivity than FTO, but shows poorer thermal stability and chemical resistivity [2]. Since sheet resistances are not (yet!) limiting the performance of PEC cells for water splitting, FTO is currently the conducting substrate of choice. For p-type photocathodes, high-work function back-contacts such as gold and platinum can be used. It should be noted that work function differences do not always correctly predict the formation of an Ohmic or Schottky contact [3]. For example, FTO appears to form an Ohmic contact with certain p-type materials such as Cu_2O [4].

The choice of conducting substrate becomes more difficult when postdeposition high-temperature treatments are necessary. This is often the case for complex oxides, such as perovskites or oxynitrides [9], which may require firing at temperatures above 600°C in order to obtain the desired crystalline phase. This prohibits the use of float glass, which softens above ~550°C. Certain types of borosilicate glass can be used up to 650°C, while fused silica or sapphire can withstand continuous exposure to temperatures up to 950°C. Unfortunately, the conductivity of ITO films quickly decreases above 350°C. FTO and ITO/FTO coatings are stable up to 600–700°C [2, 10], and may still have acceptable conductivities at higher temperatures provided the anneal treatment is kept sufficiently short (<30 min). For higher temperatures and/or longer times one needs to switch to thin noble metal films (e.g., Pt), at the price of significantly poorer optical transmission properties.

3.2.3 The Counter Electrode

To avoid performance limitations, the reaction at the counter electrode should be fast and the electrode itself should have a high catalytic activity. A common materials choice is Pt, which combines good chemical stability with a very small overpotential for hydrogen evolution (~0.1 V). Ideally, the surface area should be at least two times larger than the working electrode area, and both electrodes should face each other symmetrically to avoid inhomogeneous current densities at the working electrode. For high electrolyte concentrations (typically >0.5 M, see below), the latter requirement is less stringent due to the modest voltage drop across the electrolyte solution. A common configuration is a coiled Pt wire electrode or a piece of Pt wire mesh at about ~1 cm in front of the working electrode. A circular hole in the mesh is necessary to avoid it from blocking the light. A piece of TCO glass coated with a sputtered film of Pt is a low cost and – in most cases – perfectly acceptable alternative.

3.2.4 The Reference Electrode

When studying the properties of the working electrode – either a photoanode or a photocathode – the applied potential is a key parameter. This potential should be measured with respect to a fixed reference potential, so that any change in the applied potential reflects a change in the working electrode alone. The counter electrode cannot be used for this, since the overpotential at the counter electrode/electrolyte interface is generally unknown[5] and varies with the amount of current flowing through the cell. To avoid this current-dependence, a third electrode is added to the electrochemical cell: this is the reference electrode. The potential of the working electrode can now be measured with respect to the (fixed) potential of the reference electrode with negligible current flow through the latter.

Table 3.2 shows an overview of commonly used reference electrodes in solar water splitting research. Applied potentials are nowadays usually reported against the RHE scale for water splitting studies. Zero volt on the RHE scale reflects the

[5] Factors that may contribute to the total overpotential include slow charge transfer across the Helmholtz layer, slow reaction kinetics due to preceding or subsequent reaction steps, mass transport limitations (diffusion, convection, migration), and removal of the solvation sheet of water molecules (dipoles) surrounding each ion.

Table 3.2 Overview of reference electrodes commonly used in PEC research [12–16]

Reference electrode	Filling solution	Potential (vs. SHE)[a,b]	Remarks
Reversible hydrogen	Actual electrolyte solution, purged with H_2 gas[c]	0.0 – 0.059 × pH	RHE: This is nowadays the preferred reference scale for publications on solar water splitting.
Standard hydrogen (=normal hydrogen)	$[H^+]$ = 1.18 mol/L, $p(H_2)$ = 10^5 Pa	0.000	SHE (=NHE): Often used as reference in the literature, but now being replaced in favor of the more convenient RHE scale.
Silver/silver chloride (Ag/AgCl)	0.1 M KCl	0.289	
	1 M KCl	0.237	
	3 M KCl	0.210	
	3.5 M KCl	0.205	
	Saturated KCl	0.198	Workhorse for PEC research
	3 M NaCl	0.209	
	Saturated NaCl	0.197	
Calomel (Hg/Hg_2Cl_2)	0.1 M KCl	0.334	
	1 M KCl	0.281	NCE
	3.5 M KCl	0.250	
	Saturated KCl	0.242	SCE
	Saturated NaCl	0.236	SSCE

[a]At T = 298 K

[b]The temperature dependence is in the order of 1 mV/K

[c]RHE electrodes are commercially available, see for example, the "Hydroflex" electrode by Gaskatel GmbH

H^+/H_2 redox potential in the actual solution, irrespective of the pH. This makes it more convenient to use than the NHE scale, for which $\phi(H^+/H_2) = 0$ V only at unit activity of the H^+ ions and the dissolved H_2 gas. Silver/silver chloride electrodes are most commonly used as reference electrodes, and have largely replaced the traditional and environmentally less friendly saturated calomel electrodes (SCEs) that are based on Hg/Hg_2Cl_2. A potential measured with respect to Ag/AgCl ($\phi_{\mathrm{Ag/AgCl}}$) can be converted to the RHE scale (ϕ_{RHE}) with the following expression:

$$\phi_{\mathrm{RHE}} = \phi_{\mathrm{Ag/AgCl}} + \phi^0_{\mathrm{Ag/AgCl\ vs.\ SHE}} + 0.059 \times \mathrm{pH}. \tag{3.1}$$

Here, $\phi^0_{\mathrm{Ag/AgCl\ vs.\ SHE}}$ is the potential of the Ag/AgCl reference electrode with respect to the standard hydrogen potential (SHE). To give an example, a potential of 0.23 V vs. Ag/AgCl (satd.) in a 1 M KOH solution of pH 13.6 corresponds to a potential of 1.23 V vs. RHE.

An important selection criterion for a reference electrode is its stability in the electrolyte solution. An example of a good all-round electrode that can also be used in alkaline solutions is the XR-300 (Ag/AgCl, $\phi^0 = 0.198$ V vs. SHE) from Radiometer Analytical [11]. Other well known suppliers are Metrohm, Koslow, several potentiostat manufacturers, and many others. It should be emphasized that all reference electrodes are delicate should be well-maintained; they should not be allowed to dry out and should be stored in the appropriate solution when not in use. One should consider having at least three identical reference electrodes in the laboratory, and check their potential differences every 2 weeks or so to confirm that the deviation between any two individual electrodes is less than ± 3 mV. Larger differences may indicate that the electrode solution should be refreshed, that the frit should be declogged (by, e.g., overnight immersion in a 1 M HCl solution), or that the electrode is at the end of its life. The lifetime of a reference electrode depends on its design and manufacturer, but they typically last ~2–3 years if properly maintained. More detailed information on various types of reference electrodes can be found in Sawyer [12].

3.2.5 *The Electrolyte*

The electrolyte in an electrochemical cell consists of a solvent in which the active species to be reduced or oxidized are dissolved. For photoelectrochemical water splitting, the solvent and the active species are one and the same: water. However, since pure water is poorly conducting, supporting ions must be added to ensure that the desired current flow can be attained. The concentration of the supporting electrolyte should be sufficiently high to avoid large Ohmic voltage losses across the electrolyte. These Ohmic losses have to be compensated by increasing the applied bias potential, which adversely affects the overall efficiency of the photoelectrode – see (2.69). The voltage drop is given by $V_{\mathrm{loss}} = I \times R_{\mathrm{E}}$, where I is the total current flowing between the working and counter electrodes, and R_{E} is the electrolyte resistance. The latter is determined by the conductivity of the electrolyte, κ, and the cell constant, K_{cell}, according to:

$$R_{\mathrm{E}} = \frac{K_{\mathrm{cell}}}{\kappa}. \tag{3.2}$$

For the simple geometry shown in Fig. 3.3a, the cell constant is given by the distance between the electrodes divided by the electrode area, $K_{\mathrm{cell}} = L/A$. The cell constant depends on the geometry of the cell, and can be determined from an ac impedance measurement of R_{E} using a standard electrolyte solution of known conductivity.

Another consequence of a non-negligible voltage drop across the electrolyte is a deviation of the measured reference electrode potential, ϕ^{*}_{REF}, from the "real" reference potential, ϕ_{REF}, as illustrated in Fig. 3.3b. To avoid this, the reference electrode needs to be placed as close as possible to the working electrode. When this is not possible due to the design of the and/cell or the size of the reference electrode, a so-called *Luggin* capillary can be used. By placing the opening of this capillary close to the surface of the photoelectrode (at ~0.5–1 mm), the reference electrode "samples" the electrolyte potential at this location. While Luggin capillaries are commonplace in electrochemistry, they are rarely mentioned in publications on

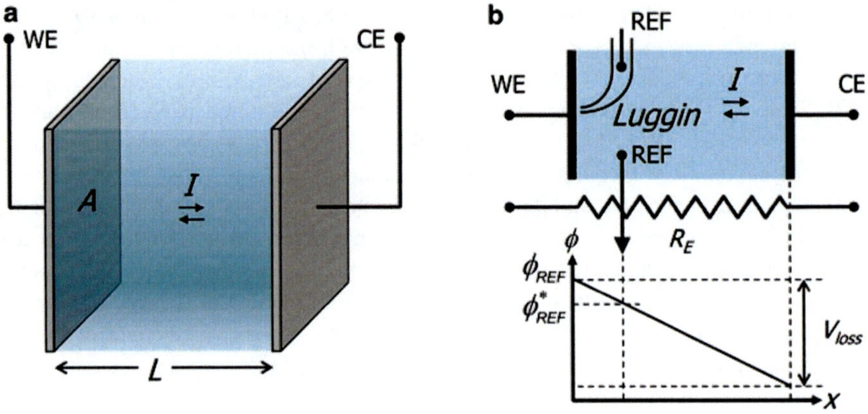

Fig. 3.3 Illustration of the potential gradient (voltage drop) across the electrolyte in an electrochemical cell. A Luggin capillary can be used to ensure that the reference electrode measures the potential close to the working electrode surface, i.e., ϕ_{REF} instead of ϕ^*_{REF}

solar water splitting. This is because the high electrolyte concentrations used in these studies usually implies that the voltage drop across the electrolyte can be neglected. However, as we shall see below this may no longer be true when photocurrent densities reach values of a few mA/cm^2, and for studies on such highly efficient photoelectrodes the use of a Luggin capillary is recommended.

The electrolyte conductivity depends strongly on the type of dissolved ions and their concentrations. It is important to realize that the relationship between the conductivity and the concentration is in general non-linear. This is due to either incomplete dissociation of the anions and cations (in the case of so-called *weak* electrolytes) and/or ion–solvent interactions.[6] Deviations from linearity may already occur at concentrations above 1 mmol/L, and will certainly play a role in practical water splitting applications. At very high concentrations (>1 M), ion-pair formation may result in a decrease of the conductivity with concentration. This is why an aqueous solution of KOH shows a conductivity maximum at a concentration of ~6 M. Table 3.3 gives an overview of the conductivities of several electrolyte solutions often used in PEC studies. To give a rough idea of the Ohmic losses that can be expected, Table 3.3 also includes the voltage drop across the electrolyte for a current density of 5 mA/cm^2. Clearly, concentrations of at least 0.5 M are required to keep Ohmic losses at an acceptable level. Moreover, potassium-based salts are preferred over sodium salts from a conductivity point of view.

In practice, aqueous H_2SO_4 or HCl solutions (0.5–1 M) are often used as an acidic electrolyte for, e.g., WO_3 [19] or TiO_2. For WO_3, NaCl has also been used in

[6] Each ion is surrounded by a solvation sheath of water molecules (see Sect. 2.6.2). The size and charge of the central ion determine the configuration of the surrounding molecules, and this in turn determines the interaction strength of the ion with neighboring ions in the solution.

Table 3.3 Electrical conductivities for various electrolytes and the corresponding electrolyte resistances and voltage drops for a current density of 5 mA/cm^2

Reference electrode	κ (Ω^{-1} m^{-1})	R_E (Ω)	V_{loss} @ 5 mA/cm^2 (mV)	T (°C)	Ionic species (in H_2O)	Limiting ionic conductivity (10^{-4} Ω^{-1} mol^{-1} m^2)
1.0 M KOH	20.1	5.0	25	20	H^+	349.8
0.5 M KOH	10.7	9.3	47	18	K^+	73.5
0.1 M KOH	2.26	44	221	18	Na^+	50.1
0.5 M NaOH	8.6	12	58	18	OH^-	197
Satd. NaCl (~5 M)	21.4	4.7	23	18	Cl^-	76.4
1.0 M NaCl	7.44	13	67	18	SO_4^{2-}	162
0.5 M NaCl	3.8	26	132	18		
0.1 M NaCl	1.07	93	467	18		
3.5 M H_2SO_4	73.9	1.4	7	18		
1.0 M H_2SO_4	36.6	2.7	14	18		
0.5 M K_2SO_4	6.2	16	81	20		
Distilled water	10^{-3} to 10^{-4}	10^5 to 10^6	∞	20		
Millipore water	~5.5×10^{-6}	~18×10^6	∞	25		

R_E and V_{loss} are calculated using $K_{cell} = 1$ cm^{-1}. Individual ionic conductivities are shown to give an impression of their relative contributions (note that listed values are for the limit of low concentrations, i.e., no interactions). Data compiled from refs. [16–18]

order to simulate saline conditions [20]. For photoelectrodes that require neutral or alkaline solutions, such as α-Fe_2O_3 [21], 0.5–1 M KOH or NaOH solutions are commonly used. Some metal oxides are only stable in fairly neutral environments. An example is $BiVO_4$, which is only stable between pH 3 and 11 and is usually studied in an aqueous solution of 0.5 M Na_2SO_4 or K_2SO_4 [22]. To avoid local pH fluctuations in this solution – which would affect the flatband potential – one should consider to add a pH buffer to the solution. An suitable choice is the well-known phosphate buffer KH_2PO_4/K_2HPO_4, which gives a pH of 6.86 at 25°C when adding 0.025 M of each species to the electrolyte [16].

To efficiently remove the evolved hydrogen and oxygen from the electrolyte solution, the electrolyte should be continuously purged with an inert gas such as nitrogen or argon. This prevents the back-reaction of dissolved H_2 and O_2 to water, and ensures that the redox potentials do not change over time (both H_2 and O_2 appear in the Nernst equations for water reduction and oxidation, see (2.52–2.53)). Continuous purging through the electrolyte solution must be avoided when the generated bubbles interfere with the light path and generate excessive noise on the photocurrent signal. If this is unavoidable, one can purge the solution thoroughly (at least 20 min) before starting the experiment, and hang the purge tube above the electrolyte to generate a blanket of inert gas above it. Stirring the solution with a magnetic stir bar greatly enhances the purging efficiency and ensures quick removal of any dissolved oxygen. Stirring to address mass transport limitations in the electrolyte is usually not necessary in view of the high ionic concentrations, and is often avoided to prevent measurement noise.

3.3 The Photoelectrochemical Measurements Setup

Photocurrent measurements are the main tool for investigating the properties and performance of photoelectrodes. Two main experimental configurations can be distinguished, based on the type of light source used: one for performance and stability measurements under high intensity white light, and the other for measuring wavelength-dependent properties using a (low-intensity) monochromatic light source. In this section, these two configurations and their components are discussed. Special attention will be given to the two main components: the potentiostat and the light source.

3.3.1 Simulated Sunlight Measurements

The ultimate test for any photoelectrochemical device or individual photoelectrode is its performance under solar irradiation. Since few places in the world have daily access to sunlight of a constant intensity, most PEC research laboratories are equipped with a solar simulator. Figure 3.4 shows an overview of the main components for a photoelectrochemical experimental test setup featuring simulated sunlight. A potentiostat is used to control the sample's potential and to measure

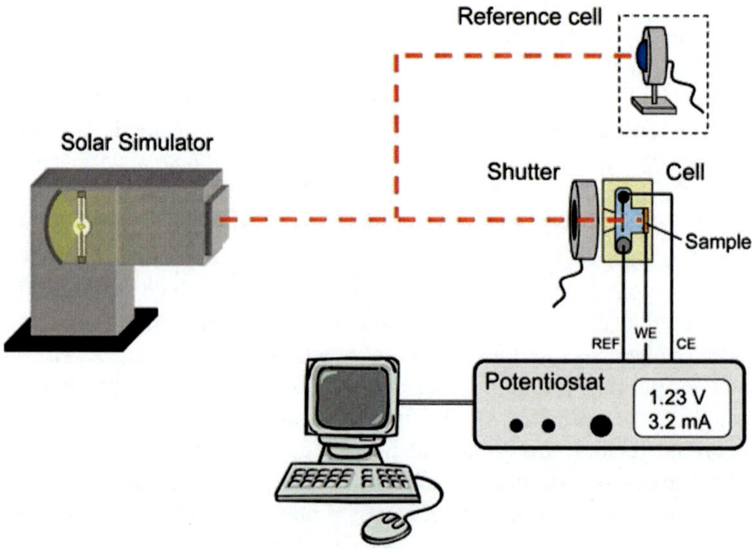

Fig. 3.4 Experimental setup for measuring the performance of a photoelectrode under irradiation with simulated sunlight

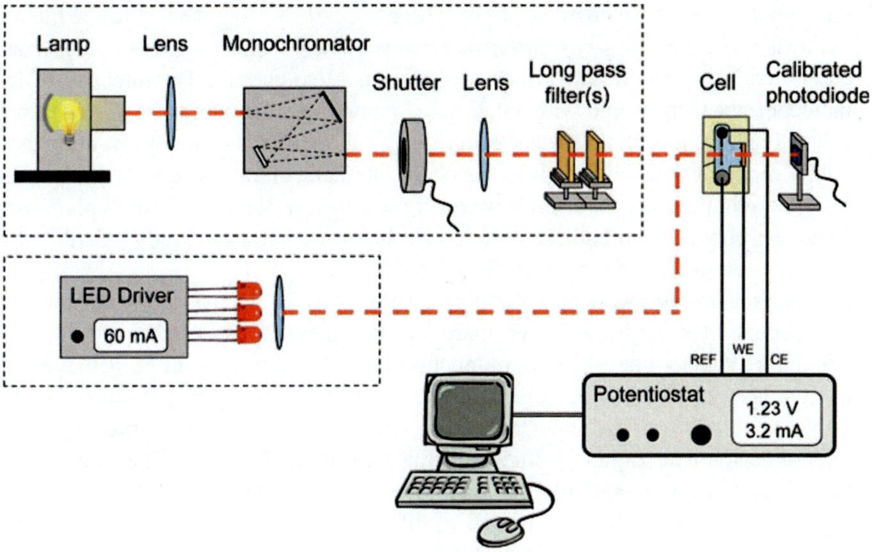

Fig. 3.5 Experimental setup for measuring the photocurrent and/or quantum efficiency as a function of wavelength

the current, and an electromechanical shutter is used to block the light for, e.g., chopped-light experiments. A calibrated solar cell or photodiode is used to verify that the intensity of the solar simulator at the sample position corresponds to the desired value – usually 1,000 W/m^2, which corresponds to 1 sun at AM1.5. A more detailed description of the individual components will be given in the paragraphs below. The setup shown in Fig. 3.4 can be used for performance measurements as well as medium-term stability measurements. The latter are usually limited to 72 h; for longer term stability measurements dedicated testing facilities that also feature temperature and humidity control may be more suitable.

3.3.2 *Wavelength-Dependent Measurements*

More detailed insights into the factors that limit the performance of a photoelectrochemical device or photoelectrode can be obtained by measuring the photocurrent as a function of wavelength. A typical setup for such measurements is shown in Fig. 3.5. The light source is a tungsten halogen or gas discharge (e.g., xenon) lamp, and a monochromator is used to filter out a narrow part of the spectrum centered around the wavelength of interest. An electromechanical shutter is used to control the exposure of the sample to the light, and optical lenses are used to properly focus the light onto the monochromator and sample. If the dispersive element inside the monochromator is a diffraction grating, the monochromator output also contains the higher orders of the diffracted light. These undesired components must be filtered out by the so-called *long-pass filters* (see below). The potential of the sample is again controlled by a potentiostat, which also measures the current. The intensity of the incident light is measured with a calibrated photodiode. In contrast to the reference cell in Fig. 3.4, this photodiode is specifically calibrated for – and meant to be used with – monochromatic irradiation. Using an internal calibration curve, it reports the incident light intensity for a single wavelength, either in Watts (=J/s) or in photons/s. From the photocurrent and the number of photons/s, one can readily calculate the external quantum efficiency or IPCE (cf. Sect. 2.8).

One aspect to be aware of with the setup of Fig. 3.5 is that the light intensities are 2–4 orders of magnitude lower than the 100 mW/cm^2 one gets with a solar simulator. Higher intensities of monochromatic illumination can be achieved by using a laser or one or more LEDs. These sources are convenient in the sense that long-pass filters are no longer necessary, and that the function of the expensive electromechanical shutter is often integrated in the LED driver. The main disadvantage, however, is that only one wavelength is available – perhaps a few more if one uses an argon ion laser or a series of different LEDs. One should also take note of the fact that the spectral half-width of LEDs is quite large, ranging between 10 and 50 nm (with 30 nm being a typical value).

3.3.3 The Potentiostat

A central component in every photoelectrochemical setup is the potentiostat. This is truly the work horse of photoelectrochemistry, but despite its important role it is often considered and treated as a black box. A basic understanding of its inner workings goes a long way in avoiding common user errors and getting the most out of one's measurements.

A simplified diagram of a potentiostat is shown in Fig. 3.6.[7] Most potentiostats have three or four test leads to be connected to the cell or device under test. The three leads that are always present are those for the working (WE), counter (CE), and reference (REF) electrodes. The optional fourth one is basically a second reference electrode which "senses" the potential of the working electrode. Depending on the manufacturer, it is labeled as "SENSE" or "RE2" or "REF2." If no fourth test lead is present, it means that it is internally connected to the working electrode connection. The potentiostat's aim in life is simple: *it measures the potential difference between the REF and SENSE inputs, and makes sure that it stays equal to the desired potential difference by adjusting the potential at the CE output.* The desired potential

[7] For a more detailed description of the working principle of the differential amplifiers, buffers, and power amplifiers shown in Fig. 3.6, the reader is referred to the excellent classic text on electronics by Horowitz and Hill [23].

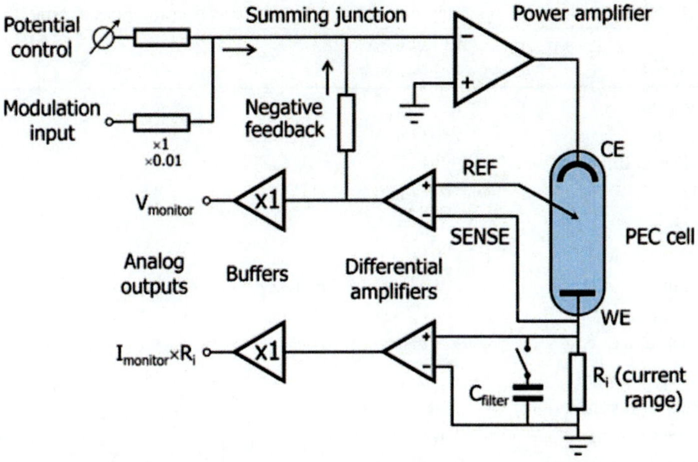

Fig. 3.6 Simplified diagram of a potentiostat

difference is the sum of the potential set by the user (either via the computer or by the potential control knob on the instrument) and the potential present at the modulation input.[8]

The potential reported by the potentiostat corresponds to the measured potential difference between the REF and SENSE electrodes. So for a simple two-electrode measurement on a practical PEC device, the REF and CE leads have to be tied together in order to measure the potential difference between WE and CE. For a three-electrode measurement, the potential of the counter electrode is unknown. In most cases this not a problem, since we are only interested in processes that take place at the working electrode. The counter electrode only serves to supply the voltage and current that is necessary to keep $V_{\text{REF-SENSE}}$ at the desired value.

The REF and SENSE inputs draw negligible current, so the entire current through the cell goes through the *current range resistor*, R_i, which converts the current to a voltage. Most potentiostats have a R_i value of 10 Ω for a 100 mA current range, 100 Ω for a 10 mA range, etc., which results in a potential difference of 1 V across R_i when the current is at the maximum of the chosen range. Parallel to this resistor is a capacitor, C_{filter}, whose value can often be selected (or altogether disconnected) by the user. Together with R_i this forms an *RC* filter that reduces the noise and gives more stable measurements. Most manufacturers specify the values of R_i and C_{filter} because their values affect the speed of the potentiostat – more on this in Sect. 3.6.5.

Standard photoelectrochemical measurements are a fairly undemanding application for modern potentiostats, and almost any commercial potentiostat will serve its purpose. It is nevertheless instructive to have a brief look at the main

[8] The modulation input is used for, e.g., impedance measurements, which will be discussed later.

Table 3.4 Specifications for a somewhat older type of research-grade potentiostat

Specifications	EG&G PAR 283 Potentiostat
Power amplifier	
Max. current	±100 mA
Compliance voltage	±80 V
Slew rate	10 V/μs (no load)
Electrometer	
Input bias current	<10 pA
Band width (−3 dB)	>9 MHz at $R_s = 1$ kΩ
Input impedance	$>10^{13}$ Ω
Input capacitance	<5 pF
I and V measurement	
Bias potential resolution	14 bits, ±8 V
Lowest current range	1 nA
Current resolution	12 bits
Current accuracy	0.2% of full scale (2% for 1–10 nA current ranges)
Frequency response	
−3 dB point	>2 MHz @ 1 mA current range and 1 kΩ source impedance >200 kHz @ 1 μA current range and 1 MΩ source impedance
Analog interface	
Analog I and V outputs	BNC, max. current corresponds to ±1 V, 50 Ω
External input	×1 (Ext. in) or × 0.02/×0.2
Digital interface	
Aux. input/output	1 DAC out, ±10 V, 12 bits
Computer interface	GPIB, RS-232
Software/drivers	Dedicated software from manufacturer, Labview drivers available on the internet

specifications and features. Table 3.4 shows some of the specifications for an EG&G PAR 283, which is a ~15-year-old research-grade potentiostat used in the author's laboratory. Its specifications are still respectable – even by today's standards – and it is more than able to handle the most demanding measurements for solar water splitting applications. The 283 has been succeeded by newer models and is no longer in production, so it simply serves to illustrate our discussion on specifications.

- The *maximum current* is one of the first specifications to consider when purchasing a new potentiostat. In practice, maximum current of 20–30 mA is more than sufficient so long as the electrode area does not exceed a few cm^2. Many potentiostats can deliver currents of 1 or 2 A, but this is overkill for most PEC research efforts.

  The *compliance voltage* is the maximum voltage that the power amplifier can deliver to the counter electrode. For poorly conducting electrolytes (e.g., organic solvents or ceramics) a high CE voltage may sometimes be necessary in order to achieve the required potential difference between the REF and SENSE inputs. For the highly conducting aqueous electrolytes in a PEC cell, however, this is never an issue and a compliance voltage as small as ±8 V is more than sufficient. In fact, a high compliance voltage can pose a serious safety problem. For example, if one accidentally disconnects the reference electrode during operation, the potential difference between REF and SENSE is undefined and the power amplifier may put the maximum compliance voltage on the counter electrode in order to compensate. Any voltage higher than 42 V (dc) is potentially lethal, so this is something to consider carefully, especially when the equipment is operated by individuals with limited experience in this field (students!).

- The *slew rate* indicates how fast the potential of the counter electrode can change to a different value. For standard voltammetry this value is not so relevant, but it may become important for, e.g., time-resolved photocurrent spectroscopy or impedance spectroscopy. Anything above 10 V/μs should be fast enough for PEC work.
- The *electrometer* refers to the differential amplifier that measures the REF and SENSE inputs. It should be able to do this while drawing negligible current from the reference electrode. The 283's 5 pA *input current* is certainly small enough – in fact, anything below a few nA is acceptable. The same goes for the input impedance, anything above 10^{10} Ω would do. The *band width* determines how fast the electrometer is. Note that the 9 MHz in Table 3.4 does not mean one can do impedance measurements up to this frequency: this value refers to the so-called "−3 dB" point, which is the frequency at which the reported signal is only ~70% of the actual signal and shifted 45° in phase. The frequency should be at least 10× lower to get reasonable accuracies: at 0.9 MHz the amplitude difference is only 0.5%, and the phase shift is reduced to ~6°. The *input capacitance* is also important, since it forms a low-pass filter with the reference electrode's impedance – more on this in Sect. 3.6.5.
- A *bias potential resolution* of 14 bits over a range of ±8 V means that the smallest potential step is $(8-(-8))/2^{14} \approx 1$ mV. This is certainly acceptable, although smaller values would allow even smoother voltammograms. The *current resolution* is where the 283 starts to show its age. To illustrate this, a 12-bit resolution for a ±2 mA current range means that the resolution is ~1 μA. Since dark currents are often more than 10^3 times smaller than the maximum photocurrents, measuring both in a single chopped-illumination experiment requires switching the current range every time the light is switched on or off. This often leads to spikes in the measured current for capacitive samples and may even cause some of the faster potentiostats to start oscillating. Such oscillations should be avoided at all cost since they may destroy the sample. Newer potentiostats have 16-bit resolution D/A and A/D converters that allow very smooth voltammograms and less current range switching.
- The *frequency response* is not so relevant for voltammetry, but it is a crucially important figure-of-merit for impedance measurements. For now, suffice it to say that practical frequencies should be 10–50× smaller than the reported -3 dB frequency at the current range used. More details on impedance measurements are discussed in Sect. 3.6.

- The *analog outputs* of the cell voltage and the cell current are essential for impedance measurements using an external FRA, or for measuring low-level photocurrents using an external lock-in amplifier. The *external input* (sometimes called the *modulation* or *polarization* input) is used to super-impose a small sinusoidal voltage on the cell during impedance measurements. It can also be used to apply the large-amplitude modulation required for, e.g., linear sweep voltammetry, although the internal sweep generators inside modern potentiostats are so good that this is almost never necessary. All potentiostats have these very convenient analog inputs and outputs, except for some models that have a built-in impedance analyzer.
- *Digital interface*: some potentiostats feature auxiliary A/D inputs and D/A outputs that can be controlled by the software. These are convenient for controlling auxiliary equipment, such as electromechanical shutters or magnetic stirrers. GPIB used to be the standard *interface to the computer* for instrument control and data exchange. However, this >30-year-old standard is now outdated and is superseded by USB and/or Ethernet interfaces. The same goes for the even older – and much slower – RS-232 interface.
- The final item is the *software* that controls the potentiostat and is used to collect and transfer the data to a computer. The importance of a user-friendly piece of software to do daily measurements cannot be overstated. Fortunately, the software packages of most manufacturers have matured into powerful and easy-to-use measurement and analysis packages. While great for standard electrochemical measurements, it should be mentioned that they offer (very) limited built-in support for controlling additional instruments, such as monochromators, shutter drivers, calibrated photodiodes, and lock-in amplifiers.[9] To get around this issue, the user has to write his own software. Today's de facto standard for doing this is Labview®, a graphical programming language developed by National Instruments. Most – but not all – potentiostat manufacturers offer Labview drivers. If not, suitable drivers may available on the internet. Some manufacturers are unwilling they provide the necessary command set information that allows researchers to write their own control software – if this is the case, one should seriously consider another brand of potentiostat for PEC studies.

When starting to work with a new potentiostat, one should be aware that there exist different sign conventions for the reported voltage and current. To avoid confusion, we strongly recommend that one adheres to the following convention for photoelectrochemistry: (a) the reported potential should reflect the potential of the working electrode with respect to that of the reference electrode, and (b) more positive potentials should give more positive currents. In other words, more positive potentials favor (anodic) oxidation of the working electrode, and anodic currents

[9] Most proprietary software packages include support for several types of frequency response analyzers, sometimes even for models made by other manufacturers.

are positive. Certain potentiostats do not follow this convention, and this can have the strange and confusing consequence that a two-electrode measurement on a simple resistor gives a negative current when the bias potential is positive. The sign convention used by a particular potentiostat can be easily determined by connecting a multimeter across the resistor in such a two-electrode measurement. We finish this paragraph with a list of well-known potentiostat manufacturers: Princeton Applied Research (PAR), Solartron Analytical, Metrohm Autolab, Ivium, Gamry, CH Instruments, and Bio-logic – and there are many others.

3.3.4 The Solar Simulator

The main figure-of-merit for a solar water splitting device is its performance under real sunlight. However, even real sunlight does not have the same intensity and spectral distribution everywhere on earth. To facilitate meaningful comparisons of device performances, the performance characteristics are usually quoted for the so-called *AM1.5G* conditions. This stands for "air mass 1.5 global," and refers to the spectral distribution and intensity of sunlight on a 37° south-facing tilted surface after it has traveled through 1.5 times the thickness of the earth's atmosphere. The latter corresponds to a solar zenith angle of 48.19°. The AM1.5G spectrum includes both the direct and the diffuse contributions of the incident sunlight, and has a total integrated intensity of 1,000 W/m^2.[10] The most recent and widely used AM1.5G reference spectrum is published by the American Society for Testing and Materials as the ASTM-G173-03 standard, and represents a reasonable average for the 48 contiguous states of the USA over a period of 1 year [24]. The corresponding international standard is IEC 60904–3 (2nd edition, 2008).

Since real AM1.5G sunlight is not readily available at all times and at all locations, solar simulators are used. Making a reasonably good solar simulator is far from trivial. One cannot use halogen lamps since their maximum color temperature is limited to ~3,200 K, which is much lower than the color temperature of the sun (5,800 K). Xenon lamps are widely accepted to give the best match to the solar spectrum. Optical filters are used to optimize the overall spectral shape and to remove (to some extent) the sharp lines in the emission spectrum of xenon. A water filter is sometimes used to remove excess spectral intensity in the IR region. Despite such efforts, many conversion efficiencies of photoelectrochemical cells reported in the literature using primitive home-made solar simulators based on, e.g., Xe lamps are overestimated, as demonstrated in an important and carefully conducted study by Murphy et al. [25]. In order to obtain reliable efficiency estimates, Murphy et al. recommend two approaches: (a) measure the performance using actual sunlight under a similar zenith angle as that of the AM1.5 spectrum, or (a) measure the

[10] The AM1.5D spectrum only contains the *direct* contributions over a 5° field of view, and has an integrated intensity of 768 W/m^2.

Table 3.5 International standards for solar simulator performance

	Standard		
Specifications	ASTM 927-05	IEC 60904-9	JIS-C-8912
Spectral match			
Class A	0.75–1.25	0.75–1.25	0.75–1.25
Class B	0.6–1.4	0.6–1.4	0.6–1.4
Class C	0.4–2.0	0.4–2.0	0.4–2.0
Irradiation nonuniformity			
Class A	≤3%	≤2%	≤2%
Class B	≤5%	≤5%	≤3%
Class C	≤10%	≤10%	≤10%
Temporal instability			
Class A	≤2%	≤2%	≤1%
Class B	≤5%	≤5%	≤3%
Class C	≤10%	≤10%	≤10%

The class of a solar simulator is indicated by three consecutive letters, e.g., "ABA." The first letter represents the spectral matching, the second the irradiation uniformity, and the third the temporal stability

IPCE as a function of wavelength and integrate this over the AM1.5 solar spectrum (using (2.74) in Sect. 2.8). Unfortunately, the first approach is not always possible in certain parts of the world. The second approach should be treated with some caution: the intensity used during monochromatic quantum efficiency measurements is usually much lower than 1,000 W/m^2, and this can give large errors for samples that show a nonlinear photocurrent response (e.g., $BiVO_4$). To avoid this, one has to either verify the linearity of the photocurrent response up to 1,000 W/m^2 (using a laser), or try to measure the IPCE under simultaneous high-intensity background illumination.

An alternative option is to purchase a good quality commercial solar simulator. Various standards have been defined to evaluate the performance of solar simulators. The three main criteria are spectral matching, uniformity of irradiation, and temporal stability. The quality with regard to each of these criteria is indicated by three consecutive letters. Table 3.5 shows the widely adopted American, European, and Japanese standards for these criteria. If a manufacturer conforms to all three standards, it is the strictest one that counts. For example, an "ABB class" solar simulator that conforms to all three standards of Table 3.5 would have a spectral irradiance (W/m^2/nm) that deviates no more than −25% to +25% from the AM1.5G intensity within the entire wavelength range (usually 400–1,100 nm). The nonuniformity of the integrated irradiation intensity (W/m^2) should be within 3% for every point inside the specified area (e.g., 2 × 2 in.), and the integrated irradiation intensity would vary less than 1% (rms) in time. A 2 × 2 in. class ABB solar simulator would do very nicely for almost all PEC studies.

From Table 3.5, it is clear that the largest uncertainties are in the spectral matching criterion. To get a feeling for the extent of the deviations, Fig. 3.7

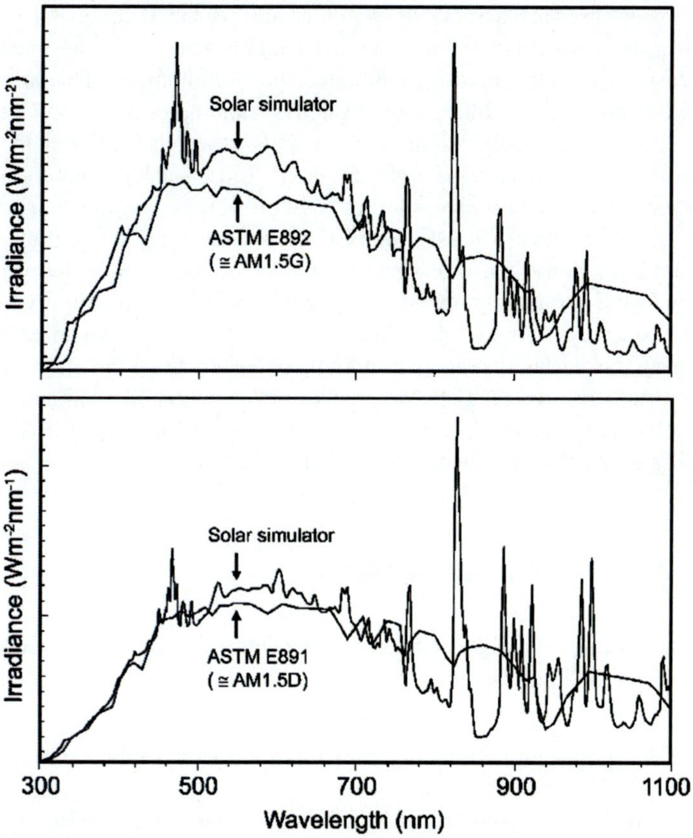

Fig. 3.7 Examples of spectral irradiance curves for a commercial solar simulator (Newport Oriel). The curves are normalized to the AM1.5G and AM1.5D spectra (*top and bottom graphs*, respectively) by matching the total power density between 250 and 1,100 nm

compares the spectrum of a commercial solar simulator to the AM1.5G and AM1.5D reference spectra. The intense xenon emission peaks around ~764, ~825, and above 880 nm can be safely ignored for PEC cells based on metal oxides, since their bandgap is usually larger than 1.7 eV (<730 nm). It is important to be aware of the series of peaks around 475 nm, even though the region between 450 and 500 nm represents only ~17% of the total spectral irradiance of a solar simulator between 300 and 700 nm. For metal oxide-based PEC cells, the most important region lies between 300 and 500 nm. For the solar simulator of Fig. 3.7, the match in this region to the AM1.5G and AM1.5D spectra is quite good. However, this should be considered an exception, and many solar simulators on the market show much poorer matching. One reason for this is that the international performance standards are only specified above 400 nm. This means that the spectral irradiance of a class ABB simulator may deviate much more than 25% below 400 nm, while still being "class A" in terms of spectral matching. The only way to make sure that a solar simulator is indeed suited for metal oxide PEC research is by inspection of the solar irradiance curves from the manufacturer, or to measure this yourself with a calibrated spectroradiometer. The latter are sold by several suppliers, e.g., International Light Technologies or Ocean Optics.

When choosing a solar simulator for PEC research, one should carefully consider the orientation of the light exit port. Most solar simulators have their exit port oriented vertically, i.e., shining "down." This is convenient for solid-state solar cells, but a bit of a hassle for PEC cells containing a liquid electrolyte – one would have to use a mirror since the photoelectrode is mounted vertically in most PEC cells. Fortunately, several manufacturers offer the possibility to have the light exit sideways. Some solar simulator manufacturers even offer integrated electromechanical shutters, with minimum exposure times as short as 200 ms and lifetimes $>10^6$ cycles. There are many manufacturers of solar simulators, and some well-known names include Oriel Instruments (Newport), Asahi, Abet Technologies, Luzchem, Wacom, Spire, Voss, etc.

3.3.5 Miscellaneous Components

As shown in Figs. 3.4 and 3.5, a photoelectrochemical setup contains several other components in addition to the potentiostat and solar simulator. We briefly discuss them below.

Lamp – For wavelength-dependent photocurrent measurements, a combination of a lamp and a monochromator is commonly used as a monochromatic light source (Fig. 3.5). The lamp should have a broad and homogeneous emission spectrum to ensure reasonable light intensities from the UV up to the near-IR. Xenon and tungsten-halogen lamps are most often used. Xe lamps have higher intensities, especially in the UV, and can be nicely collimated due to their small spot size. The downside is that Xe contains sharp emission lines around 475, 765, 825, and 880 nm, and that the short-term temporal stability is significantly worse than that of halogen lamps. Xe lamps are gas-discharge lamps and have to be ignited, which can cause large electromagnetic spikes during ignition. Without proper shielding by the manufacturer, this can interfere with sensitive electronic equipment and bring a potentiostat into oscillation, possibly destroying the sample.[11] This is not a problem for tungsten–halogen lamps. These lamps give very smooth spectra without specific emission peaks, and still give reasonable output power above 300 nm when a "quartz" (i.e., fused silica) lamp envelope is used. The tungsten filament is usually coiled and has a certain size, typically 7×4 mm^2 for a 250 W lamp. The highest light intensity is achieved when the projected image of the filament exactly fills the

[11] In exceptional cases, ignition of a high-power gas discharge lamp can even destroy electronic equipment. This has happened once in the author's laboratory to a Solartron 1286 potentiostat after switching on an older type 450 W Xe lamp.

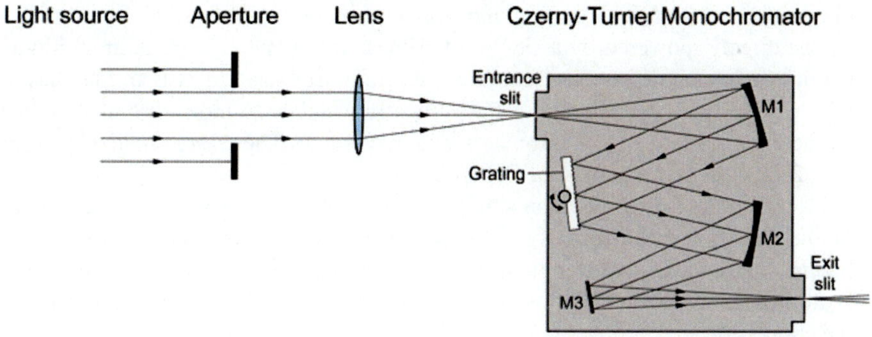

Fig. 3.8 Schematic diagram of a Czerny–Turner monochromator illuminated by a focused beam of light. The diffraction grating consists of a highly reflecting planar surface in which sawtooth-shaped grooves are etched. Mirrors M1 and M2 are concave-shaped, mirror M3 is a planar mirror that directs the light to the exit slit. The aperture in front of the lens prevents overfilling of the mirror, which would result in stray light inside the monochromator

area of the sample. Choosing a lamp with a higher power rating does not give higher light intensities at the sample, since the filament temperature is limited to ~3,200 K. Despite their somewhat lower irradiation intensity, quartz tungsten-halogen lamps are often preferred for PEC research due to their low cost, temporal stability, and ease of use. Manufacturers of Xe and halogen light sources include Oriel (Newport), OBB Corp, Sutter, Cermax, Hamamatsu, PTI, etc.

Monochromator – The monochromator filters the desired wavelength out of the incident broad-band irradiation. The dispersive element is usually a grating, which consists of a series of closely spaced sawtooth-shaped grooves that are etched in a highly reflecting mirror surface. The desired output wavelength is selected by rotating the grating with a stepper motor. Figure 3.8 shows a simplified diagram of a Czerny–Turner monochromator, one of the most common designs. For PEC research, one of the main selection criteria is to get as much light as possible into the monochromator. This can be achieved by selecting a modest focal length (0.1–0.3 m) and the largest diameter grating one can afford. This translates into a small *F-number* (*F*/#), which is defined as the focal length divided by the aperture (diameter) of the optical element. For example, a 150 mm monochromator with a 36 mm grating has an *F*-number of *F*/4. An *F*/2 monochromator has more light gathering power, but will also be more expensive due to the larger grating and mirrors. To ensure optimal filling of the grating with the incident light, the lens in front of the monochromator entrance should have a matching *F*/#. A lens with a larger *F*/# does not fully illuminate the grating and will lead to lower intensities at the output. A smaller *F*/# may overfill the grating and lead to stray light inside the monochromator. The latter should be avoided to prevent undesired wavelength components reaching the output. An *aperture stop* (variable-size aperture) in between the lamp and the lens is sometimes used to avoid overfilling of the grating (Fig. 3.8). A very convenient tool to check for stray light at the output is a miniature spectrometer. These spectrometers, offered by, e.g., Ocean Optics, have an optical fiber input and contain a grating and a CCD camera in a very small package that can be directly powered by a USB port. The easiest way to ensure optimal illumination of the monochromator entrance and minimize stray light is by purchasing light sources with outputs that are specifically matched to monochromator inputs (sometimes called "monochromator illuminators"). These are available from, e.g., Princeton Instruments and Oriel/Newport.

At the output side of the monochromator, the light exits through a slit. The width of the slit is variable (usually between 10 μm and 3 mm) and determines the intensity and wavelength range of the light. An important parameter in this regard is the linear dispersion of the monochromator. Its value can be estimated with the expression

$$\frac{d\lambda}{dx} = \frac{10^6 \cos\beta}{Nkf}, \tag{3.3}$$

where β is the angle between the grating normal and the diffracted light, m is the diffraction order (monochromators always use the first order, so $m = 1$), N is the number of grooves/mm of the grating, and f is the effective focal length in mm. The value of β depends on the wavelength and varies between ~17° and 35°. As an example, consider a $f = 150$ mm monochromator with a 600 g/mm grating. Equation (3.3) predicts a linear dispersion of ~10 nm/mm. Even a slit opening as large as 1 mm[12] (to obtain sufficient output intensity) would still give sufficient optical resolution for PEC studies. A 1200 g/mm grating would improve the resolution by a factor of two, but also results in a narrower wavelength range. In practice, the wavelength range is limited by the efficiency of the grating (i.e., not by the mechanical range of the monochromator). A wavelength range between 300 and 800 nm is well suited for PEC work, and most manufacturers offer gratings that cover this range. It is important to realize that grating monochromators also transmit higher order diffracted light. So when the monochromator is set to 600 nm, it also contains 300, 200, and 150 nm components that need to be filtered out (see below). Again, a miniature spectrometer is a convenient tool to quickly check the presence of second order components, and also to verify the spectral width of the emitted light. Gratings can be *blazed*, which means that the angles of the grooves are optimized to give the highest reflection at a specific wavelength. As a rule of thumb, gratings can be used between 2/3 and 2 times the blaze wavelength. So for most PEC work, a grating blazed at 400 or 500 nm would be optimal. For more detailed information on monochromators and gratings, the reader is referred to general optics textbooks [26–28] and the extensive information available on several manufacturer's websites [29–31].

Lenses – The quality of the lenses does not need to be as good as, e.g., best form laser lenses, but they do need to be made from fused silica to ensure transparency at UV wavelengths. Lens diameters of 50 mm (2 in.) are generally sufficient to capture all the light from the lamp and from the monochromator exit.

[12] Monochromators have both entrance and exit slits. It is recommended to use the same width for both slits.

Long pass filters (sometimes called *high pass* filters) only transmit light above a certain wavelength. They are essential for removing higher order diffracted light from grating-type monochromators. They are available as colored glass filters that are either square or round, with sizes ranging from 12 to 50 mm and a thickness between 1 and 3 mm. The type of filter is indicated as, e.g., WG280 or GG495, where the number represents the wavelength at which the transmission is 50% ($\lambda_{50\%}$). The transition is very steep: for a 3 mm filter, the transmission can change from <1% to >85% within 20–30 nm. The main manufacturer of these filters, Schott AG, offers an Excel-based tool that describes the spectral characteristics in detail [32]. Another manufacturer of long pass filters is Hoya.

Electromechanical shutters are used to quickly switch the illumination on and off. They are offered with diameters ranging between 1 and 90 mm, and with switching times down to a few milliseconds for the smaller models. The electromagnetic coils are driven by dedicated shutter drivers that have to be purchased separately. Depending on the model, these drivers can be controlled externally via a TTL-compatible input and/or they have their own timing electronics, usually with a TTL-compatible sync output. The shutter blades are often coated with a PTFE (Teflon®) coating to minimize the resistance. To avoid excessive heating of the shutter blades when using lasers or other high-intensity sources, certain models can be ordered with a reflecting metallic coating on one side of the shutter blades. For setups with a monochromator, a small-size shutter with an aperture somewhat bigger than the slit size can be placed directly after the monochromator (Fig. 3.5). Alternatively, the shutter can be placed in between the lens and the sample, at the focus point of the lens where the spot is at its smallest size. A shutter aperture of 5–6 mm in diameter is a suitable choice in most cases. Research-grade shutters of this size can have opening times as short as ~1–2 ms and continuous operating frequencies as high as 20 Hz. For setups with a solar simulator, shutters with a larger area are usually required to ensure full illumination of the sample. As mentioned in Sect. 3.3.4, some solar simulator manufacturers offer internal shutters with opening times of ~200 ms. If this is too slow, research-grade shutters with a clear aperture of 45 mm are available with opening times as small as 14 ms and continuous repetition rates of 2.5 Hz. Vincent Associates (Uniblitz® shutters) and CVI Melles Griot are well-known manufacturers of electromechanical shutter systems.

Calibrated photodiodes fall in two categories: those used to calibrate white light sources (e.g., solar simulators) and those calibrated for measuring one wavelength at a time for a certain range of wavelengths. The output of a solar simulator is usually measured with a calibrated PV solar cell or Si photodiode. An important – but often overlooked – assumption is that the spectrum of the solar simulator corresponds to the spectrum used during calibration. Because of this, it is probably best to purchase the photodiode and solar simulator from the same manufacturer. Alternatively, one can use a thermopile for calibrating white light sources. A thermopile contains a black body material that converts the incident radiation into heat, and the temperature is measured by large number of thermocouples connected in series. It gives an output voltage that is proportional to the absolute power of the incident light (in W or in W/m^2) and its response covers a wide range of wavelengths – depending on the type from the infrared (> 5 μm) down to the UV (200 nm). It is almost insensitive to the spectral shape of the light source, but it has a much slower response time (>1 s) than a Si-based detector.

The intensity of monochromatic light sources can be determined with calibrated Si photodiodes that come with a wavelength-dependent response curve. The response is defined in units of A/W. Calibrated photodiodes are available from many manufacturers, but finding types that can be used in the UV (down to 300 nm for metal oxide PEC research) can be challenging. The PD300-UV from Ophir Optics is a convenient model used in the author's laboratory which covers a wavelength range of 200–1,100 nm and powers ranging from 20 pW to 300 mW. The calibration curve is stored within a separate controller, which is connected to the measurement computer via a USB or RS-232 interface. The program sends the current wavelength to the controller, after which the measured output (in Watts) can be read back.

3.4 Photocurrent–Voltage Measurements

Current–voltage measurements are the most important technique to determine the performance characteristics of photoanodes or photocathodes for water splitting. It allows one to quickly obtain the following information:

- The photocurrent density and energy conversion efficiency as a function of applied potential.
- The dark current as a function of the applied potential.
- The sign of the photocurrent (anodic or cathodic).
- The photocurrent onset potential.
- Electron- vs. hole-transport limitations (front- and back-side illumination).
- Transient effects that indicate recombination.

The measurements are done by measuring the (photo)current while sweeping the potential at a constant scan rate – usually between 1 and 50 mV/s. This can be done in the dark, under constant illumination, or by switching the illumination on and off at a fixed frequency.

3.4.1 Dark Current and Photocurrent Onset Potential

Figure 3.9a shows a linear sweep voltammogram of a 0.2% Si-doped α-Fe_2O_3 photoanode deposited on FTO glass. At potentials more positive than ~1.7 V_{RHE}, the dark current starts to increase, indicating the electrochemical oxidation of water at the surface. This is due to an increase in the concentration of free holes in the valence band as the Fermi level is pulled down (more positive potential) and starts to approach the valence band in the region near the surface. A positive photocurrent is observed upon illumination, as indeed expected for an n-type photoanode.

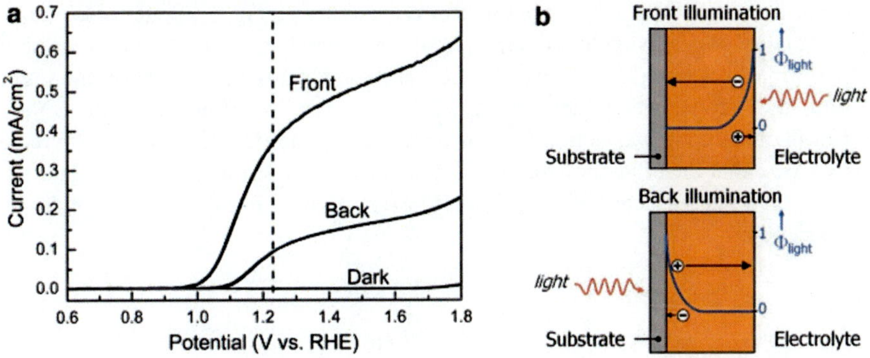

Fig. 3.9 (**a**) Current–voltage curve of a spray-deposited film of 0.2% Si-doped α-Fe_2O_3 in the dark and under continuous front- or back-side illumination with 80 mW/cm^2 simulated sunlight. The scan rate was 10 mV/s. The substrate is FTO-coated glass on which a thin (~5 nm) interfacial layer of undoped SnO_2 was sprayed prior to depositing the Fe_2O_3 [33]. (**b**) Illustration of the difference between front- and back-side illumination of a thin film photoanode or photocathode. The curved line represents the exponentially decaying light intensity in the film (Lambert–Beer)

The photocurrent onset potential is at ~0.95 V vs. RHE. This is much more positive than the flatband of α-Fe_2O_3, which is at ~0.3 V_{RHE} [34, 35]. The difference is due to slow oxidation kinetics (catalysis) [36] and/or to recombination in the space charge region – see Sect. 2.6.6.

3.4.2 *"Water Splitting Efficiencies"*

If the incident light intensity is known, the energy conversion efficiency can be calculated from the photocurrent density. If one assumes that 100% of the photocurrent is used to actually split water (a reasonable assumption for a concentrated KOH or NaOH electrolyte), this also gives the solar-to-hydrogen conversion efficiency. The question is then at which applied bias potential one should take the photocurrent. Ideally, this should be at zero bias for a two-electrode measurement, and presumably also at zero bias vs. RHE for a three-electrode measurement.[13] However, at these bias potentials the photocurrent is often very

[13] The situation is a bit complicated in a three-electrode system since the potential of the counter electrode is not known. One can argue that for a counter-electrode with a low overpotential for H_2 evolution, the potential of the counter electrode should be close to that of the reversible hydrogen potential – provided that the concentration of dissolved hydrogen is high enough. This can be easily verified by a separate measurement of V_{WE}–V_{CE} during a three-electrode linear sweep voltammogram. One particular test in the author's laboratory showed that at a current density >0.5 mA/cm^2, the potential of a coiled Pt wire electrode approached that of the reversible hydrogen electrode to within ~0.1 V. At lower current densities, however, deviations as large as 1 V were observed.

small for a cell with a single photoelectrode (see, e.g., Fig. 3.9) due to unsuitable band edge positions or low overpotentials/slow catalysis. To allow a meaningful performance comparison between different photoelectrodes reported in the literature, the photocurrent densities at 1.23 V_{RHE} (for photoanodes) and 0 V_{RHE} (for photocathodes) are often used as performance metrics. However, it is crucial to realize that the efficiency for an individual photoanode is actually *zero* for a bias of 1.23 V_{RHE}, as can be seen from the expression for the solar-to-hydrogen efficiency:

$$\eta_{STH} = \frac{j_{photo}(V_{redox} - V_{bias})}{P_{light}}. \tag{3.4}$$

This expression is the so-called "applied bias photon-to-current efficiency," or *ABPE* [1]. Any claim of a certain solar-to-hydrogen efficiency for a single photoanode at a nonzero bias potential should therefore be understood as an extrapolation toward a tandem cell device, in which the required bias voltage is generated by a photovoltaic cell that is placed *behind* the photoanode [37, 38]. This PV cell should of course have an equally large area as the photoanode, so that it only absorbs that part of the solar spectrum which is not absorbed by the photoanode. This important distinction is often overlooked in the literature. Moreover, it should be realized that efficient tandem architectures can be difficult to achieve in practice. This is particularly true for nanoporous photoelectrodes, in which extensive scattering may prevent the light from actually reaching the PV device [39]. For these reasons, claims of high "water splitting efficiencies" for any photoelectrode that requires a bias potential should be strongly discouraged. The only way in which a realistic value for the water splitting efficiency can be obtained is to measure the current for a two-electrode device under short circuit conditions.

3.4.3 Front- and Back-Side Illumination

An interesting observation in Fig. 3.9a is that the photocurrent under front-side illumination (via the electrolyte) is much larger than that under back-side illumination (via the substrate). As first pointed out by Lindquist et al. [40], a dependence of the photocurrent on the illumination direction indicates either electron or hole transport limitations in the bulk of the film. This is illustrated in Fig. 3.9b. Under front-side illumination, most light is absorbed near the semiconductor/electrolyte interface. This means that the photogenerated electrons have to travel a larger distance before reaching the interface than the photogenerated holes. For back-side illumination, the situation is reversed. Since surface kinetics, band bending, and other aspects are not (or hardly) affected by the direction of illumination, the smaller back-side illumination photocurrent observed in Fig. 3.9a implies that hole transport is more difficult than electron transport in Fe_2O_3. This is indeed

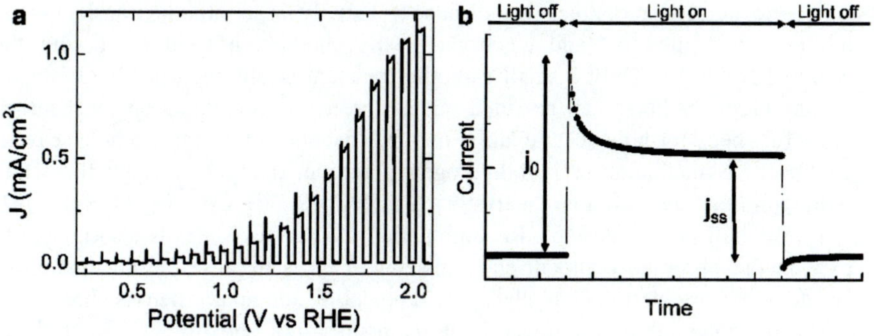

Fig. 3.10 (**a**) Voltammogram for a spray-deposited $BiVO_4$ photoanode on FTO glass under chopped AM1.5 illumination. (**b**) Current vs. time curve for $BiVO_4$ under high-intensity illumination with 364 nm light from a continuous-wave Ar^+ laser at a potential of 1.23 V_{RHE}. In both cases a 0.15 M K_2SO_4 aqueous electrolyte solution was used

consistent with the very small hole diffusion length of 4–20 nm reported for hematite – see Chap. 4 for a more extensive discussion on this. For undoped Fe_2O_3 films the situation is reversed: the photocurrent under front-side illumination is lower (not shown), indicating that the material behaves as an insulator and that slow electron transport limits the photocurrent. Note that differences in front- and back-side illumination due to slow carrier transport can only be observed when a substantial light intensity gradient is present in the film. A good rule of thumb is that (αL) should be >1, where α is the absorption coefficient and L the film thickness.

3.4.4 Photocurrent Transients

The photocurrent–voltage curve is sometimes recorded while switching the light on and off with a shutter. An example of such a "chopped" AM1.5 voltammogram is shown in Fig. 3.10a for a $BiVO_4$ photoanode. Small differences between the dark current and the photocurrent can be easily observed with this technique, and this can help to accurately determine the photocurrent onset potential. Note that Fig. 3.10a shows pronounced transients when the light is switched on. Such transients are often observed and indicate recombination of the photogenerated charge carriers. To explain this, consider the transient photocurrent response shown in Fig. 3.10b for a $BiVO_4$ photoanode illuminated with 364 nm light from an argon ion laser (cw) at a fixed potential of 1.23 V_{RHE} (the x-axis in this figure represents the time instead of the potential, but this does not affect the discussion). When the light is switched on, the photocurrent instantaneously rises to its peak value (j_0). After that, it decays to a steady-state value (j_{SS}) as part of the photogenerated electrons and holes ($j_0 - j_{SS}$) recombine. This recombination can be due to accumulation of holes near the surface, accumulation of electrons in the bulk (i.e., slow electron transport), or trapping of electrons or holes at surface states.

The accumulation of holes would cause an equally large cathodic transient when the light is turned off and electrons in the conduction band react with the accumulated holes. Only a small cathodic transient is observed in Fig. 3.10b, so accumulation of holes can be ruled out as the main recombination mechanism. A useful check for hole accumulation in a photoanode is by adding a hole scavenger, such as methanol [22] or hydrogen peroxide, or by catalyzing the water oxidation reaction with a co-catalyst such as IrO_x [9, 36], $Co^{II/III}$ [21], RuO_2 [41, 42], or cobalt phosphate [43]. Recombination via surface states is also unable to explain the observed photocurrent transients. This is because the occupation of surface states is determined by the applied potential, and no transient is observed at the same potential but at lower light intensities (not shown). For an in-depth discussion on surface state recombination the reader is referred to two interesting papers by Peter [44] and Salvador [45], who pioneered the detailed analysis of photocurrent transients in the early 1980s. The origin of the transients in Fig. 3.10 is believed to be due to the poor electron transport properties of $BiVO_4$. This is indeed consistent with the observation that the photocurrent of $BiVO_4$ for front-side illumination is always lower than for back-side illumination.

3.5 Photocurrent Action Spectra

The spectral response of a photoelectrode can be investigated by measuring so-called *photocurrent action spectra*, in which the photocurrent is recorded as a function of the wavelength at a constant applied potential. One usually also records the intensity of the incident light as a function of wavelength in a separate measurement using a calibrated photodiode. The following information can be obtained from photocurrent action spectra:

- The incident photon-to-current conversion efficiency (IPCE).
- An indication in which part of the film more/less recombination occurs.
- An estimate of the photoelectrode performance under AM1.5 conditions.
- The sign of the photocurrent.
- Electron- vs. hole-transport limitations (front- and back-side illumination).
- Transient effects that indicate recombination.

While there is clearly some overlap between the information that can be obtained from a photocurrent voltammogram (Sect. 3.4) and an action spectrum, it should be realized that action spectra are generally recorded at much lower light intensities. This means that the measured currents under illumination are relatively small, and the dark current may no longer be negligible. The photocurrent therefore always has to be reported as the difference between the current under illumination and the dark current ($i_{photo} = i_{light} - i_{dark}$). Since most samples show some degree of current drift over a time scale of 5–10 min, i_{light} and i_{dark} are best recorded in a single measurement by chopping the light on and off. Care should be taken to ensure that the current has reached a stable value before it is recorded. The preferred scan direction is from high wavelengths to low wavelengths, i.e., from low-energy photons to high-energy photons. This minimizes the chance that the sample itself changes during the measurement, e.g., through the creation of surface states during illumination. Photocurrent action spectra are usually recorded in two parts to allow changing the long pass filters that block the second order of the diffracted light from the monochromator. For example, a scan between 300 and 700 nm can be divided into a scan from 700 $\rightarrow$ 530 nm using a GG495 filter, and from 530 $\rightarrow$ 300 nm using a WG280 filter. If the region around 530 nm contains important information, one can also switch filters at shorter wavelengths (e.g., 430 nm using a GG395 filter) to avoid distracting discontinuities in the photocurrent.[14]

3.5.1 *Quantum Efficiencies*

The *IPCE* – also called the external quantum efficiency – is obtained by dividing the measured photocurrent by the number of photons using (2.71). The difference between the measured IPCE and the maximum value of 100% reflects the sum of the incident photons that are reflected or transmitted (instead of absorbed), and the number of photogenerated electron–hole pairs that recombine before reaching the outer circuit. The *absorbed photon-to-current conversion efficiency* (APCE) only reflects losses due to charge carrier recombination within the semiconductor, i.e., it is corrected for reflection and transmission losses. This is also referred to as the *internal* quantum efficiency. The APCE is a more useful parameter than the IPCE when trying to optimize the material in terms of purity, crystallinity, and defect density. It can be obtained from the IPCE by dividing it through the optical absorption of the sample using (2.72).

An example of an IPCE curve for a (rather poor) $BiVO_4$ photoanode is shown in Fig. 3.11. Less than 10% of the incident photons end up contributing to the photocurrent. The other 90% is lost due to recombination and reflection + transmission of the incident light. Fig. 3.11 also shows that front-side illumination gives lower photocurrents than back-side illumination, consistent with the poor electron transport in $BiVO_4$ (cf. Sect. 3.4.4). The decrease in IPCE below 330 nm for front-side illumination is consistent with electron transport being a limiting factor; at this wavelength the absorption coefficient is so high that all the light is absorbed near the semiconductor/electrolyte interface, and the electrons have to travel through the entire film to reach the back contact. The decrease below 350 nm for back-side illumination is due to the absorption of the FTO substrate.

[14] In contrast to photocurrent measurements, discontinuities in the ICPE spectrum can be easily avoided by using the same combination of wavelength range and long pass filters while measuring the light intensity with the calibrated photodiode.

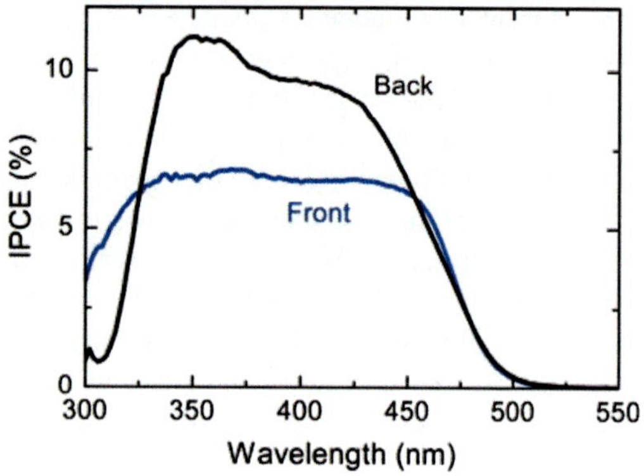

Fig. 3.11 IPCE spectrum for a spray-deposited $BiVO_4$ photoanode on FTO glass, recorded at a potential of 1.23 V_{RHE} in a 0.15 M K_2SO_4 aqueous electrolyte solution

Several examples of IPCE and APCE spectra for another material, α-Fe_2O_3 (hematite), are shown in Chap. 4. This chapter discusses the various features of these spectra in detail, and nicely illustrates how these spectra can help in determining the regions of the film in which recombination is most pronounced.

3.5.2 Extrapolation to AM1.5 Performance

From the IPCE spectrum, an estimate of the total photocurrent under AM1.5 illumination can be obtained by multiplying the IPCE value with the AM1.5 photon flux for each wavelength, followed by integration over all wavelengths. The expression for this total solar photocurrent is given by (2.74) in Sect. 2.8. The value obtained should match the current measured under actual AM1.5 conditions provided that the sample shows a linear response. As outlined in Sect. 2.8, the presence of surface traps or slow surface reaction kinetics may cause nonlinear behavior, and can result in large deviations between the predicted solar photocurrent and the actual value under AM1.5 conditions. The effect of slow surface reaction kinetics has recently been observed in the author's laboratory for $BiVO_4$. The poor water oxidation kinetics for this (un-catalyzed) material under high light intensities results in an AM1.5 photocurrent that is ~8 times smaller than that predicted from the IPCE measured at low light intensities using (2.74) [46].

3.6 Impedance and Mott–Schottky Measurements

Of the techniques discussed in this chapter, impedance measurements are arguably the most complicated measurements to perform and interpret. The discussion below is limited to the most common form of impedance analysis used in solar water splitting research: measuring a Mott–Schottky plot to determine the donor density and flatband potential of a thin film photoelectrode. The basic principles of the measurement are discussed, and the necessary precautions to ensure a valid data set and correct interpretation are outlined.

3.6.1 The Impedance of a Thin Film Photoelectrode

The donor density (N_D) in a semiconductor film can be determined by measuring the space charge capacitance, C_{SC}, as a function of the applied potential, ϕ_A. Since the derivation of this relationship was already given in Sect. 2.5.2, we only repeat the final result, which is the *Mott–Schottky equation*:

$$\frac{1}{C_{SC}^2} = \frac{2}{\varepsilon_0 \varepsilon_r e N_D A^2}\left(\phi_A - \phi_{FB} - \frac{kT}{e}\right). \tag{3.5}$$

A plot of the slope of $1/C_{SC}^2$ vs. ϕ_A should give a straight line with a slope that is proportional to $1/N_D$, while the intercept with the potential axis gives the flatband potential, ϕ_{FB}.

The challenge is to distinguish C_{SC} from other contributions of the photoelectrochemical cell. These can be either capacitive or resistive in nature.[15] Other capacitive elements are the Helmholtz capacitance and surface state capacitances. Examples of resistive elements are the semiconductor bulk, the electrolyte, contacts, and charge transfer resistances. The contributions of all these elements can be modeled by a so-called *equivalent circuit*. Figure 3.12a describes most of the contributions that can be found in metal oxide semiconductor/electrolyte systems. Extracting the space charge capacitance from the overall response of such a complicated circuit can be difficult. Fortunately, many contributions can be ignored, and in the most simple case this leads to the circuit shown in Fig. 3.12b. The overall impedance of this simple *RC* series circuit is given by

$$Z^* = R - j\frac{1}{\omega C_{SC}} \tag{3.6}$$

Here, ω represent the angular frequency at which the impedance is measured. It is equal to $2\pi f$, with f the frequency in Hz. The presence of j ($=\sqrt{-1}$) in (3.6) indicates that Z^* is a complex number – the asterisk serves as a reminder of this. It can

[15] There are several other, more complicated elements available to describe the various processes that can occur in a photoelectrochemical cell, such as the Warburg element (to model diffusion), the Constant Phase Element (CPE, used to describe processes that have a distribution of time constants or activation energies), and transmission lines (to model porous electrodes [47]). Porous electrodes and CPE elements that represent nonideal capacitive elements are briefly discussed below. For more detailed information, the reader is referred to the literature [48, 49].

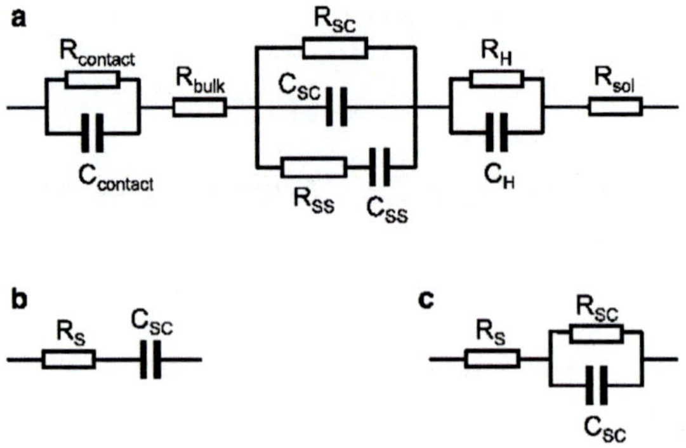

Fig. 3.12 Equivalent circuit diagrams for photoelectrochemical cells. (**a**) Extensive circuit that contains contributions from the semiconductor/back contact interface ($R_{contact}$, $C_{contact}$), the semiconductor bulk resistance (R_{bulk}), the space charge region (R_{SC}, C_{SC}), surface states (R_{SS}, C_{SS}), the Helmholtz layer (R_H, C_H), and the electrolyte resistance (R_{sol}). (**b**) Simplest circuit possible in which only the space charge capacitance and an overall series resistance is present. (**c**) Circuit that is often encountered in practice for photoelectrodes

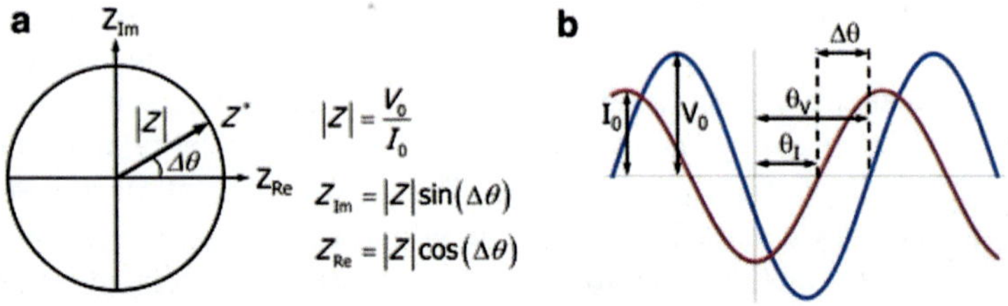

Fig. 3.13 (**a**) Representation of the complex impedance. (**b**) Sinusoidal voltage and current signals from which the complex impedance is calculated

therefore be represented as a vector in the complex plane, as shown in Fig. 3.13a. For this simple *RC* series circuit, it can be easily seen that the real part of the impedance, Z_{Re}, corresponds to the resistance *R*, and the imaginary part, Z_{Im}, is equal to $(-1/\omega C_{SC})$. This means that the space charge capacitance can be directly determined from the imaginary part of the impedance. Thus, by measuring the imaginary part of the impedance as a function of the applied potential, one can obtain a Mott–Schottky curve in which $1/C_{SC}^2$ is plotted against ϕ_A.

3.6.2 The Mott–Schottky Plot: An Example

Before going into the details of the measurement principles, equipment, and precautions, we briefly illustrate a few basic features of a Mott–Schottky plot using the example shown in Fig. 3.14 [50]. The Mott–Schottky curves in this figure

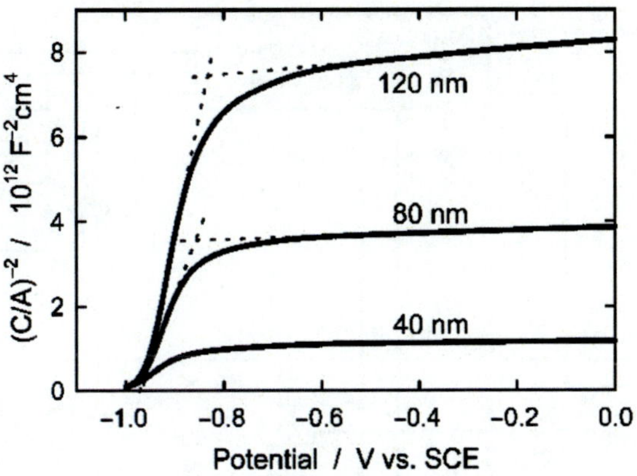

Fig. 3.14 Mott–Schottky curves for 40, 80, and 120 nm dense anatase TiO_2 films on ITO, made by electron beam evaporation of TiO_{2-x}. The curves are recorded at a modulation frequency of 10 kHz and an amplitude of 5 mV, using a scan rate of 1 mV/s. The electrolyte was a 0.5 M KOH solution (pH 13.7). The electrode surface area is 7.0 mm^2 for all three samples, and the capacitance values are normalized with respect to surface area. Reproduced from ref. [50] with permission from The Electrochemical Society

are for dense, electron-beam evaporated anatase TiO_2 films with a thickness of 40, 80, and 120 nm. The material is n-type, which implies that the depletion layer width increases when the potential is made more positive. Extrapolation of the steep slope between -1.0 and -0.8 V vs. SCE gives the flatband potential of TiO_2. The three curves give nearly identical values for ϕ_{FB}, with an average value of (-0.98 $\pm$ 0.04) V vs. SCE. From the slopes, donor densities between 4.1×10^{16} and 2.9×10^{17} cm^{-3} are found for the films.

An interesting feature of the Mott–Schottky curves in Fig. 3.14 is the capacitance plateau found at potentials positive of -0.7 V vs. SCE. At these potentials, the depletion layer width actually exceeds the TiO_2 film thickness, and further extends into the underlying (n-type) ITO film. Clearly, such a plateau will only be observed for low donor densities and/or very thin films. The slope of the curve at the plateau is determined by the donor density of the ITO, for which a value of 1×10^{20} cm^{-3} is found in this particular case [50, 51]. Moreover, from the capacitance C_L at the intersection of both curves (dashed lines in Fig. 3.14) and the known thickness (L) of the films, it is possible to determine the dielectric constant of the TiO_2 using the expression for a parallel plate capacitor ($C_L = \varepsilon_0\varepsilon_r A/L$). A value of 55 was found for the dielectric constant of polycrystalline anatase TiO_2 (this value was also used to calculate the donor densities mentioned in the previous paragraph).

3.6.3 The Frequency Response Analyzer

The real and imaginary parts of the impedance of the sample can be measured with a FRA. This is a fairly complicated machine that can be purchased from several companies. Some are sold as stand-alone machines (Solartron, Princeton Applied

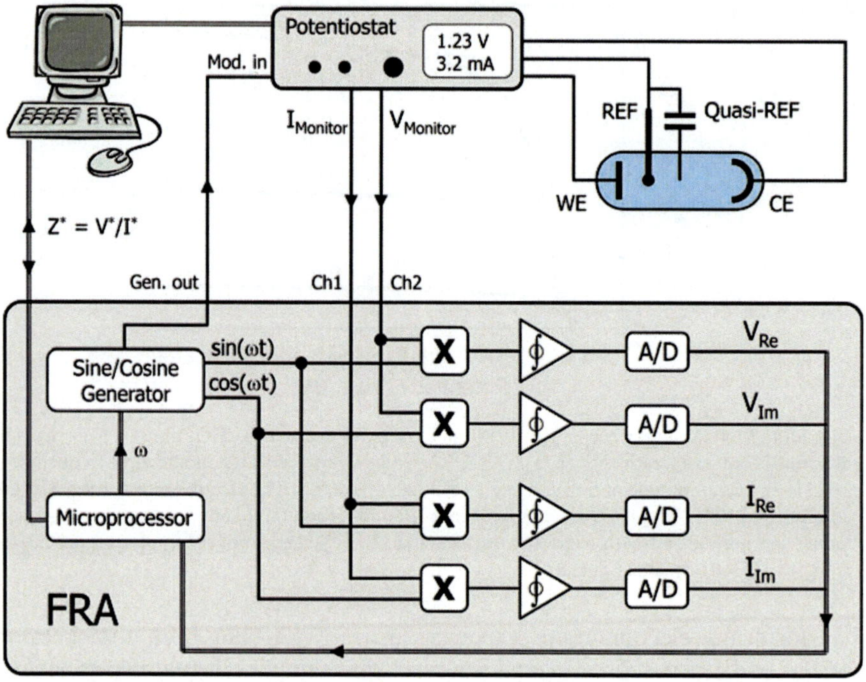

Fig. 3.15 Schematic diagram of a frequency response analyzer (FRA), showing the proper connections to the potentiostat for impedance spectroscopy and Mott–Schottky measurements. The function of the quasi-reference electrode is explained in Sect. 3.6.5

Research), while others are sold as an integral part of a potentiostat or as a modules that can be separately purchased (Princeton Applied Research, Gamry, Metrohm-Autolab, Ivium).

While the FRA is often treated as a black box by the photoelectrochemist, a basic understanding of its inner workings may help to avoid measurement errors. We briefly describe its operating principles using the diagram shown in Fig. 3.15. The FRA generates a small sinusoidal voltage V_{ref} of a certain frequency (ω_{ref}) and amplitude (A_{ref}), which is used to modulate the potential of the sample via the modulation input of the potentiostat. When the system behaves linearly, this results in a sinusoidal response of the current that only differs in amplitude and phase from the voltage perturbation (Fig. 3.13b). The measured potential and current are then fed back into the FRA via the analog outputs of the potentiostat, and are further processed inside the FRA in order to calculate the impedance.

To ensure a linear response of the system, the amplitude of the perturbation voltage needs to be small, usually 5–10 mV for photoelectrochemical devices. Such a small perturbation also gives a very small current response. The first job of the FRA is to isolate these signals from the background signal. The latter consists of the steady-state (dc) current and voltage plus the unavoidable electrical noise. This seems an impossible task, since the signals that we are interested in can easily be 1,000× smaller than the background signals. Fortunately, the FRA can do this very well using a so-called *phase sensitive detector* (PSD). To understand how this works, consider a sinusoidal input signal that can be described by $V_{sig} = A_{sig} \sin(\omega_{sig}t + \theta_{sig})$. After some preconditioning,[16] the PSD multiplies this signal with its internally generated reference signal, $V_{ref} = A_{ref} \sin(\omega_{ref}t + \theta_{ref})$. Some trigonometric manipulation shows that the resulting signal can be written as

$$\begin{aligned} V_{mult} &= A_{ref}A_{sig} \sin(\omega_{ref}t + \theta_{ref}) \sin(\omega_{sig}t + \theta_{sig}) \\ &= \tfrac{1}{2}A_{ref}A_{sig} \cos([\omega_{ref} - \omega_{sig}]t + \theta_{ref} - \theta_{sig}) - \\ &\quad \tfrac{1}{2}A_{ref}A_{sig} \cos([\omega_{ref} + \omega_{sig}]t + \theta_{ref} + \theta_{sig}). \end{aligned} \tag{3.7}$$

The resulting signal has two frequency components, one at $(\omega_{ref} + \omega_{sig})$ and one at $(\omega_{ref} - \omega_{sig})$. The PSD then proceeds to integrate this signal during an integer number of cycles of the reference signal. If the signal frequency is the same as the reference signal's frequency ($\omega_{ref} = \omega_{sig}$), the PSD give a dc output voltage that is proportional to the signal's amplitude:

$$V_{PSD} = \tfrac{1}{2}A_{sig}A_{ref} \cos(\theta_{ref} - \theta_{sig}). \tag{3.8}$$

More importantly, *the result of this integration is zero for all other signal frequencies*. This means that the PSD very effectively rejects noise and other undesired signal components; it only detects the signal of interest, i.e., any signal that has the same frequency as the reference signal. The output of the PSD is further amplified and converted into a digital value by an A/D converter in the FRA.

As shown in Fig. 3.15, there are actually *four* phase-sensitive detectors in a FRA: two for the voltage and two for the current. Each signal is multiplied by the reference signal and by a signal that is shifted 90° in phase from the reference signal. The latter is given by $A_{ref} \sin(\omega_{ref}t + \theta_{ref} + 90°) = A_{ref} \cos(\omega_{ref}t + \theta_{ref})$. This results in two values that represent the real and imaginary components of each signal. To see this, remember that the voltage and current signals can also be written in complex notation:

$$\begin{aligned} V_{sig} &= V_0 \,\mathrm{Exp}\big(j(\omega_{sig}t + \theta_V)\big) = V_0 \cos(\theta_V) + jV_0 \sin(\theta_V) = V_{Re} + jV_{Im}, \\ I_{sig} &= I_0 \,\mathrm{Exp}\big(j(\omega_{sig}t + \theta_I)\big) = I_0 \cos(\theta_I) + jI_0 \sin(\theta_I) = I_{Re} + jI_{Im}. \end{aligned} \tag{3.9}$$

The in-phase signal of the voltage is equal to V_{Re}, while the 90° out-of-phase signal is the imaginary part, V_{Im} (and similarly for the current). As a final step, the

[16] Preconditioning steps are, e.g., removal of the dc component of the signal by passing it through a high pass filter (a capacitor), or amplification/attenuation of the signal by a certain fixed factor. These steps are often necessary to ensure that the signal falls within the range that the FRA can handle internally. These steps do not (and should not) influence the final measurement results.

microprocessor of the FRA divides the voltage by the current to give the complex impedance of the system in the form of numerical values for Z_{im} and Z_{re}:

$$Z^* = \frac{V_{Re} + jV_{Im}}{I_{Re} + jI_{Im}} = Z_{Re} + jZ_{Im}. \tag{3.10}$$

As shown in Sect. 3.6.1, we can now calculate the space charge capacitance by multiplying Z_{im} with $-\omega$.[17]

3.6.4 *Measurement Parameters*

To measure a Mott–Schottky plot, the potential is swept at a constant scan rate while continuously measuring the imaginary part of the impedance at a fixed modulation frequency. The starting potential is a potential where a large depletion layer is present. The scan then proceeds in the direction of the flatband potential, and is reversed (or stopped) at – or just before – ϕ_{FB}. When approaching the flatband potential, one has to be careful that the current does not rise too high and causes damage to the sample. The scan rate is usually chosen based on the integration time of the FRA and the required spacing between the data points. Typical values range between 1 and 20 mV/s. The longer the FRA integration time, the more reliable the data is and the less noise one sees in the measurement. For modulation frequencies of a few hundred Hz and more, an integration time between 0.5 and 1 s generally gives good results. The amplitude of the modulation signal generated in the FRA is often chosen to be relatively large (0.1–1 V), and is then attenuated by a factor of 10, 50, or 100 inside the potentiostat. This is done to minimize noise pick-up in the cable between the FRA's generator output and potentiostat's modulation input. The modulation amplitude imposed on the sample should be as low as possible to avoid nonlinearities. Since the current–voltage characteristics of semiconductors typically vary as Exp(eV/kT), the amplitude should be much smaller than 25 mV.[18] Amplitudes higher than 10 mV should be avoided, and a value as small as 5 mV often gives good results. To ensure that the potentiostat is fast enough, its internal filters – such as the capacitor across the range resistor in Fig. 3.6 – should be switched off. This is crucially important, and easy to forget when writing one's own program to do these measurements. In some potentiostats the bandwidth of the power amplifier itself can be changed.

[17] When writing one's own measurement software, it should be realized that the FRA simply divides the voltage signals at both input channels – it does not know what these signals represent. Since the (voltage) signal that represents the current is given by $I_{monitor} \times R_i$ (Fig. 3.6), the FRA reports the impedance as $V/(I \times R_i)$ – it does not "know" the value of the range resistor R_i. The software should therefore multiply the reported value by R_i to obtain the actual impedance.

[18] The value of kT/e at room temperature is ~25 mV.

For impedance measurements, one should select the fastest bandwidth model. The performance of the FRA can be optimized by using *ac coupling* for the channel 1 and channel 2 inputs. This inserts a capacitor in the signal line and thereby subtracts the dc component of the signal and improves the FRA's ability to analyze small signals in a large background. Note that this setting should only be used for frequencies that are at least 10× higher than the -3 dB point of the ac-coupled input. The −3 dB point is usually below 10 Hz and can be found in the specifications.

3.6.5 Recognizing Instrumental Limitations

The most critical parameters to choose are the FRA's modulation frequency and the potentiostat's current range. A more sensitive current range gives higher quality data (higher accuracy, less noise), but it also makes the potentiostat slower. This can be understood from the diagram in Fig. 3.6, where the range resistor R_i is seen to limit the speed at which the sample's capacitance can be charged – essentially forming a low-pass filter. To ensure that the measured impedance does not depend on the potentiostat's settings (the data should only reflect the properties of the sample), the potentiostat needs to be fast enough to follow the modulation. The best way to verify this is to measure a quick series of impedance spectra at different values of the potentiostat's current range. As explained below, these measurements should be carried out at potentials not too far from the flatband potential. Fig. 3.16a shows simulated impedance spectra for the *R*(*RC*) circuit shown in Fig. 3.12c for three values of the range resistor, whose values correspond to current ranges of 10 mA, 1 mA, and 100 μA. For the 100 μA current range (R_i = 10 kΩ), the imaginary part of the impedance drops steeply at high frequencies due to the limited bandwidth of the potentiostat. This can be avoided by choosing a higher current range, but this also results in smaller signals and higher noise levels. If the signals are not too small this can be remedied by choosing a longer integration time. Longer times than ~2 s are, however, not practical as they result in very long measurements or very widely spaced data points.

The effect of bandwidth limitations on a Mott–Schottky measurement is illustrated in Fig. 3.17a. The data shown in this figure are simulated for a current range of 100 μA (R_i = 10 kΩ). The other parameters are the same as those used to simulate the 10 kΩ curve of Fig. 3.16a. At 100 Hz and 10 kHz, the latter figure shows a perfectly linear relation between Log($-Z_{Im}$) and Log(*f*), with a slope of −1 that corresponds to an ideal capacitive response ($-Z_{Im} = (\omega C)^{-1}$). The resulting Mott–Schottky plots for these frequencies are therefore identical. At 50 kHz, however, the Mott–Schottky plot is significantly different, and at 100 kHz the data are entirely unreliable. This shows that one should stay well away from the frequency region at which the slope of Log($-Z_{Im}$) vs. Log(*f*) starts to deviate from −1.

In addition to the value of the range resistor R_i, other parameters also affect the effective bandwidth of the system. As shown in Fig. 3.16b, the bandwidth improves when the width of the depletion layer (*W*) increases. This is because a larger value

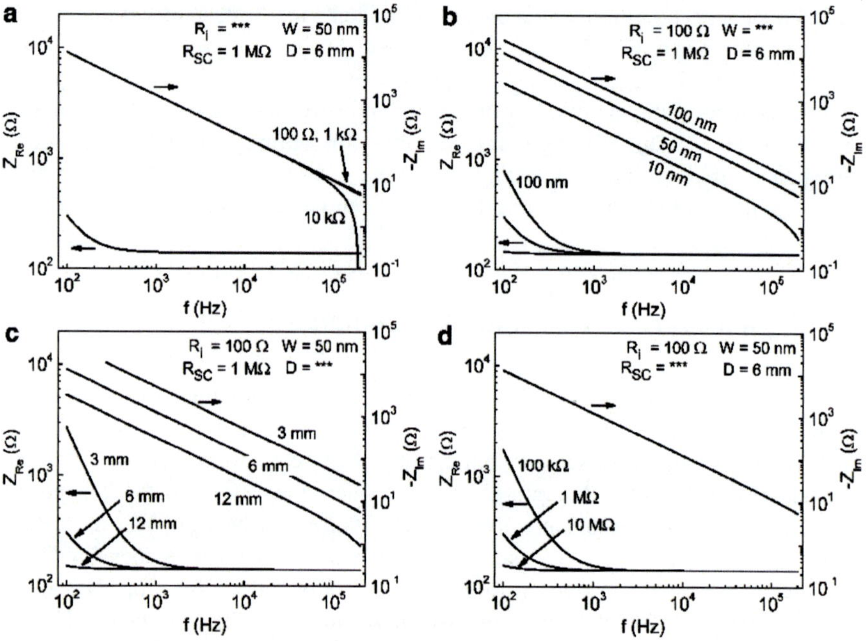

Fig. 3.16 Simulated impedance spectra (Bode plots) for the *R*(*RC*) circuit from Fig. 3.12c. The simulation includes a simple model that describes the bandwidth limitation of a (hypothetical) potentiostat. Several parameters are varied: (**a**) the range resistor, R_i; (**b**) the depletion layer thickness *W*; (**c**) the sample diameter *D*; (**d**) the resistance parallel to the space charge capacitance, R_{SC}. The series resistance R_S is fixed at 140 Ω and C_{SC} is calculated using $C_{SC} = (\varepsilon_0\varepsilon_r A/W)$, where ε_r is the dielectric constant (25) and the surface area *A* is given by $1/4\pi D^2$. Note that the simulation in (**a**) does not show the increase in noise that one expects at less sensitive current ranges

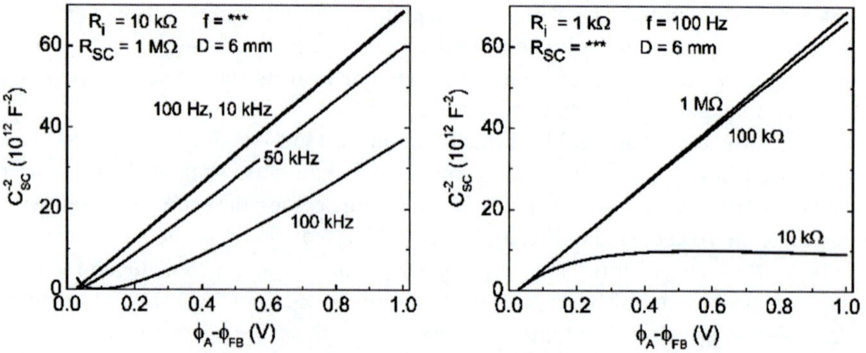

Fig. 3.17 Simulated Mott–Schottky plots for an n-type semiconductor. The same model is used as in Fig. 3.16, except that the depletion layer thickness is now calculated from (2.41) using a donor density of 10^{18} cm^{-3}. *Left*: Calculated curves using parameters that correspond to the $R_i = 10$ kΩ curve in Fig. 3.16a, showing the deviations that can occur at high frequencies when a (too) low current range limits the bandwidth of the potentiostat. *Right*: Deviations at low frequencies due to the presence of R_{SC} parallel to C_{SC}, based on the parameters used in Fig. 3.16d

of W gives a smaller space charge capacitance, and a concomitantly smaller time constant for the low pass filter formed by R_i and C_{sample}. This also implies that one should check the bandwidth limitations of the system by measuring the impedance spectra at potentials close to the flatband potential, i.e., the potential regime where C_{SC} has its highest value.

A very effective method to increase the available frequency window is by using smaller samples. The effect of the effective surface area is shown in the simulation of Fig. 3.16c. A reduction in sample diameter by a factor of two decreases the capacitance by a factor of 4, i.e., it scales quadratically.

3.6.6 The Reference and Quasi-Reference Electrodes

In addition to the potentiostat and the sample, the reference electrode may also affect the effective bandwidth of the system. This is because the impedance of the reference electrode forms a low pass filter with the input capacitance of the potentiostat's electrometer. For example, a 100 kΩ reference electrode combined with a 5 pF input capacitance (cf. Sect. 3.3.3) gives −3 dB point of ~0.3 MHz ($f = (2\pi RC)^{-1}$), which would limit the frequency for accurate measurements to ~30 kHz. The impedance of the reference electrode mostly depends on the type of diaphragm or membrane that is used.[19] A highly permeable membrane gives a low impedance, but also risks contamination of the reference electrode's internal solution with ions from the electrolyte of the PEC cell. It should also be realized that the impedance may increase over time when the membrane get clogged.

One can measure the impedance of a reference electrode using impedance spectroscopy. This can be done using a simple two-electrode configuration, in which the reference electrode is connected as the working electrode (!), and the counter electrode is a regular Pt counter electrode. The electrolyte should be highly conducting, e.g., a 1 M KOH or NaCl solution. One should be careful to keep the applied bias potential at 0 V at all times to avoid large dc current flows through the reference electrode – this may destroy it. The impedance spectrum is measured with a 5 or 10 mV modulation amplitude, over a frequency range of, e.g., 100 Hz to 10 kHz. The imaginary part of the impedance should be very small, and the real part should be constant over a fairly large range. If this is indeed the case, the real part of the impedance is the impedance of the reference electrode.

For high-impedance reference electrodes, or for potentiostats with a high input capacitance, one should consider using a so-called *quasi-reference* electrode. This is a piece of Pt wire that is immersed in the electrolyte next to the normal reference electrode, and connected to the reference electrode via a small capacitor, as shown in Fig. 3.15. The capacitor blocks any dc signals, so for constant or slowly varying

[19] This membrane forms the junction between the internal solution of the reference electrode and the electrolyte of the PEC cell.

potentials the potentiostat only "sees" the normal reference electrode. At high frequencies, however, the impedance of the capacitor decreases and the potentiostat only senses the potential via the Pt wire. The impedance of the capacitor is given by $Z = (2\pi fC)^{-1}$, and C should be dimensioned so that the impedance is less than 1–10 kΩ at the frequencies of interest. A value of 100 nF gives good results in almost all cases. Using a quasi-reference electrode for impedance measurements is always a good idea, even if the reference electrode has an acceptably low impedance. It has no undesired side-effects, and it ensures that the high-frequency data remains reliable if the resistance of the reference electrode membrane increases over time (due to, e.g., clogging).

3.6.7 Sample Limitations and Complications

Even if there are no instrument-related limitations, obtaining a representative Mott–Schottky plot can still be a major challenge – or even impossible – due to the nature of the sample itself. For example, if the equivalent circuit of Fig. 3.12b is too simple to model the sample's behavior, additional circuit elements are necessary which introduce cross-terms in the expressions for the real and imaginary parts of the impedance. The space charge capacitance can then no longer be obtained by measuring Z_{Im} (and/or Z_{Re}) at a single frequency. Instead, one has to measure a series of impedance spectra at various applied potentials, and then identify a single equivalent circuit that adequately describes all these spectra. The values of all the elements in this circuit can then be determined as a function of the applied potential from a nonlinear least-squares fit of each individual impedance spectrum. The final step is to construct a Mott–Schottky plot from the fitted values of C_{SC} vs. the potential. Clearly, this is a complicated procedure that one usually tries to avoid if the goal is simply to measure the donor density or the flatband potential. There are several possible reasons why a simple *RC* series circuit may not be sufficient to describe the sample's behavior. A few of the most important ones are described below.

Parallel resistance R_{SC} – If the semiconductor film contains a large number of defects, or has pin-holes that expose the underlying back-contact to the electrolyte, this is often accompanied by a relatively large steady-state background current. This can be modeled by a resistance R_{SC} in parallel with the space charge capacitance, as illustrated in Fig. 3.12c. The presence of R_{SC} mainly affects the real part of the impedance at low frequencies. While it does not seem to influence the imaginary part of the impedance all that much, the simulation in Fig. 3.17b shows that significant deviations of the Mott–Schottky curve can still occur at low frequencies for sufficiently small values of R_{SC}. Deviations of this kind can be avoided by avoiding low frequencies and by improving the quality of the sample.

Surface States – The presence of surface states introduces one or more additional *RC* series elements in the equivalent circuit, as illustrated in Fig. 3.12a. If the energy level of the surface states is far away from the band edges, surface state charging will be slow and ignored by choosing sufficiently high modulation frequencies. While the physical or chemical origin of surface states is often difficult to determine and poorly understood, their contribution to the impedance of the photoelectrode is well established [52–54]. Specifically, the presence of surface states is known to cause frequency dispersion in the Mott–Schottky plot. When considering only the space charge and surface state capacitances and resistances, the influence of the surface state capacitance on the overall capacitance of the sample can be estimated with [53]:

$$C_{\mathrm{eff}} = C_{\mathrm{SC}} + \frac{C_{\mathrm{SS}}}{1 + (\omega R_{\mathrm{SS}} C_{\mathrm{SS}})^2}. \tag{3.11}$$

The density of surface states at the Fermi level, $N_{\mathrm{SS}}(E_{\mathrm{F}})$, can be determined by measuring an impedance spectrum and extracting the value of C_{SS}, which is given by $C_{\mathrm{SS}} = eAN_{\mathrm{SS}}(E_{\mathrm{F}})$ [53]. By doing this for a range of potentials, one can determine the energetic distribution of surface states over the energy gap. This may help to design passivation strategies for surface states in order to suppress recombination. Depending on their nature and energetic position, surface states can sometimes be passivated by suitable surface treatments, e.g., by depositing a few atomic layers of Al_2O_3 or SiO_2 on top of the photoelectrode's surface.

Helmholtz capacitance – For samples with very high donor densities, the space charge capacitance may become so large that it is no longer possible to ignore the Helmholtz capacitance, C_{H}, which has a constant value of 10–20 μF/cm^2. The effect on the Mott–Schottky curve is a shift of its slope to more negative potentials (for photoanodes). This means that the slope itself is unaffected and can still be used to determine the donor density. The flatband, however, shifts by an amount given by [52]

$$\Delta\phi_{\mathrm{FB}} = -\frac{\varepsilon_0 \varepsilon_{\mathrm{r}} e N_{\mathrm{D}}}{2C_{\mathrm{H}}^2}, \tag{3.12}$$

It should be noted that extremely high donor densities ($>10^{20}$ cm^{-3}) are required for an appreciably shift of the (apparent) flatband potential.

Low carrier concentrations – In some cases, the background (or *bulk*) concentration of free carriers is so low that the entire film is depleted at all values of the applied potential. This means that the semiconductor film effectively behaves as an insulator with a constant (potential-independent) capacitance. This is not often observed for metal oxide photoelectrodes, which usually have free carrier concentrations between 10^{17} and 10^{19} cm^{-3} due to either impurities, intrinsic defects, or intentionally introduced dopants. However, there are exceptions. Unpublished results from the author's laboratory show that TiO_2 films deposited by spray deposition of high-purity TTIP (99.99%) dissolved in ethanol yield insulating films. A different situation occurs if the donor or acceptor level is relatively far away from the band edge, and ionization of these species is slow.[20] Under these conditions, the Mott–Schottky plot shows a horizontal curve for high frequencies, similar to what is observed for an insulator. At low frequencies, however, the carriers can follow the modulation voltage and a normal Mott–Schottky curve is found. This situation has been observed for spray-deposited $InVO_4$ photoanodes [55].

Nanostructured electrodes – a final important example is that of nanostructured porous electrodes. Such mesoporous morphologies are used to increase the active surface area (better catalysis), improve light absorption through scattering, and/or to provide short diffusion paths for the photogenerated minority carriers to the electrolyte. The latter is essential in the case of, e.g., α-Fe_2O_3 due to its very small minority carrier diffusion length (see Chap. 4). The impedance of mesoporous semiconductor photoelectrodes can be described by the so-called transmission lines, which are basically ladder networks of resistive and capacitive elements. The analysis of such systems is a fairly complicated topic. Much progress has recently been made in this area, most notably by Bisquert et al., and for a more in-depth discussion the reader is referred to the literature [47, 56].

Under certain conditions it may still be possible to determine the donor density of a porous photoelectrode from a Mott–Schottky measurement. This is the case for high donor densities, when the space charge width is small and is therefore still able to track the surface contours. An example of this for nanostructured Si-doped α-Fe_2O_3 photoanodes is shown in Fig. 3.18, reported by Cesar et al. [57]. The actual surface area was estimated by dye absorption experiments, and the estimated donor density for this system was ~10^{20} cm^{-3}. The concave shape of the Mott–Schottky plot is consistent with a gradual decrease in the effective surface area (cf. (3.5)) as the potential is increased and the depletion layer progressively penetrates into the bulk [58].

3.6.8 *Nonideal Space Charge Capacitance: The Constant Phase Element*

Even if the depletion layer contribution can be successfully separated from the other contributions in the system, one may find that it does not behave as an ideal capacitor. A clear indication of this is the observation of frequency dispersion in the Mott–Schottky plot, or a somewhat depressed semicircle instead of a perfectly round one in a Nyquist plot of the impedance spectrum.[21] In such cases, it is often

[20] The donors/acceptors are then no longer shallow, but deep.

[21] Note that a semicircle in the Z' vs. Z'' (Nyquist) plot is only observed in the presence of a resistive element in parallel to the space charge capacitance.

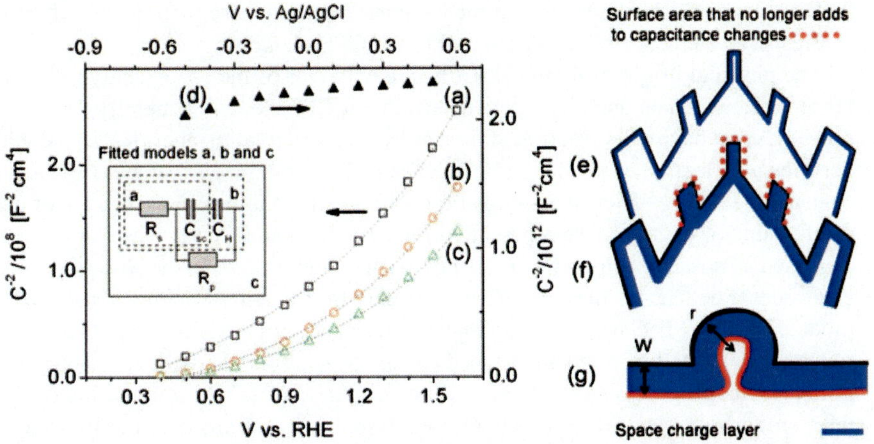

Fig. 3.18 Mott–Schottky plots of Si-doped (curves *a–c*) and undoped (curve d) mesoporous hematite photoanode. The capacitances for curves **a**, **b**, and **c** are obtained from an Si-doped sample and models a, b, and c, respectively, shown in the inset of the left-hand plot. Curve d is obtained from the undoped film and model a (series *RC*). The dashed lines connecting the data points represent the variable active surface area fit. Sketches **e–g** depict the development of the space-charge layer in a mesoporous semiconductor as function of applied potential, illustrating a decrease in active surface area at advancing space-charge layer width in two dimensions. (**e**) Near flat band potential with maximum surface area. (**f**) Total depletion of smaller feature at increased bias potential. (**g**) Decreased active surface area in concave curved surface. Reprinted with permission from ref. [57], copyright, 2009 American Chemical Society

possible to describe the space charge response by a so-called Constant Phase Element (CPE). The impedance of a CPE is given by[22]

$$Z_Q = Y_Q^{-1}(j\omega)^{-n}. \tag{3.13}$$

Depending on the value of n, the electrical behavior of a CPE can range between purely resistive ($n = 0$) and purely capacitive ($n = 1$). Y_Q is often interpreted as the numerical value of the capacitance of the element. This is only true for $n = 1$, since for lower values of n the units are (Ω^{-1} s^n) instead of Farads (Ω^{-1} s). This means one cannot directly calculate the characteristic time constant, τ, for an RQ circuit as one would normally do for an RC circuit ($\tau = RC$). Hsu and Mansfeld showed that the equivalent capacitance of a CPE can be written as [59]:

$$C_Q = Y_Q(\omega_{\max})^{n-1}. \tag{3.14}$$

Here, $\omega_{\max}$ is the frequency at which the imaginary value of the impedance has a maximum, and C_Q has the proper dimensions of Farads (Ω^{-1} s). It should be noted

[22] CPE elements are usually indicated by the letter Q, analogous to the letters R and C used for resistive and capacitive elements, respectively.

that ω_{max} can only be determined if the Nyquist plot shows a semicircle, i.e., if there is a resistive element present in parallel to the CPE element.

The physical origin of nonideal, CPE-like behavior of the space charge capacitance is often unclear and difficult to determine. A CPE element generally describes processes that have a distribution of time constants or activation energies, so it could describe a collection of surface or bulk states with energy levels that are distributed over the bandgap.[23] Dielectric relaxation phenomena in a thin surface layer of the semiconductor or at the semiconductor/electrolyte double layer have also been proposed as possible causes for frequency dispersion [60, 61]. In the case of porous electrodes (Sect. 3.6.7), the capacitive element of the transmission line model can be modeled as a CPE element [47]. Finally, current inhomogeneities due to surface roughness effects are often mentioned as an explanation for CPE-like behavior. For such electrodes, however, the frequency dispersion has been reported to disappear at the high electrolyte concentrations (>0.5 M) that are commonly used in PEC research [61].

Acknowledgments The author gratefully acknowledges the NWO-ACTS Sustainable Hydrogen program (project 053.61.009) and the European Commission's Framework 7 program (NanoPEC, Project 227179) for support.

References

1. Chen, Z.B., Jaramillo, T.F., Deutsch, T.G., Kleiman-Shwarsctein, A., Forman, A.J., Gaillard, N., Garland, R., Takanabe, K., Heske, C., Sunkara, M., McFarland, E.W., Domen, K., Miller, E. L., Turner, J.A., Dinh, H.N.: Accelerating materials development for photoelectrochemical hydrogen production: Standards for methods, definitions, and reporting protocols. J. Mater. Res. **25**, 3–16 (2010)
2. Gordon, R.G.: Criteria for choosing transparent conductors. MRS Bull. **25**, 52–57 (2000)
3. Sze, S.M.: Physics of Semiconductor Devices. Wiley, New York (1981)
4. De Jongh, P.E., Vanmaekelbergh, D., Kelly, J.J.: Photoelectrochemistry of electrodeposited Cu_2O. J. Electrochem. Soc. **147**, 486–489 (2000)
5. Minami, T.: New n-type transparent conducting oxides. MRS Bull. **25**, 38–44 (2000)
6. Ferekides, C.S., Mamazza, R., Balasubramanian, U., Morel, D.L.: Transparent conductors and buffer layers for CdTe solar cells. Thin Solid Films **480**, 224–229 (2005)
7. Fortunato, E., Ginley, D., Hosono, H., Paine, D.C.: Transparent conducting oxides for photovoltaics. MRS Bull. **32**, 242–247 (2007)
8. Minami, T., Miyata, T., Yamamoto, T.: Stability of transparent conducting oxide films for use at high temperatures. J. Vac. Sci. Technol. A **17**, 1822–1826 (1999)
9. Abe, R., Higashi, M., Domen, K.: Facile fabrication of an efficient oxynitride TaON photoanode for overall water splitting into H_2 and O_2 under visible light irradiation. J. Am. Chem. Soc. **132**, 11828–11829 (2010)

[23] A bulk or surface state with a single, discrete energy level is probably better described by a single additional *RC* component instead of a CPE element.

10. Goto, K., Kawashima, T., Tanabe, N.: Heat-resisting TCO films for PV cells. Solar Energy Mater. Solar Cells **90**, 3251–3260 (2006)
11. Radiometer Analytical. http://www.radiometer-analytical.com/. Accessed 29 Aug 2011
12. Sawyer, D.T., Sobkowiak, A., Roberts, J.L.: Electrochemistry for Chemists. Wiley-Interscience, New York (2010)
13. Rodgers, B.: Research Solutions & Resources LLC. http://www.consultrsr.com/resources/ref/index.htm. Accessed 28 Dec 2010
14. Compton, R.G., Sanders, G.H.W.: Electrode Potentials. Oxford University Press, Oxford (1996)
15. Compton, R.G., Banks, C.E.: Understanding Voltammetry. World Scientific Publishing, Oxford (2007)
16. Hamann, C.H., Hamnett, A., Vielstich, W.: Electrochemistry. Wiley-VCH, Weinheim (2007)
17. Gilliam, R.J., Graydon, J.W., Kirk, D.W., Thorpe, S.J.: A review of specific conductivities of potassium hydroxide solutions for various concentrations and temperatures. Int. J. Hydrogen Energy **32**, 359–364 (2007)
18. Bertrand, G.L.: Conductivity of ionic solutions. http://web.mst.edu/~gbert/conductivity/cond.html. Accessed 28 Dec 2010
19. Santato, C., Odziemkowski, M., Ulmann, M., Augustynski, J.: Crystallographically oriented mesoporous WO_3 films: synthesis, characterization, and applications. J. Am. Chem. Soc. **123**, 10639–10649 (2001)
20. Alexander, B.D., Kulesza, P.J., Rutkowska, L., Solarska, R., Augustynski, J.: Metal oxide photoanodes for solar hydrogen production. J. Mater. Chem. **18**, 2298–2303 (2008)
21. Kay, A., Cesar, I., Grätzel, M.: New benchmark for water photooxidation by nanostructured alpha-Fe_2O_3 films. J. Am. Chem. Soc. **128**, 15714–15721 (2006)
22. Sayama, K., Nomura, A., Arai, T., Sugita, T., Abe, R., Yanagida, M., Oi, T., Iwasaki, Y., Abe, Y., Sugihara, H.: Photoelectrochemical decomposition of water into H_2 and O_2 on porous $BiVO_4$ thin-film electrodes under visible light and significant effect of Ag ion treatment. J. Phys. Chem. B **110**, 11352–11360 (2006)
23. Horowitz, P., Hill, W.: The Art of Electronics. Cambridge University Press, Cambridge (1989)
24. Solar Irradiance Data, ASTM-G173-03 (AM1.5, global tilt): http://rredc.nrel.gov/solar/spectra/am1.5/. Accessed 11 Aug 2010
25. Murphy, A.B., Barnes, P.R.F., Randeniya, L.K., Plumb, I.C., Grey, I.E., Horne, M.D., Glasscock, J.A.: Efficiency of solar water splitting using semiconductor electrodes. Int. J. Hydrogen Energy **31**, 1999–2017 (2006)
26. Jenkins, F.A., White, H.E.: Fundamentals of Optics. McGraw-Hill, New York (1981)
27. Hobbs, P.C.D.: Building Electro-Optical Systems – Making It All Work. Wiley, New York (2000)
28. Hecht, E.: Optics. Addison Wesley, Reading (2001)
29. Lerner, J.M., Thevenon, A.: Optics Tutorial. http://www.horiba.com/scientific/products/optics-tutorial/. Accessed 10 Jan 2011
30. Oriel Monochromators: http://www.newport.com/. Accessed 10 Jan 2011
31. Acton Research Monochromators: http://www.princetoninstruments.com/. Accessed 29 Aug 2011
32. Schott Optical Filters: http://www.schott.com/advanced_optics/english/filter/index.html. Accessed 10 Jan 2011
33. Liang, Y.Q., Enache, C.S., van de Krol, R.: Photoelectrochemical characterization of sprayed alpha-Fe_2O_3 thin films: influence of Si doping and SnO_2 interfacial layer. Int. J. Photoenergy (2008). doi:10.1155/2008/739864
34. Kennedy, J.H., Frese Jr., K.W.: Flatband potentials and donor densities of polycrystalline alpha-Fe_2O_3 determined from Mott–Schottky plots. J. Electrochem. Soc. **125**, 723–726 (1978)
35. Sanchez, C., Hendewerk, M., Sieber, K.D., Somorjai, G.A.: Synthesis, bulk, and surface characterization of niobium-doped Fe_2O_3 single crystals. J. Solid State Chem. **61**, 47–55 (1986)

36. Tilley, S.D., Cornuz, M., Sivula, K., Gratzel, M.: Light-induced water splitting with hematite: improved nanostructure and iridium oxide catalysis. Angew. Chem. Int. Ed. **49**, 6405–6408 (2010)
37. Grätzel, M.: Photoelectrochemical cells. Nature **414**, 338–344 (2001)
38. Grätzel, M.: Mesoscopic solar cells for electricity and hydrogen production from sunlight. Chem. Lett. **34**, 8–13 (2005)
39. Brillet, J., Cornuz, M., Le Formal, F., Yum, J.H., Grätzel, M., Sivula, K.: Examining architectures of photoanode-photovoltaic tandem cells for solar water splitting. J. Mater. Res. **25**, 17–24 (2010)
40. Lindquist, S.E., Finnstrom, B., Tegner, L.: Photoelectrochemical properties of polycrystalline TiO_2 thin film electrodes on quartz substrates. J. Electrochem. Soc. **130**, 351–358 (1983)
41. Inoue, Y., Asai, Y., Sato, K.: Photocatalysts with tunnel structures for decomposition of water. 1. $BaTi_4O_9$, a pentagonal prism tunnel structure, and its combination with various promoters. J. Chem. Soc. Faraday Trans. **90**, 797–802 (1994)
42. Sato, J., Saito, N., Yamada, Y., Maeda, K., Takata, T., Kondo, J.N., Hara, M., Kobayashi, H., Domen, K., Inoue, Y.: RuO_2-loaded beta-Ge_3N_4 as a non-oxide photocatalyst for overall water splitting. J. Am. Chem. Soc. **127**, 4150–4151 (2005)
43. Kanan, M.W., Nocera, D.G.: In situ formation of an oxygen-evolving catalyst in neutral water containing phosphate and Co^{2+}. Science **321**, 1072–1075 (2008)
44. Peter, L.M., Li, J., Peat, R.: Surface recombination at semiconductor electrodes. 1. Transient and steady-state photocurrents. J. Electroanal. Chem. **165**, 29–40 (1984)
45. Salvador, P., Gutierrez, C.: Analysis of the transient photocurrent time behavior of a sintered n-$SrTiO_3$ electrode in water photoelectrolysis. J. Electroanal. Chem. **160**, 117–130 (1984)
46. Abdi, F.F., van de Krol, R.: Efficient cobalt phosphate-catalyzed $BiVO_4$ photoanodes for photoelectrochemical water splitting (Submitted)
47. Bisquert, J., Garcia-Belmonte, G., Fabregat-Santiago, F., Ferriols, N.S., Bogdanoff, P., Pereira, E.C.: Doubling exponent models for the analysis of porous film electrodes by impedance. Relaxation of TiO_2 nanoporous in aqueous solution. J. Phys. Chem. B **104**, 2287–2298 (2000)
48. MacDonald, J.R.: Impedance Spectroscopy: Emphasizing Solid Materials and Systems. Wiley-Interscience, New York (1987)
49. Orazem, M., Tribollet, B.: Electrochemical Impedance Spectroscopy. Wiley-Interscience, New York (2008)
50. Van de Krol, R., Goossens, A., Schoonman, J.: Mott–Schottky analysis of nanometer-scale thin-film anatase TiO_2. J. Electrochem. Soc. **144**, 1723–1727 (1997)
51. Van de Krol, R., Goossens, A., Schoonman, J.: Mott–Schottky analysis of nanometer-scale thin-film anatase TiO_2 (Erratum vol. 144, pg 1723, 1997). J. Electrochem. Soc. **145**, 3697–3697 (1998)
52. Gomes, W.P., Cardon, F.: Electron-energy levels in semiconductor electrochemistry. Prog. Surf. Sci. **12**, 155–215 (1982)
53. Morrison, S.R.: Electrochemistry of Semiconductor and Oxidized Metal Electrodes. Plenum, New York (1980)
54. Goossens, A., Schoonman, J.: The impedance of surface recombination at illuminated semiconductor electrodes – a nonequilibrium approach. J. Electroanal. Chem. **289**, 11–27 (1990)
55. Enache, C.S., Lloyd, D., Damen, M.R., Schoonman, J., Van de Krol, R.: Photo-electrochemical properties of thin-film $InVO_4$ photoanodes: the role of deep donor states. J. Phys. Chem. C **113**, 19351–19360 (2009)
56. Bisquert, J., Grätzel, M., Wang, Q., Fabregat-Santiago, F.: Three-channel transmission line impedance model for mesoscopic oxide electrodes functionalized with a conductive coating. J. Phys. Chem. B **110**, 11284–11290 (2006)
57. Cesar, I., Sivula, K., Kay, A., Zboril, R., Grätzel, M.: Influence of feature size, film thickness, and silicon doping on the performance of nanostructured hematite photoanodes for solar water splitting. J. Phys. Chem. C **113**, 772–782 (2009)

58. Schoonman, J., Vos, K., Blasse, G.: Donor densities in TiO_2 photoelectrodes. J. Electrochem. Soc. **128**, 1154–1157 (1981)
59. Hsu, C.H., Mansfeld, F.: Technical note: concerning the conversion of the constant phase element parameter Y_0 into a capacitance. Corrosion **57**, 747–748 (2001)
60. Dutoit, E.C., VanMeirhaeghe, R.L., Cardon, F., Gomes, W.P.: Investigation on frequency-dependence of impedance of nearly ideally polarizable semiconductor electrodes CdSe, CdS and TiO_2. Ber. Bunsenges. Phys. Chem. **79**, 1206–1213 (1975)
61. Oskam, G., Vanmaekelbergh, D., Kelly, J.J.: A reappraisal of the frequency-dependence of the impedance of semiconductor electrodes. J. Electroanal. Chem. **315**, 65–85 (1991)

Part II
Materials Properties and Synthesis

Chapter 4
Nanostructured α-Fe_2O_3 Photoanodes

Kevin Sivula

4.1 Introduction

Since the seminal demonstration of photocatalytic hydrogen production from water splitting with TiO_2 [1], scientists and engineers have sustained the search for a photoelectrochemical (PEC) water-splitting material combining efficient solar light harvesting, high quantum efficiency, practical durability, and low cost. However to date, no single semiconducting material which meets all of these requirements has been found [2]. While systems delivering high solar-to-hydrogen (STH) efficiency have been demonstrated using III–V semiconductor materials with optimized band gaps and energy levels straddling the hydrogen and oxygen redox potentials their cost and stability are a major disadvantage [3]. In contrast, many transition metal oxide semiconductors offer suitable stability to the chemically harsh environment of water, oxygen, and light inherent to PEC water splitting. For example, TiO_2 and WO_3 have been thoroughly investigated and optimized to deliver high single-wavelength quantum efficiencies [4, 5]. Unfortunately the stability of these oxides in aqueous environments, aided by an energetically low valence band which is defined largely by O 2p energy levels, also brings a large bandgap which renders these materials transparent to most of the incident solar illumination. This limits the maximum possible STH conversion efficiency. Specifically, considering the bandgap energies, E_g, alone, TiO_2 (anatase, $E_g = 3.2$ eV) and WO_3 ($E_g = 2.6$ eV) could only convert 3.4 and 10.2%, respectively, of the sun's energy into hydrogen. A more accurate evaluation of the potential efficiencies, including expected kinetic losses, suggests that only 5% STH is possible with WO_3 [6].

In addition to sufficient light absorption and stability, a semiconducting material used for PEC water splitting must be made from raw materials with ample availability

K. Sivula (✉)
Institut des sciences et ingénierie chimiques, Ecole Polytechnique
Fédérale de Lausanne, CH-1015 Lausanne, Switzerland
e-mail: kevin.sivula@epfl.ch

R. van de Krol and M. Grätzel (eds.), *Photoelectrochemical Hydrogen Production*,
Electronic Materials: Science & Technology 102, DOI 10.1007/978-1-4614-1380-6_4,

to be relevant for energy conversion on a global scale. Iron(III) oxide is a promising material in light of these requirements. With a potential to convert 16.8% of the sun's energy into hydrogen [6], it has been extensively examined for application to solar water splitting. In this chapter the advantageous properties of Fe_2O_3 as well as the challenges it presents are examined in the context of PEC water splitting. The most recent efforts at improving photon harvesting by nanostructuring are reviewed, and continuing efforts to control the surface catalysis, and enhance the electronic properties by doping, are also addressed by examining recent literature reports in the field.

4.2 Iron(III) Oxide: A Promising Material for Solar Hydrogen Production

As iron is a readily oxidized metal, and iron ions can exist in different charge states (mainly 2+ and 3+) separated by relatively small energy differences, there are many forms of iron oxide both found in nature and that can be synthesized easily in the lab. In addition, because iron is a common element in the earth's crust (6.3% by weight) iron oxide is everywhere around us, giving, for example, the geological formations of the southwest USA their brown-red color. Commonly (and perhaps unfortunately), iron oxide is better known as a nuisance material, and by its common name – rust – the red-brown corrosion product of iron in the presence of oxygen and moisture. However, its chromatic characteristics also imply the ability to absorb solar irradiation. This coupled with its abundance and nontoxicity make iron oxide a particularly attractive material for use in solar energy conversion. The distinct properties of all of the iron oxides are comprehensively presented in a recent book by Cornell and Schwertmann [7]. Here, a brief summary is presented of the important and unique properties of iron(III) oxide, α-Fe_2O_3, or hematite, the most frequent iron oxide employed for solar hydrogen production. These properties highlight both the promising aspects and the challenges that hematite presents as a material for solar energy conversion.

4.2.1 Structural and Material Properties

For iron oxide in its fully oxidized form, each iron atom is in the +3 oxidation state. There are thus two O^{2-} anions for every three Fe^{3+}. Even considering this stoichiometry, there are many types of Fe_2O_3 including amorphous structures. The α (hematite) and γ (maghemite) forms are two most common crystalline structures found in nature. Maghemite can be considered to be a form of magnetite (Fe_3O_4), a ferrimagnetic spinel iron(II–III) oxide, with cation vacancies. Indeed, natural maghemite arrives from the oxidation of magnetite, and retains strong (ferri) magnetic properties. While this form of iron oxide has found wide-spread use as

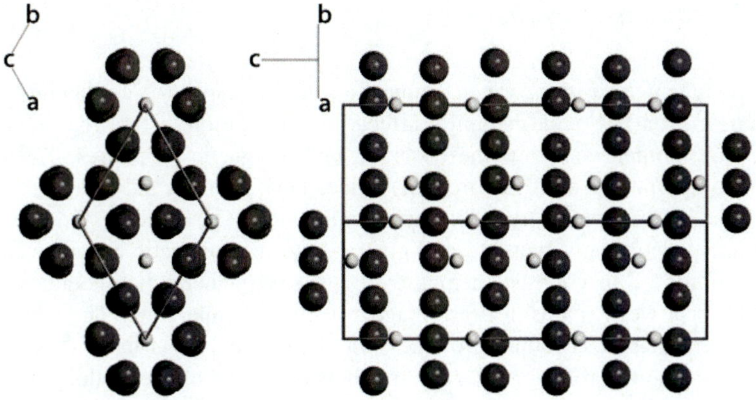

Fig. 4.1 The crystal structure of hematite as seen looking down the [001] direction (*left*) and the [8] direction (*right*). Oxygen anions are *dark gray* and iron cations are *light gray*. The unit cell is outlined, and the orientation map is shown for each view

a magnetic recording medium (giving floppy disks and cassette tapes their brown color) the physical properties have not been found to be suitable for photoelectrochemistry. In addition, maghemite, as with all of the other iron oxides converts to hematite, the most thermodynamically stable form of iron oxide (under standard conditions) upon treatment with heat.

Hematite, alpha-Fe_2O_3, is the most common form of iron oxide. Due to its prevalence, hardness (6.5 on Mohs scale), and chemical and thermal stability (decomposition temperature of 1,566°C), it has found historic use as a polishing powder known as jeweler's rouge. In addition, its ability to absorb light has made it an attractive choice for pigments in paints and even cosmetics. Alpha-Fe_2O_3 has the same crystal structure of corundum, Al_2O_3, which is trigonal (hexagonal scalenohedral, symbol $\bar{3}$ 2/m) with space group $R\bar{3}c$ with lattice parameters a = 5.0356 Å, c = 13.7489 Å, and six formula units per unit cell [7]. It is easy to understand hematite's structure based on the packing of the anions, O^{2-}, which are arranged in a hexagonal closed-packed lattice along the [001] direction. The cations, Fe^{3+}, occupy the two-thirds of the octahedral interstices (regularly, with two filled followed by one vacant) in the (001) basal planes. The tetrahedral sites remain unoccupied. Two views of the hematite crystal structure are shown in Fig. 4.1. Here the layered structure of iron oxide is evident; layers of closed-packed oxygen anions alternate with iron cations bilayers in the [001] direction. The arrangement of cations can also be thought of producing pairs of FeO_6^{9-} octahedra which share edges with three neighboring octahedra in the same plane and one face with an octahedron in an adjacent plane in the [001] direction. This face sharing is responsible for a trigonal distortion of the octahedron. In the FeO_6^{9-} octahedra which share faces, iron atoms are repelled along the direction normal to the [001] axis. Consequently, the O–O distance in shared faces is shorter than in an unshared face. However, the electronic structures of these distorted FeO_6^{9-} octahedral are quite similar to undistorted clusters [7].

4.2.2 Magnetic Properties

The arrangement of the oxygen anions and the high-spin (d^5 configuration) iron cations in iron(III) oxide naturally affects the orientation of the iron atoms' spin magnetic moment and thus the observed bulk magnetic properties. The diverse magnetic properties of all the iron(III) oxides have been recently reviewed [9]. In brief, the inverse spinel structure of maghemite (γ-Fe_2O_3) results in the presence of two interpenetrating sublattices and gives rise to ferrimagnetic behavior and high susceptibilities. In contrast, hematite is antiferromagnetic (temperatures below 260 K) or a weak (parasitic) ferromagnet at room temperature. The latter is due to the ferromagnetic coupling of the spins within the (001) basal (iron bilayer) planes and antiferromagnetic coupling between iron bilayers along the [001] direction [10]. Here the trigonal distortion of the $FeO_6{}^{9-}$ octahedra causes the spins to become slightly canted (about 5°) causing the destabilization of their perfectly antiparallel arrangement (parasitic ferromagnetism). Above the Néel temperature (950 K), the iron spins become thermally disordered and a paramagnetic material results [9]. While the magnetic properties of hematite are not particularly pertinent to its photoelectrochemical performance, the iron spin configuration does influence the optoelectronic and carrier transport properties of hematite. In addition, exploiting magnetic characteristics with Mössbauer spectroscopy to identify phases in iron oxides prepared for electrochemical application has proven to be a useful technique [11–13].

4.2.3 Optoelectronic Characteristics

The dense (hexagonal close) packing of oxygen combined with the interstitially positioned iron yields a very dense structure (5.26 g/cm^3) exhibiting a high polarizability and a high refractive index (3.15) [14]. This layered structure also generates complex behavior when interacting with photons and electrons. These attributes of hematite should be properly understood for application as a semiconductor for solar water splitting. Decades of research have passed scrutinizing the mechanisms of charge transport and light absorption, and while hematite is still not completely understood, much progress has been made.

The absorption of photons by hematite is shown in Fig. 4.2 (left side) and begins in the near infrared spectral region. This initial weak absorption is due to d–d transition states between electron energy levels of the d^5 Fe^{3+} ion, which are split by an intrinsic crystal field [16]. Two absorption bands in this spectral region centered around $\lambda = 860$ and 640 nm with absorption coefficients, α, of the order 10^3 cm^{-1} have been accurately assigned using ligand field theory to spin-forbidden transitions from $^6A_{1g}$ to $^4T_{1g}$ and $^4T_{2g}$, respectively [15, 16]. While photoexcitation of hematite at these wavelengths has been shown to increase its conductivity [15], sustained photocurrent is not observed in a PEC system (see Sect. 4.2.4) upon

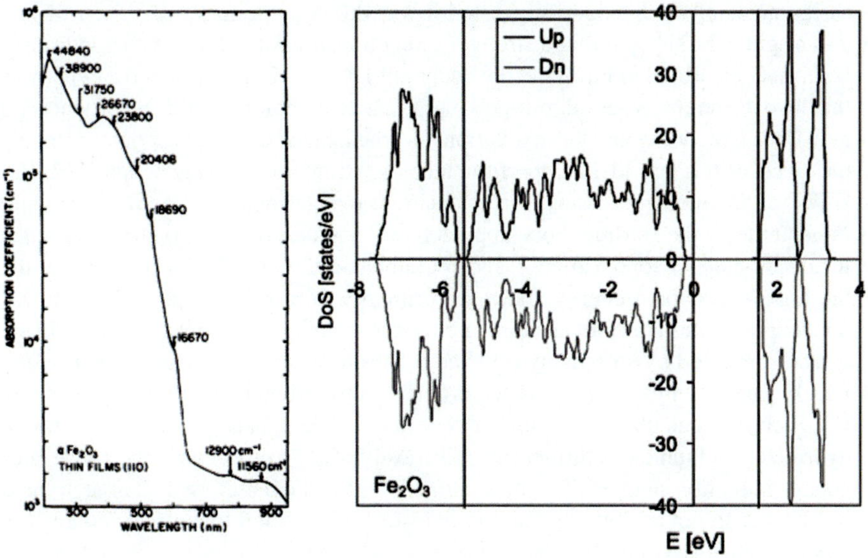

Fig. 4.2 Optical and electronic properties of hematite: The left graph shows the absorption coefficient as a function of photon wavelength in hematite thin films prepared by r.f. sputtering (from [15] used with permission). The *right graph* shows the calculated density of electron states (spin up and down) in hematite. The Fermi energy is at 0 eV and two lines at about −5.5 and 1.8 eV divide the plot into three regions: the lowest energy region contains the occupied Fe-d states; the next energy region contains the occupied O-p states; the region above the gap contains the empty Fe-d levels (Reprinted figure with permission from Velev, J., Bandyopadhyay, A., Butler, W.H., Sarker, S.: Phys. Rev. B: Condens. Matter **71**, 205–208 (2005). Copyright 2005 by the American Physical Society)

irradiation below the bandgap energy, E_g (which, depending on the method of preparation of hematite, is usually reported to be between 1.9 and 2.2 eV or 650–560 nm [14]). The absorption coefficient of pure hematite increases abruptly at the band gap energy and further up the electromagnetic spectrum α continues to increase with additional absorption peaks having been observed centered on 2.4, 3.2, and 5.8 eV (with α on the order of 10^5 cm^{-1}) in samples of polycrystalline hematite [17]. Hematite's strong absorption of yellow to ultraviolet photons in the visible region and transmission of orange to infrared photons gives it a characteristic red color. However, it is fairly uniform reflectivity as a function of visible light wavelength gives polished bulk samples of hematite a metallic look.

Since the electronic nature of the band gap in hematite is of great interest to understand its performance as a material for solar energy conversion, much work has focused on this aspect. Tauc analysis of the bandgap absorption onset, which assumes that the energy state bands are parabolic with respect to the crystal momentum, most frequently indicates an indirect (phonon-assisted) band gap transition [18]. However, a few recent reports of a direct band gap in hematite have been attributed to quantum size-effects [19, 20]. The initial orbital assignments of the bandgap suggested it was due to an indirect transition of Fe^{3+} d–d origin [15, 21], and that a stronger direct transition involving a charge transfer from an O 2p orbital to Fe 3d did not occur until 3.2 eV [21]. This led to a hypothesis that two different types of p-type charge carriers (holes) could be produced in hematite, depending on the excitation mechanism, and were responsive for the observed difference in PEC performance as a function of wavelength [22, 23]. However, the most recent ab initio calculations to determine the electronic structure of hematite by the Hartree–Fock approach [24] and density functional theory with a local spin-density approximation and coulomb correlation [25, 26] both predict that the highest occupied energy states are primarily O p in character and the lowest unoccupied states are from an empty Fe-d band (see Fig. 4.2, right). This conclusion is also supported by soft-X-ray (O *K*-edge) absorption and emission spectroscopy, which when compared to photoemission spectra from configuration-interaction FeO_6 cluster calculations, confirm that the valence band is at least strongly hybridized and indicates further that it is mostly of O 2p character [27]. These latest results not only suggest that pure stoichiometric hematite is a charge transfer insulator, and not a Mott–Hubbard type insulator – a detail that affects conduction models – but also contradicts the suggestion that two types of holes, originating from different transitions, could cause different PEC behavior.

Finally, as discussed in the next section, much effort has been placed on doping hematite with various elements to increase charge carrier conduction. These elements are typically present at 1 atom% or less, and while substitutional transition metal impurities have been predicted to introduce interband gap energy states [25], the same dopants have not been reported to significantly change the bandgap absorption energy or the absorption coefficient [28–31].

4.2.4 Charge Carrier Transport and Dynamics

During an electromagnetic energy absorption event in hematite, a photon with energy greater than the bandgap will excite an electron from an occupied to an unoccupied state. For this process to be useful for water splitting the excited electron and remaining electron vacancy (hole) must spatially separate and live long enough to reach points of their collection or reaction. Luminescence studies on nanoparticles show that the excitation of hematite with photons of 3.2 eV results in fluorescence quantum yields of about 1×10^{-5} [32]. This extremely low value of radiative carrier recombination suggests that fast nonradiative processes such as carrier trapping and phonon coupling are limiting the excited state lifetime [33]. Indeed, the ultra fast dynamics of excited states in hematite have been studied in nanoparticles, thin films, and single crystals using femtosecond laser spectroscopy [32, 33]. In nanoparticles it was found that the excited state decay profiles were independent of the pump power and the probe wavelength, and were not affected by lattice doping or surface modification. However, the lifetime of the excited state was determined to be very short – 70% of the transient absorption had disappeared after just 8 ps and could not be detected after 100 ps [32]. Studies on epitaxially grown thin films (100 nm using an oxygen plasma-assisted molecular beam) reduced the influence of defect states in the bulk or on the surface compared to the nanoparticle case, but found similar excited state dynamics, which were described by the following scenario: Initially, hot electrons relax to the conduction band edge within 300 fs. Then their recombination with holes and trap states occurs within 3 ps. The resulting trap states can exist for hundreds of picoseconds or longer [33]. In addition, the dominant carrier trapping mechanism was ascribed to mid-gap Fe^{3+} d–d states. These carrier trap states have been reported to be 0.5–0.7 eV below the conduction band edge giving them an optical transition of about 1.5 eV [34, 35] and are thus likely the same spin-forbidden transitions intrinsic to hematite which were discussed in the previous section.

A common metric sought by researchers in the field of semiconductor solar energy conversion is the excited carrier diffusion length which describes how far an average excited state can travel before recombination, and is estimated by taking the square root of the product of the carrier lifetime and the diffusion coefficient. Thus an ultra short lifetime alone does not prevent high efficiency in a semiconductor for PEC energy conversion if the carriers have a high mobility through the crystal lattice. Accordingly, understanding the transport of carriers in hematite is equally important to understanding the lifetime.

The earliest work by Morin on pure sintered polycrystalline hematite describes very low electrical conductivities (ca. $10^{-14}\ \Omega^{-1}\ cm^{-1}$) [36], conduction electron concentrations of $10^{18}\ cm^{-3}$ at 1,000 K, and an activation-energy type electron mobility on the order of $10^{-2}\ cm^2/V/s$ [37]. Studies on pure single crystals showed conductivities below $10^{-6}\ \Omega^{-1}\ cm^{-1}$ and were not suitable for further electronic characterization [31]. These unusually small values obliged electrical conduction to be explained by Fe^{3+}/Fe^{2+} valence alternation on spatially localized 3d orbitals. Modeling studies have had success matching empirical data by describing the conduction mechanism with a small polaron model which includes the effect of the larger size of the Fe^{2+} ion and the associated lattice distortion (polaron) [38, 39]. Conduction of electrons or holes can then be best described by hopping of polarons with an activation energy. This conduction mechanism causes the mobility of carriers to increase with increased temperature, as transport is assisted by lattice vibrations [40]. The localized electron, small polaron model has also been further supported by experimental results including pulse radiolysis [41] and the previously mentioned femtosecond laser spectroscopy on nanoparticles [32].

Further studies of hematite single crystals identified a highly anisotropic electron transport with conduction within the iron bilayer (001) basal plane four orders of magnitude greater than perpendicular directions (parallel to [001]) [42, 43]. This discrepancy cannot be explained by the proximity of iron cations alone, as the shortest Fe–Fe distances are actually along the [001] direction and nevertheless they are quite similar to Fe–Fe distances in the basal planes [44]. However, the anisotropic conductivity can be classically explained considering Hund's rule and the magnetic structure of hematite. The ferromagnetic coupling of the spins in the (001) basal planes and antiferromagnetic coupling along the [001] direction create an environment where electrons can move (n-type conductivity) within the iron bilayers (an environment of parallel spins) but are forbidden to hop across the oxygen plans to an iron bilayer with the opposite spins. Thus conduction in the [001] direction could only involve the movement of holes in the form of $Fe^{3+} \rightarrow Fe^{4+}$ electron transfer (p-type conductivity). This process is significantly slower in hematite [40]. While this classical explanation is generally accepted, recent ab initio electronic structure calculations combined with electron transfer (Marcus) theory have correctly predicted the large transport anisotropy in hematite but suggested it arises from the slowness of both hole and electron mobility across basal oxygen planes [10]. This is contrast to the classical conception as it does not consider electron transport to be forbidden in the [001] direction, but instead identifies the most important factor that influences the carrier mobility to be the electronic coupling – a quantity found to depend on both a superexchange interaction between the bridging oxygen atoms and the d-shell electron spin coupling.

Regardless of the specific details of the conduction mechanism it is clear that the conduction properties of pure hematite are not appropriate for use as a PEC semiconducting material. However, the conductive properties of hematite can be significantly enhanced by adding impurities to act as electronic dopants. Indeed, it is possible to increase conductivities and obtain both p-type and n-type alpha-Fe_2O_3 by doping with atoms such as (but not limited to) Mg^{2+}, Cu^{2+} (p-type) or with Ti^{4+}, Sn^{4+}, Zr^{4+}, Nb^{5+} (n-type) [40]. This behavior can be easily explained if we consider hematite to be a small polaron conductor with n- and p-type charge carriers assumed to be Fe^{2+} and Fe^{4+}, respectively. Then, if the impurity atoms substitute for a Fe^{3+} ion in the crystal lattice, they would cause the creation of Fe^{2+} (n-type) or Fe^{4+} (p-type) to balance the lattice charge. By doping at sufficient levels, high carrier conductivities can be attained. For example, Zr^{4+} was doped into single crystals to give donor densities on the order of 10^{19} cm^{-3}, conductivities around 0.1 Ω^{-1} cm^{-1} , and increased electron mobility (perhaps due to a increase in dielectric constant) of 0.1 $cm^2/V/s$ [31]. A review by Shinar and Kennedy outlines the dopants used for n-type iron oxide and their affect on electronic and PEC properties [45].

In addition to foreign impurities acting as electronic dopants, oxygen vacancies play an important role in the electronic behavior of hematite, and are the most commonly present intrinsic defects found in hematite [17]. In fact, samples of natural specular hematite have been found to be degenerate semiconductors due to oxygen deficiencies [34]. This is because oxygen vacancies act as n-type dopants by creating Fe^{2+} ions. Elevated temperature treatment (above 800°C) in air or anaerobic environments has been shown to create oxygen vacancies in hematite [17]. In this case, the oxygen vacancies may result in the formation of a Fe_3O_4 on the surface, distorting conduction measurements [38], and causing extensive recombination of carriers and reduced photoconduction that can be regained by oxygen treatment [46].

Finally, based on the lifetime and charge carrier mobility found in hematite, the diffusion length of excited carriers is expected to be small. Kennedy and Frese calculated the diffusion length for holes in n-type (Ti^{4+} doped) polycrystalline samples using the experimental photocurrent efficiency and a band-bending semiconductor liquid interface model to determine the mean diffusion length for holes, L_p, outside the space charge region as 2–4 nm [18], and a similarly indirect method gave the same order of magnitude for pure Fe_2O_3 [47]. It is important to note that this value is significantly smaller compared to other oxides used for PEC water splitting. For example, values of L_p up to 10^4 and 150 nm have been reported for TiO_2 [48] and WO_3 [49], respectively. The ramifications of such a low value for L_p on the performance of hematite for PEC water splitting are discussed in the next section.

4.2.5 Photoelectrochemical Properties: Challenges for Solar Water Oxidation

Hardee and Bard [50] first turned to Fe_2O_3 as a material for water photolysis in 1976 seeking a photoanode material that was both stable under anodic polarization and capable of absorbing light with wavelengths longer than 400 nm. Before this, materials like GaAs, CdS, and Si had been found to absorb an appreciable amount of solar irradiation, but also to be unstable in aqueous environments under anodic condition. In contrast, materials like TiO_2, $SrTiO_3$, and SnO_2 had been identified as stable, but had bandgaps over 3 eV. Hardee and Bard made thin films of Fe_2O_3 on Ti and Pt substrates by chemical vapor deposition (CVD) of $Fe(AcAc)_3$ and observed photocurrent from their electrodes under illumination of 500 nm light at a potential of 0.8 V vs. SCE in 0.5 M KCl as electrolyte. Soon after Yeh and Hackerman [51] simply heated iron foils in air to make n-type iron oxide electrodes that gave photocurrents with no observable corrosion problems from pH 4.0 to 14 and Quinn et al. [52] showed that the (012) face of flux grown single crystals had collection efficiencies of around 20% at 0.5 V vs. SCE in 2 M NaOH (1.57 V vs. RHE) with 475 nm light. Further work by Kennedy et al. showed that p-type doping (with Pb^{2+} and Ca^{2+}) produced electrodes with high resistivity, but Si^{4+}, Sn^{4+}, and Ge^{4+} doped n-type hematite performed better than Ti^{4+}-doped hematite [28]. In addition, these initial studies showed that hematite had a flatband potential, V_{fb}, too low in energy to reduce water, and an external bias had to be applied to complete the water-splitting reaction (the electron energy scheme for a n-type hematite photoanode water-splitting system is shown in Fig. 4.3, left). Despite this, hematite was still considered to be a promising n-type photoanode material for solar water splitting due to its capability to absorb a significant portion of solar irradiation (29% of AM 1.0 [52]) and superior stability. Over the next decade numerous reports were published describing continuing studies with pure and doped Fe_2O_3 made by various routes. Notable extensive PEC studies were performed on Ti-doped polycrystalline-sintered hematite pellets [18, 23], and Nb-doped single crystals [53] – the latter reporting champion performance with an incident photon conversion efficiency (IPCE) of about 37% at 370 nm and 1.23 V vs. RHE in 1 M NaOH (27% at 1.06 V, see Fig. 4.3, right).

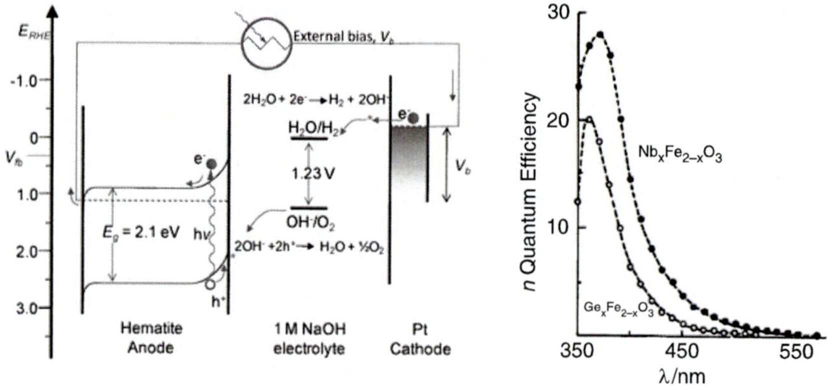

Fig. 4.3 Challenges presented by hematite: at left is an energy diagram showing a typical value of the flatband potential, V_{fb}, of n-type hematite, and the PEC water-splitting operation of a hematite photoanode under illumination and with an applied external bias, V_b. The right graph (from [53], with permission) shows the quantum efficiency vs. wavelength for Nb-doped and Ge-doped hematite single crystals at 0 V vs. SCE in 1 M NaOH (1.06 V vs. RHE). The efficiency is expressed in units of percent

A good review of these seminal studies on hematite has been given by Lindgren et al. [14] and in summary, the first decade of work identified the many challenges of employing this material for PEC water splitting:

1. Flatband potential too low in energy for water reduction [23, 52]
2. Large overpotential and slow water oxidation kinetics [23]
3. Low absorption coefficient, requiring relatively thick films [54, 55]
4. Poor majority carrier conductivity [35, 53]
5. Short diffusion length of minority carriers [18]

These limitations deterred much further interest in using hematite as a photoanode for water oxidation, as it appeared that the conversion of solar energy with high efficiency with it was not possible. Specifically, the small polaron transport and the deep inter-bandgap traps in hematite, which were discussed in Sect. 4.2.3, result in challenge numbers 4 and 5 above. Drawback 4 requires high doping levels to increase the ionized donor concentration and thus the conductivity. This in turn reduces the width of the space charge layer [56]. For example, donor dopant concentrations in the champion Nb-doped single crystals were found to be 5×10^{19} cm^{-3} by Mott–Schottky analysis [53]. Assuming classical depletion layer theory [56], this would result in a depletion layer width, W, of 7 nm at a band bending of 0.25 V and $\varepsilon = 100$. The low absorption coefficient in hematite (challenge 3) then implies that most of the incident photons will not be absorbed in the space charge region if the geometry of the photoanode is planar because the absorption depth (defined as α^{-1} = depth at which 63% of the photons are absorbed according to the Beer–Lambert law) for hematite ranges from 120 to 46 nm for photon wavelengths from 550 to 450 nm [17]. This would not be problematic if the diffusion length of the minority carriers was large compared to the absorption depth, but it is only 2–4 nm (challenge 5). The effect of these limitations emerges as low quantum efficiencies, especially for the longer wavelengths. This is shown in Fig. 4.3 (right) with the highly efficient Nb-doped single crystals [53]. Even though peak quantum efficiency is close to 30% at $\lambda =$ 370 nm, it drops to below 5% at 450 nm and even though all of the incident irradiation is absorbed at 550 nm in the 0.1-mm thick films used, the quantum efficiency is less than 1%.

The above challenges and limitations suggest that the standard planar single crystal or sintered disk electrode geometries are not suitable for hematite, and indeed recent efforts with hematite have shown that challenges 3–5 can be overcome by optimizing the electrode morphology. These recent and exciting advances in hematite structuring are described in detail in the next section. In addition, the renewed efforts with structuring hematite have instigated further research to overcome challenges 1 and 2 listed above. These efforts are reviewed in Sect. 4.4.

4.3 Morphology Control of Fe_2O_3

Both the small (ca. 10 nm) space charge layer present in highly doped hematite (resulting from the necessity for conductivity) and the short (ca. 4 nm) diffusion length of minority carriers discussed in the previous section suggest that the ideal morphology of a hematite photoanode would be one where all of the hematite is within 10–20 nm of the semiconductor/liquid junction (SCLJ). This would give photogenerated holes a high probability of arriving at the SCLJ to oxidize water. Of course, one could simply deposit a 10 nm layer of hematite on a transparent substrate using a traditional deposition method. Then the poor absorptivity of such a film could be overcome by stacking multiple films in tandem. Indeed, this solution was first proposed by Itoh and Bockris in 1984 [55]. While this could fundamentally resolve the issue, it is cumbersome and expensive to implement, and in practice the thin iron oxide films were found to have poor performance due to the increased recombination of the photogenerated holes [54]. Fortunately, the recent development of tools to control the dimensions and morphology of materials at the nanometer length scale has offered more practical morphological solutions for hematite. Of course, the application of nanostructuring techniques to PEC solar hydrogen production has included more materials than simply hematite, with materials including WO_3 and TiO_2 also seeing noteworthy development. Two recent review articles from the research groups of both Augustynski [57] and Van de Krol [58] present a general overview of this topic. Currently, using new nanostructuring techniques, research groups are designing hematite with the proper morphology, feature size and dopants in order to enhance its ability to split water from solar irradiation. This section examines how directing the nanostructure of hematite has led to significant advances in its use as a material for photoelectrochemcial water splitting. This is done by analyzing how different deposition techniques have allowed the morphology control of hematite and affected its performance as a photoanode.

4.3.1 Colloidal Hematite

The nanoscopic manipulation of hematite was first mastered by Matijević's group in the early 1980s [59]. They created well-defined hematite colloidal particles through a hydothermal route and demonstrated morphological control by changing the concentration and the oxidative environment present during nucleation and growth. Spheres, disks, and plates of hematite were observed with lengths in the 100 nm range. However, no photochemistry was investigated with materials of this type until Moser and Grätzel reported halide photooxidation with colloidal dispersions of α-Fe_2O_3 2 years later [60]. In this work, 60 nm particles dispersed in aqueous solution were irradiated with 347 nm laser light and the extremely rapid (~10 ns) formation of I_2^- from iodide at high quantum efficiency (>80%) was observed. This result suggested an advantage of employing nanosized hematite where a larger fraction of the photogenerated carriers are produced closer to the surface as compared to the single crystal hematite or sintered pellets which had been reported at the time. Similar studies following shortly thereafter used various electron relays (*e.g.*, methyl viologen) and hole scavengers (e.g., methanol) to investigate the ability of Fe_2O_3 colloids to produce hydrogen from water [61–63]. While the conversion efficiency remained quite low even in these sacrificial systems, the optimum particle size, balancing bulk recombination and light harvesting, was estimated to be 100 nm in this work [63].

To date there have been numerous reports investigating the photocatalytic properties of colloidal α-Fe_2O_3. Most recently, layered "nanoflowers" [64] and hollow nanoparticles [65] synthesized via a simple solvothermal approach have shown enhanced photocatalytic activities for the degradation of Rhodamine B and diethyl phthalate, respectively, due to their large surface area. While these studies can give insight to the relative catalytic activity of different structures, given the band edge energies of hematite and the sluggish oxygen evolution reaction (OER) kinetics, it is likely that Fe_2O_3 colloids will not find application for water splitting but remain at the degradation of organics. However, the continued development of the precise control over hematite nanoparticle shape and size, like that recently demonstrated by Hu and Yu [66], should allow access to ideal electrode morphologies in the future.

4.3.2 Porous Thin Films from Colloids

Even if hematite nanoparticle dispersions could split water efficiently, the problems of gas separation and the reverse reactions (hydrogen oxidation and oxygen reduction) occurring in the heterogeneous mixture would remain. A more practical approach to water splitting is arguably a PEC cell with the anode and cathode evolving oxygen and hydrogen separately. This configuration provides the challenge of creating hematite with large surface area-to-volume ratio on the surface of

Fig. 4.4 Scanning electron micrograph (*left*) of the edge of a porous electrode made from sintered Fe_2O_3 sols on a conducting cassiterite (SnO_2) support (magnification 5,800 times). Absorption spectrum of the hematite electrode and quantum efficiency as a function of wavelength (*right*) obtained in 0.1 M NaOH at 1.4 V vs. RHE when illuminating the electrode through the substrate (SE) and directly onto the interface with the electrolyte (EE). Note that the scale of the IPCE values differs with a factor of 100 in the two cases. From [67] used with permission

a conducting support. The most straightforward way to create nanostructured photoelectrodes of hematite is to coat a sol onto an electrode and sinter to enforce interparticle connections – a method first perfected for solar energy conversion with TiO_2 in the dye-sensitized solar cell. This method was first attempted with hematite by Bjorksten et al. in 1994 [67]. Here, Fe_2O_3 sols created by the hydrolysis of $FeCl_3$ were concentrated and combined with surfactant before they were doctor-bladed onto transparent conducting oxide substrates (F-doped SnO_2) and sintered at 560°C. Micron-thick, porous thin films of necked hematite were observed to have good adhesion to the substrate and a primary particle size in the 25–75-nm range (see Fig. 4.4, left) However, the quantum efficiency (IPCE) of these photoanodes toward water splitting was quite low – on the order of 1% at 400 mV vs. SCE in 0.1 M NaOH (1.4 V vs. RHE) with 400 nm incident irradiation (see Fig. 4.4, right). The observations that the IPCE was 100 times lower when the anode was illuminated from the hematite/electrolyte interface as compared to the substrate/hematite interface and that the quantum efficiency did not improve significantly (only to 1.7%) when LiI was added as a hole scavenger led the authors to conclude that charge carrier recombination was the controlling factor for the photocurrent. The higher quantum efficiency of similar particles when dispersed in electrolyte [60] pointed to grain boundaries between the particles to be the cause of the recombination and poor performance. This limitation was addressed by Qain et al. [68] by altering the film thickness to optimize the light absorption/carrier transport issue. However, no significant improvement was obtained. Cathodic spikes observed during transient measurements in this work suggested that no space charge layer was present, and that electron recombination at the hematite/electrolyte interface was also an important limitation.

Recent work on Fe_2O_3 films prepared by colloidal methods has identified a possible application of sensing organics in gas streams due to a change in electrical resistivity [69, 70] and suggested that the inclusion of metal nanoparticles [71, 72] or the presence of a spinel phase (γ-Fe_2O_3) [73] may increase the catalytic activity of these electrodes. However, a fundamental issue of particle size recognized in the seminal work [67] has yet been unresolved: Despite attempts to use particle sizes commensurate with the hole diffusion length (<10 nm), the temperatures required to sinter and make the films photoelectrochemically active result in primary particle sized too large to expect high photocurrents. Incorporating dopant atoms [74] and controlling feature size with templating techniques [75, 76] may result in an increase in performance for hematite films prepared by this inexpensive and scalable technique.

4.3.3 Fe_2O_3 Nanowire Arrays

An elegant solution to the problem of grain boundaries in hematite films created from the colloidal approach is to use nanosized rod or wire arrays. An array of single crystalline nanorods with diameters in the 10-nm range, attached and oriented orthogonally to the conducting substrate would eliminate grain boundaries, enhancing conduction to the substrate while still allowing photoproduced holes to efficiently reach the semiconductor/electrolyte interface. A simple method to create hematite arrays on a variety of substrates via a metastable akaganeite phase produced from the controlled precipitation of Fe(III) in aqueous solution was first reported by Vayssieres et al. [77], and investigated for water photoelectrolysis soon after [19]. Bundles of individual 5 nm nanorods with an average diameter of 50 nm and a length of 0.1–1.5 μm were investigated in perpendicular and parallel orientations to the substrate (Fig. 4.5, left). While the authors were able to show a small improvement when controlling the orientation (5% IPCE at 360 nm for the perpendicular nanorods vs. 3% for the parallel) the overall photocurrents remained low under white light illumination, even with hole scavenging iodide present. The large difference between the electrode performance when illuminating from different sides of the semitransparent photoanode observed in this work [79], and a recent report [80] examining the surface photovoltage on electrodes prepared the same way, suggest that bulk defects and the slow oxygen evolution kinetics of hematite are the major factors limiting the performance of this system.

Despite the poor performance of the nanowire arrays prepared by this technique it remains an intriguing morphology for hematite photoanodes. Another facile method to produce hematite nanowires by the simple thermal oxidation of iron foils has been reported by many groups [78, 81–84]. Due to the increased volume of the oxide over the metal, when foils of iron are thermally oxidized under the right

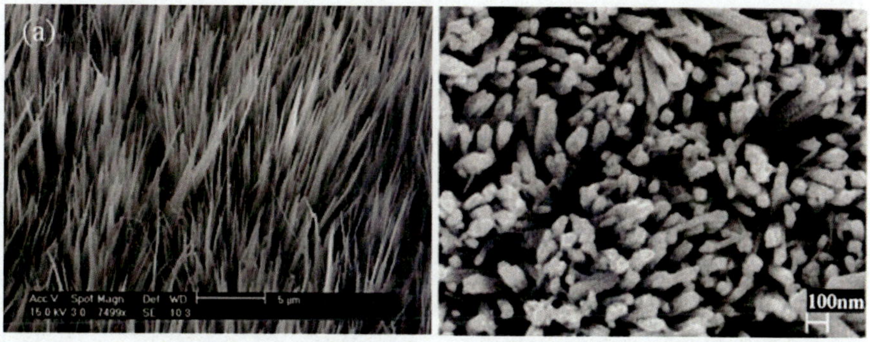

Fig. 4.5 Examples of hematite nanowire morphologies: SEM images of hematite prepared by the thermal oxidation method (*left*) from [78] and aqueous method (*right*) from [77] used with permission

conditions, arrays of Fe_2O_3 nanowires spontaneously grow from the surface of the iron. With diameters of 20–40 nm and lengths of up to 5 μm (see Fig. 4.5, right) these nanowire arrays would have large surface area, sufficient light absorption and a direct path for the conduction of electrons to the substrate (basal planes are oriented perpendicular to the substrate [83]), making them a very attractive morphology for hematite. This thermal oxidation method has been demonstrated on multiple (even transparent) substrates [84], and methods for doping the particles have been employed [78]. However, no realistic reports of activity toward water oxidation have appeared. This may be due to the defects present in nanowires prepared this way or the presence of suboxides near the interface of the substrate which greatly enhance recombination [85].

4.3.4 *Electrochemical Fe_2O_3 Nanostructuring*

The development of nanostructuring techniques using potentiostatic anodization has provided another possible route to create structured hematite photoelectrodes. While it has long been known that the potentiostatic anodization of certain metals can create structuring on small length scales, only recently have methods been developed to prepare ordered arrays of nanosized tubes or pillars with this technique. For example, ordered nanotube arrays of TiO_2 created using potentiostatic anodization have been successfully employed in multiple device applications including the solar photolysis of water [86]. Like the nanowire approach, these nanotube arrays have the potential to reduce crystal grain boundaries and enhance electron conduction to the current collector. In addition, the small thickness of the tube walls should allow for efficient hole collection. Recently, Grimes and coworkers showed that iron foils could be nanostructured using anodization in a glycerol-based electrolyte containing 1% NH_4F + 1% HF + 0.2% HNO_3 [87]. Ordered nanopores

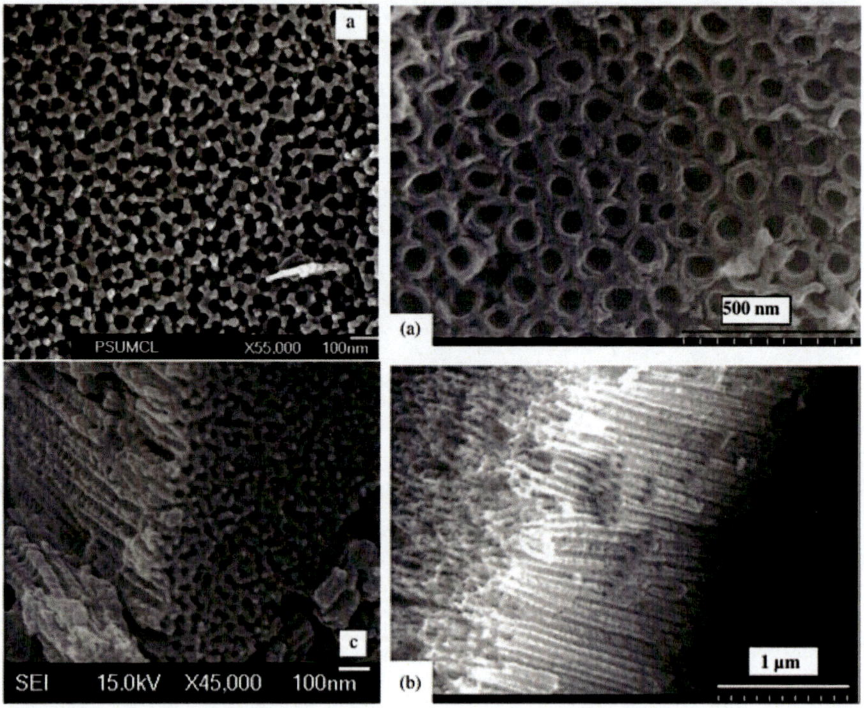

Fig. 4.6 Hematite morphologies accessible by potentiostatic anodization: The *left* (*top* and *bottom*) shows SEM images of structures from the work of Prakasam et al. [87] and the *right* (*top* and *bottom*) shows images of tubes from Rangaraju et al. [88]. SEM images used with permission from [87] and [88]

were observed with pore size ranging from 20 to 250 nm and depths up to 600 nm depending on the anodization voltage and time (see Fig. 4.6, left). Under simulated solar illumination these photoanodes produced a photocurrent of 0.05 mA/cm^2 at 0.4 V vs. SCE in 1 M NaOH. This current increased tenfold with the addition of 0.5 M H_2O_2 indicating good electron transport in these films, and suggesting the surface kinetics of the water oxidation reaction were limiting the water-splitting photocurrent. The same research group improved on this result by alloying the iron with titanium before anodization [89]. Water-splitting photocurrents were obtained from films that were 6.6% Fe, based on metal, and were largely anatase TiO_2. They observed reasonable photoactivity with a quantum efficiency of ca. 4% with 400 nm light at 1.5 V vs. RHE. In contrast to this Fe–Ti–O work, very well-defined nanotube arrays (see Fig. 4.6, right) created from pure iron foils have been subsequently reported by a different research group [88]. In this work, a single anodization step with 0.1 M NH_4F + 3 vol% water in ethylene glycol created nanotubes with walls less than 50 nm and lengths of about 1.5 μm. After an optimized annealing treatment these electrodes were found to be a mixture of both hematite and maghemite by XRD and had only small photocurrents (160 μA/cm^2, 1.23 V vs.

RHE in AM 1.5 light compared to 120 μA/cm^2 dark current). However, a double step anodization procedure with the addition of sodium tripolyphosphate in the electrolyte, and the subsequent annealing created a dendrite-like morphology situated over the tube array. Electrodes prepared this way exhibited a photocurrent of about 1 mA/cm^2 at AM 1.5 and 1.23 V vs. RHE. This is quite remarkable given the presence of maghemite and as the films were not intentionally doped to increase the carrier concentration. However, a large carrier concentration (10^{21} cm^{-3}) was found via Mott–Schottky indicating a large amount of dopant or impurity present. The authors attribute the high photocurrent to the double-layered structure which includes both a large surface area for water oxidation and a vertically orientated nanotube array for electron transport.

In subsequent work by the same group, a sono-electrochemical anodization method was employed to obtain either nanoporous or nanotubular Fe_2O_3 depending on the anodization conditions [90]. The optimization of conditions gave nanotubes with walls an impressive 5–6-nm thick and microns in length. Annealing these films resulted in pure hematite which gave similar photocurrents (ca. 1 mA/cm^2 at AM 1.5 and 1.23 V vs. RHE) and similar carrier densities to the previous work. While the initial tube size presented in this work is remarkable and an important step for obtaining hematite with the ideal morphology, the tubes were initially amorphous Fe_2O_3. The annealing step necessary for crystallinity and photoactivity probably demolishes the meticulous nanostructuring through a sintering process like has been observed for fine-structured hematite at high temperatures [67].

In addition to the nanostructuring of Fe_2O_3 with potentiostatic anodization, hematite photoanodes have been prepared by electrodepostion of precursors from solution. Recently, McFarland and coworkers reported an electrodeposition method of iron hydroxides from $FeCl_3$ solutions. The subsequent annealing at high temperature (700°C) then resulted in porous hematite films [20, 91]. This method readily allows for the incorporation of dopants which were found to have an effect on the morphology of the sintered electrodes. With feature sizes of 30–40 nm the observed IPCEs of 8% at 400 nm at 1.2 V vs. RHE in 1 M NaOH with 15% Mo doping are a notable result. However, the large particle size remains a limitation to higher photocurrents. Another example of electrodeposition has been recently reported by Spray and Choi [92] using an anodic method as opposed to the cathodic method investigated by the McFarland group. They were able to demonstrate impressive morphology control ranging from wire arrays to porous films by varying the solution pH. The films were photoactive in an electrolyte-containing iodide, but water oxidation photocurrents were not reported.

In general, it is clear that major advances have been made recently in the nanostructuring of hematite by electrochemical methods. However, the potential to oxidize water with hematite prepared in the methods described above has been limited by the crystallinity and ultimate feature size after the requisite annealing.

4.3.5 Spray Pyrolysis Techniques

The colloidal, nanowire, and anodization approaches to deliberately nanostructure hematite have provided great leaps in hematite nanostructuring but only moderate success so far for water-splitting photocurrents. However, from its first report [93] the spray pyrolysis (SP) method has continuously produced thin films of photoactive hematite. Typically, an aqueous or ethanolic solution of Fe^{3+} is simply sprayed onto a hot (~400°C) substrate and photocurrents in the 0.5–1.0 mA/cm^2 range can be obtained under simulated solar illumination (1 M NaOH, 1.6 V vs. RHE) [94–97] with the exception of one report [98] that is improbable in view of the action spectrum reported by the authors [99]. Surprisingly, modest photocurrents are obtained with this method even without intentionally adding dopants to increase the conductivity of the Fe_2O_3. However, it is likely that the commonly used precursors like $FeCl_3$ or $Fe(AcAc)_3$ leave chlorine or carbon residues behind that could act as electron donors. Typically, this technique can produce films with nanometer crystalline domain size, but the films are generally compact (not porous) and the surface area limits the conversion efficiency. However, the ease and reproducibility of the technique allow for fundamental studies on dopants [95, 100] and substrate effects [96], and careful optimization has produced films with a notable quantum efficiency (IPCE = 25% at 400 nm and 1.42 V vs. RHE) [100]. In a recent report from Duret and Grätzel, the spray pyrolysis technique was modified to create nanostructured films [94]. Here an ultrasonic spray nozzle was used to create very fine droplets of the precursor which were entrained in a flow of dry air and carried rapidly and horizontally over a heated substrate. The low concentration of the precursor in the gas stream allowed the slow growth (~100 nm/h) of mesoscopic hematite with a leaflet-type structure consisting of 100-nm-sized platelets with thickness of 5–10 nm bundled into 50 nm sheets oriented perpendicular to the F-doped SnO_2 support (see Fig. 4.7, top). These ultrasonic spray pyrolysis (USP) films were found to perform impressively for water splitting, producing 1.3 mA/cm^2 under AM 1.5 (100 mW/cm^2) simulated solar irradiation in 1 M NaOH with an applied potential of 1.23 V vs. RHE. The IPCE was found to be 16% with 375 nm illumination at the same applied potential, and over 30% at 1.6 V vs. RHE. This was a threefold increase over control films created by a conventional spray pyrolysis method in the same work, and even an improvement over optimized photoanodes created by SP [100]. The advantage of the nanostructured morphology could be directly observed by comparing normalized IPCEs of the UPS and SP electrodes (Fig. 4.7, bottom). The IPCE is higher for longer wavelengths (400–600 nm), indicating the more efficient conversion of holes due to the morphology. Subsequent work on these films revealed the importance of Si doping, and showed an enhanced (110) reflection from XRD as compared to powder samples indicating a preferred orientation of the (001) basal plane normal to the substrate [101, 102]. The vertical orientation of the (001) planes is thought to be very important to the conduction of photogenerated electrons to the conducting substrates, as this is the least resistive direction of conduction in hematite.

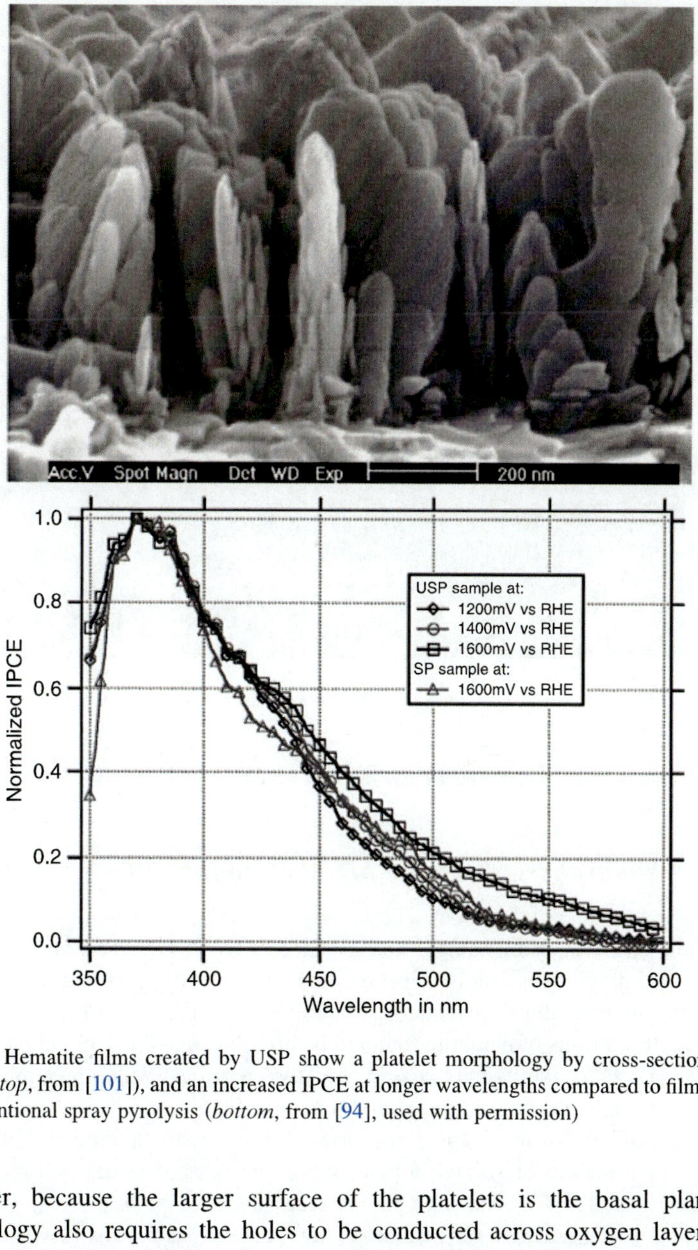

Fig. 4.7 Hematite films created by USP show a platelet morphology by cross-sectional SEM analysis (*top*, from [101]), and an increased IPCE at longer wavelengths compared to films created by conventional spray pyrolysis (*bottom*, from [94], used with permission)

However, because the larger surface of the platelets is the basal plane, this morphology also requires the holes to be conducted across oxygen layers. This detail and the poor kinetics of water oxidation were thought to be limiting the performance in this system.

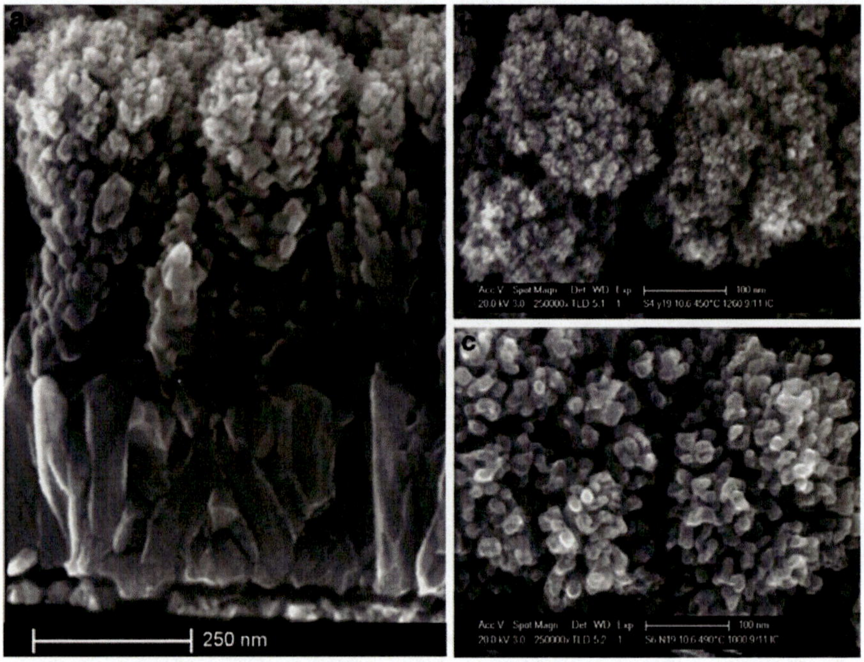

Fig. 4.8 Morphology of Si-doped hematite films created by the APCVD technique as observed by SEM. The left side shows a cross-sectional view of the Fe_2O_3, prepared at a substrate temperature of 420°C (heater setpoint of 470°C), on top of the $F{:}SnO_2$ substrate. The right shows top-down views of films prepared with heater set points of 450°C (*above*) and 490°C (*below*). From [13] used with permission

4.3.6 Atmospheric Pressure Metal Organic CVD

Continuing efforts from Grätzel and coworkers to perfect the nanometer length scale morphology of hematite thin films for water splitting beyond the USP films have led to the development of a simple, yet effective process to make nanostructured films of hematite found to be highly active for water photooxidation (Fig. 4.8). This method is based on the thermal decomposition of iron pentacarbonyl in an atmospheric pressure chemical vapor deposition (APCVD) reactor. $Fe(CO)_5$ vapors are transported, along with tetraethyl orthosilicate (TEOS) as a source of silicon dopant, by an air stream vertically to a substrate maintained at 420°C. This high temperature and the poor thermal stability of the $Fe(CO)_5$ (the half-life for decomposition at 300°C is 5.3 ms [103]) causes the homogenous nucleation of nanoparticles in the gas phase [104]. These particles are then subjected to a thermophoretic force which limits their approach to the hot substrate and results in fractal-like cauliflower structures suggesting a diffusion-limited aggregation (rather than surface reaction limited) mechanism [13]. Similar to the films prepared by the USP method, the silicon dopant was found to affect the

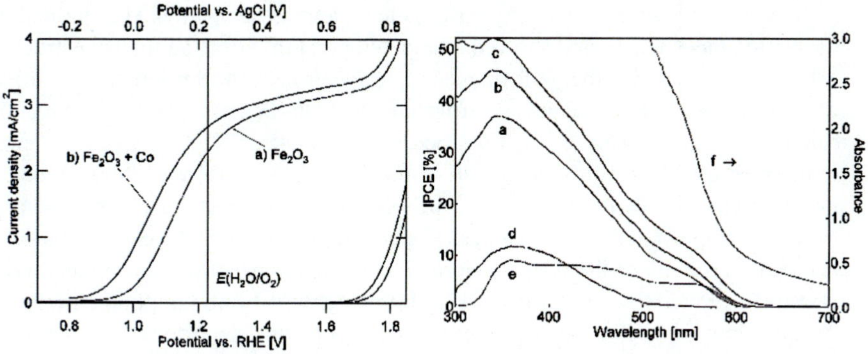

Fig. 4.9 Photoelectrochemical characterization of Si-doped Fe_2O_3 electrodes prepared by the APCVD technique. The *left graph* shows the current-voltage characteristics in darkness and under simulated sunlight (100 mW/cm^2) at a scan rate of 50 mV/s. (**a**) Unmodified Fe_2O_3; (**b**) the same electrode after cobalt treatment. The right graph shows the incident photon-to-current efficiency (IPCE) spectrum of the same electrode at different conditions. (**a**) Unmodified Fe_2O_3 at 1.23 V_{RHE}; (**b**) after cobalt deposition, at 1.23 V_{RHE}, (**c**) at 1.43 V_{RHE}, (**d**) at 1.03 V_{RHE}, (**e**) at 1.23 V_{RHE} but illuminated through the SnO_2 substrate; (**f**) absorption spectrum of this electrode. From [105] used with permission

nanostructure, producing a smaller feature size in both cases and an alignment of the basal crystal plane normal to the substrate [102, 105]. While this latter observation is considered to be a major factor improving the performance of these electrodes by facilitating electron transport to the collector, another report found the crystalline orientation of hematite unimportant to the performance [106].

The optimization of the APCVD deposition technique led to films producing a photocurrent of 1.8 mA/cm^2 at 1.23 V vs. RHE under simulated solar illumination (AM 1.5, 100 mW/cm^2) in 1 M NaOH. This was further increased to 2.2 mA/cm^2 by treating the surface with Co^{2+} ions, which act to lower the energy barrier for water oxidation and reduce the applied potential needed [105] (see also Sect. 4.4.2). With IPCEs of 42% at $\lambda = 370$ nm and 1.23 V vs. RHE these electrodes represent the most efficient Fe_2O_3 photoanodes for water oxidation to date –surpassing even the single-wavelength quantum efficiencies of Nb-doped α-Fe_2O_3 single crystals (IPCE = 37% at 370 nm and 1.23 V vs. RHE) [53]. This benchmark performance corresponds to 3.3% STH conversion efficiency when used in a tandem configuration (see Sect. 4.4.1). The PEC characterization of this champion electrode is shown in Fig. 4.9. Particularly notable is the increased IPCE of the photoanodes at wavelengths between 500 and 600 nm as compared to the Fe_2O_3 film prepared by the USP method [94] and especially the single crystal electrodes shown in Fig. 4.3. Because the high surface area and small feature size of the APCVD films the photocurrent action spectrum is much broader than the single crystal results: photons with longer wavelength have a higher chance for being absorbed within the hole diffusion length, resulting in much higher solar photocurrents than the single crystal electrodes.

In continuing work with this deposition technique Cesar et al. [13] examined the deposition mechanism, and showed a large temperature effect on the nanostructure which corresponded to the performance of the electrodes. In addition, the electrodes were studied by impedance spectroscopy; a Mott–Schottky analysis gave the calculated donor density of the Si-doped Fe_2O_3 films as 1.7×10^{20} cm^{-3} – close to the predicted 6×10^{20} cm^{-3} if all of the 1.6 at% Si [105] was substitutionally incorporated into the lattice as Si^{4+}. This observation suggests that a space charge layer of ~5 nm could be present and aiding hole transport the SCLJ even in the small features on the top of the APCVD film. The high donor density also created good electron conductivity and films up to 1 μm delivered similar photocurrents as thinner films when illuminated from the semiconductor/electrolyte interface even though photons were absorbed further from the conducting substrates. This suggests that electron percolation through the film is not a limiting factor.

Recently another group has reported the application of the APCVD technique for Fe_2O_3 photoanodes using both ferrocene and iron pentacarbonyl as iron sources, and different sources of Si for doping [107, 108]. First, ferrocene as a precursor was shown to be effective in preparing photoactive Fe_2O_3 films without a separate dopant species [108]. Photocurrents were ca. 0.6 mA/cm^2 at 1.23 V vs. RHE in 1 M NaOH under simulated solar irradiation. Furthermore, the larger feature size observed in the ferrocene-prepared electrodes (50–100 nm) as compared to the Fe $(CO)_5$ indicates the importance of the decomposition of the carbonyl precursor and the nucleation of nanoparticles before reaching the hot substrate. Next, the silicon doping was studied by varying the Si precursor species and feed rate [107]. In addition to TEOS, tetramethyl orthosilicate (TMOS) was also found to be an effective source of Si. XPS studies verified the +4 oxidation state of Si in both cases. In addition, trimethyl orthoformate (TMOF) was used as a silicon-free control, and resulted in electrodes with significantly less photocurrent. Although their findings did not result in a definitive explanation of the silicon's role, it did suggest that a combination of factors including the formation of lattice defects and nanoparticle cluster formation caused by the Si precursor's decomposition and incorporation, additional to the classical doping effect, may be contributing to the high performance observed in these electrodes.

Despite the impressive performance of the APCVD iron oxide films the solar photo current remains low compared to the theoretical value of 12.6 mA/cm^2 predicted for a material with this bandgap [6]. This is because, despite absorbing almost all of the light (see Fig. 4.9, left, curve f) the quantum conversion efficiencies are relatively low and are especially poor (<20%) in the region where hematite has an indirect bandgap transition (λ = 610–450 nm). Importantly, there is over 10 mA/cm^2 of solar photocurrent available in this wavelength range. The poor performance can be primarily attributed to the cauliflower-type morphology. While this morphology has an appropriate feature size of 10 nm at the SCLJ, the individual cauliflower structures consist of thick stems which increase the number of photons absorbed far from the SCLJ.

4.3.7 Extremely Thin Absorber–Host-Scaffold Approach

In an extension of the stacked thin film approach proposed by Itoh and Bockris [55] an increased photoresponse should also be obtainable with hematite by coating a thin film conformally on a suitable nanostructured collector with a large specific surface area, in analogy to the dye-sensitized solar cell (DSC) [109] and similar extremely thin absorber (ETA) photovoltaics [110]. In the case of the DSC, the photoanode is constructed from a light-absorbing dye molecule anchored to a high surface area mesoporous TiO_2 scaffold, which collects and transports the photoexcited electrons. The DSC's host-scaffold/guest-absorber approach effectively decouples light harvesting and charge transport while maximizing the IPCE by means of the thin (mono)layer of absorber on the high surface area, transparent collector. Given the success of this concept for photovoltaic energy conversion as well as the light absorbing and photogenerated hole transport limitations of α-Fe_2O_3, this is a promising direction for increasing the quantum efficiency of hematite.

A difficulty in applying this method for hematite is finding a suitable host material which satisfies the following criteria: First, the conduction band of the scaffold material must lie lower in energy than the conduction band of hematite to allow efficient electron transport across the host/guest interface and second, the scaffold material should have a larger band gap than hematite so as to not compete with the light absorption. Few oxide semiconductors meet both of these requirements. However, Grätzel and coworkers have recently demonstrated this concept for hematite photoanodes using nanostructured WO_3 as the host scaffold [8]. In this work the host scaffold was coated with a thin layer (ca. 60 nm) of Fe_2O_3 by the APCVD method (see Fig. 4.10, top) and a 20% increase in the photocurrent was observed over control samples prepared on the flat F:SnO_2 substrates. This increase was attributed to an increase in the absorbed photon-to-current efficiency (APCE), especially in the wavelength range from 500–600 nm (Fig. 4.10, bottom).

While this demonstration of the host/guest approach indicates it is a promising and accessible methodology to fully exploit the light harvesting capabilities of hematite for solar water splitting, the best results were obtained with iron oxide coatings 60 nm thick over the host scaffold. This is too thick based on the hole transport limitations already discussed. Attempts to use thinner films of iron oxide resulted in poor performance, similar to other reports of thin films Fe_2O_3 [54]. Further understanding and control of the interface between the iron oxide and the conducting substrate as well as the electronic properties and crystallinity of ultra-thin films of Fe_2O_3 are necessary to advance this promising nanostructuring approach.

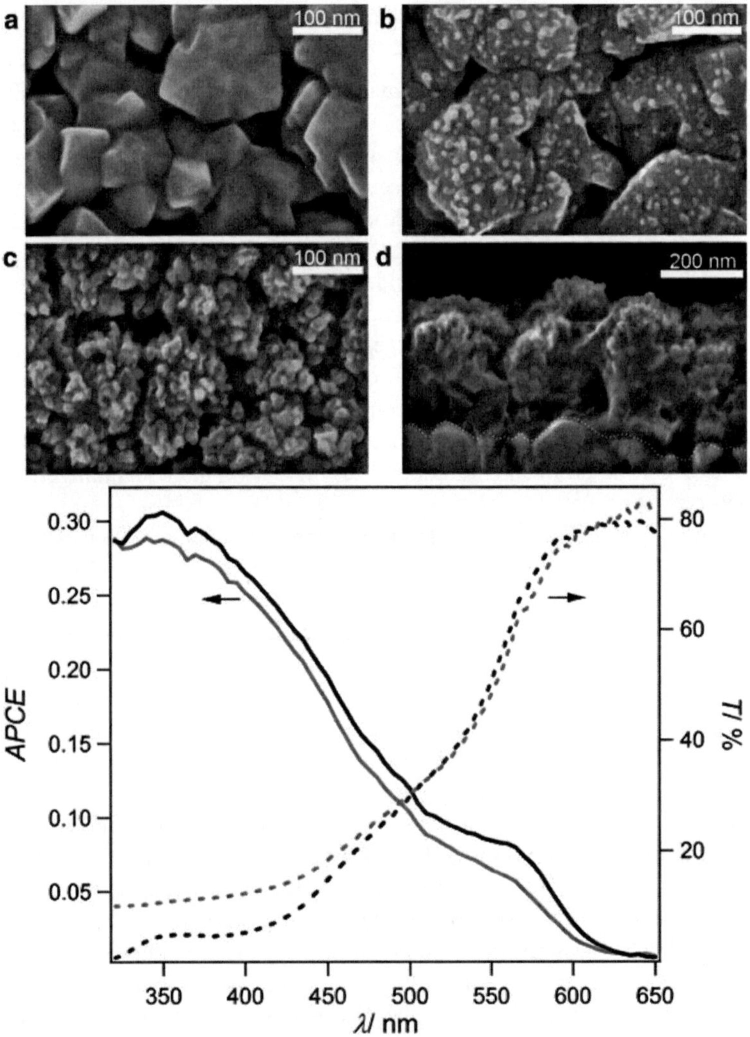

Fig. 4.10 (*Top*) Scanning electron micrographs of the WO_3/Fe_2O_3 host/guest system. Image (**a**) shows the morphology of the WO_3 host scaffold before iron oxide deposition. After 1 min of Fe_2O_3 deposition at the standard deposition rate (~100 nm/min) the WO_3 film is covered in iron oxide nanoparticles (**c**) while at a lower deposition rate (~10 /min, edge of growth spot) the iron oxide nanoparticles are seen just starting to cover the WO_3 scaffold (b). The cross-section of the film, corresponding to (**c**), is shown in (**d**). (*Bottom*) The APCE (*solid lines*) at 1.43 V_{RHE} and the transmittance, *T* (*broken lines*) of photoanodes prepared with a 1-min Fe_2O_3 deposition time plotted against the incident light wavelength, λ. The host/guest and the control electrode responses are indicated by *dark* and *light lines*, respectively

Table 4.1 Comparison of hematite photoanode structuring methods

Technique	Feature size	Dopant	Performance	References
Single crystal	Micron	Nb	IPCE 37% at 370 nm and 1.2 V_{RHE}	[53]
Sintered polycrystalline disk	Micron	Si	IPCE 34% at 400 nm and 1.23 V_{RHE}	[45]
Colloidal deposition of porous films	50 nm	None	IPCE 1% at 400 nm and 1.4 V_{RHE}	[67]
Solution-based nanowires	50 nm × 500 nm	None	IPCE 7% at 400 nm and 0.9 V_{RHE} (in KI)	[19]
Potentiostatic anodization (nanotubes)	6 nm × 1 μm	None	Before annealing: none reported After annealing[a]: 1.0 mA/cm^2 at 1.23 V_{RHE} and AM 1.5 (87 mW/cm^2)	[90]
Electrodeposition	20 – 50 nm	Mo	IPCE 8% at 400 nm and 1.2 V_{RHE}	[20]
Ultrasonic spray pyrolysis	10 nm × 300 nm	Si	1.5 mA/cm^2 at 1.23 V_{RHE} and AM 1.5 (100 mW/cm^2) and IPCE 14% at 400 nm and 1.23 V_{RHE}	[94, 101]
Atmospheric pressure CVD	5 nm × 300 nm	Si	2.2 mA/cm^2 at 1.23 V_{RHE} and AM 1.5 (100 mW/cm^2) and IPCE 36% at 400 nm (42% at 370 nm) and 1.23 V_{RHE}	[105]
Extremely thin absorber	50 nm film	Si	1.3 mA/cm^2 at 1.23 V_{RHE} and AM 1.5 (100 mW/cm^2) and IPCE 26% at 400 nm and 1.43 V_{RHE}	[8]

[a]Feature size not reported after annealing

4.3.8 *Nanostructuring Method Comparison*

Many different approaches have been taken to nanostructure Fe_2O_3 for use in PEC water splitting, and at least as many methods of characterizing the electrodes have also appeared. As such it can be difficult to compare the results in the literature, especially when variations in light sources give large over-estimations of photocurrents due to superfluous UV light. A recent and important contribution by Murphy et al. addresses these issues and suggests reliable methods for evaluating electrode conversion efficiencies [6]. While not all of the reports discussed in this section have followed these suggestions it can still be helpful to compare the top performances from the different structuring techniques and the resulting performance. This is done in Table 4.1.

It is clear, by looking at Table 4.1, that while many methods of structuring hematite have been reported, the IPCE values at ca. 400 nm rarely come close that of the single crystal or sintered polycrystalline electrodes prepared during the first decade of work. Even the champion electrodes made by APCVD merely match the IPCE of the single crystal work at ca. 400 nm without the enhanced surface treatment. The key to the increased photocurrents in the APCVD electrodes has, of course, been the nanostructuring, which allows for the more efficient conversion of the longer wavelengths of light. Further work on the nanostructuring will, no doubt, see solar photocurrents rise, but the increase in IPCE beyond 40% at 400 nm and 1.23 V_{RHE} will require research to tackle hematite's other remaining challenges.

4.4 Overcoming Hematite's Remaining Challenges

Many techniques have been used to nanostructure hematite, and while clear advances have been made over planar, single crystal approaches, in general the performance of nanostructured electrodes has remained poor compared to the quantum efficiencies found in optimized photoanodes prepared from TiO_2 or WO_3. While this is partly due to the absence of a method to prepare a "perfect" nanostructure, there are still limitations in (1) the understanding of the critical material properties affecting efficiencies and (2) the surface oxidation reaction kinetics. In addition, based on the flatband potential of hematite it cannot be used to split water unassisted by a bias voltage. In this section, these limitations are clearly detailed and new promising approaches to overcome these remaining challenges are reviewed.

4.4.1 Flatband Potential and Tandem Cell Approach

As mentioned in Sect. 4.2.5, one of the major drawbacks of hematite is that its flat band potential is below the water reduction potential. This means that the electrons exiting the conduction band of hematite and traveling to the cathode do not have sufficient energy to reduce water and create hydrogen; a small external bias must be added to hematite to complete the water-splitting reaction. The overall efficiency of solar water splitting then must account for the power input by this external bias. However, a clever solution to this problem was invented by Augusttynski and Grätzel [111]. Since hematite only absorbs solar irradiation up to wavelengths of 600 nm, transmitted photons ($\lambda > 600$ nm) can be converted by another photosystem to provide the necessary potential to complete the water-splitting reaction and accomplish unassisted solar hydrogen production from water. This second photosystem is most practically a photovoltaic device as photocathode materials typically suffer from stability issues. A promising candidate for this photovoltaic device is the DSC as these devices can be produced for one-third to one-fifth the cost of a conventional silicon solar cell and the absorption range can be tuned by selecting the appropriate dye.

Ideally in this hematite/DSC water-splitting tandem cell, only one photovoltaic system would provide the necessary potential to complete the water-splitting reaction. However, due to the large overpotential needed for water oxidation (see Sect. 4.4.2) and other loss processes, two solar cells are needed to provide enough voltage to the current state-of-the-art Fe_2O_3. For example, saturation (plateau) photocurrents from Fe_2O_3 prepared from APCVD are not reached until 1.4 V vs. RHE. While a traditional solution can still be envisioned with two solar cells with half the active area each behind the hematite photoanode, the recent development of new dyes for the DSC opens up new possibilities for optimizing light harvesting in these water-splitting tandem cells. Different architectures for the hematite/DSC system were recently investigated by Brillet et al. [112] and it was found that a three-level tandem (hematite/suqarane dye DSC/black dye DSC) was the most efficient at splitting water with a calculated overall efficiency of 1.4% solar to hydrogen. This is much lower than the predicted 3.3% with this system based on iron oxide photocurrent alone, and losses were attributed to light scattering and the large overpotential for water oxidation by hematite.

A tandem cell configuration would not be needed if the flat band potential of hematite could be shifted so that water could be directly split by solar light. While numerous groups have reported a V_{fb} 0.3 – 0.4 V lower than the hydrogen evolution potential [14], Hu et al. recently suggested that a CoF_3 treatment of Ti-doped Fe_2O_3 could actually shift the flat band potential and allow the direct hydrogen production by hematite [113]. While additional investigation of this claim is necessary to verify sustained hydrogen production at zero bias, it is indeed a promising route to reduce or eliminate the necessary applied bias.

4.4.2 *Oxygen Evolution Overpotential and Surface Catalysis*

Based on the flat band potential of hematite an external bias of only 0.3 0.4 V vs. RHE should be necessary to initiate the water-splitting reaction. Once the applied bias is greater than the flatband potential, the band bending will drive photogenerated holes to the surface for reaction. However, photocurrent onset potentials are usually not observed until 0.8–1.0 V vs. RHE even for the single crystal electrodes. The remaining potential of ca. 0.5 V is due to the overpotential for water splitting caused by unfavorable surface properties. These unfavorable properties have two origins. First, there is evidence that oxygen vacancies [17, 23, 46] and crystalline disorder [13] can be present at the surface. This would create additional Fe^{2+} recombination centers for holes, and can even result in Fermi level pinning in some electrodes [94]. Second, the reaction kinetics for water oxidation are slow, not just due to the complicated four-electron mechanism that must occur, but also compared to other oxide semiconductors [79]. This may be due to the more pronounced metal 3d character of the valence band compared to other oxides [23].

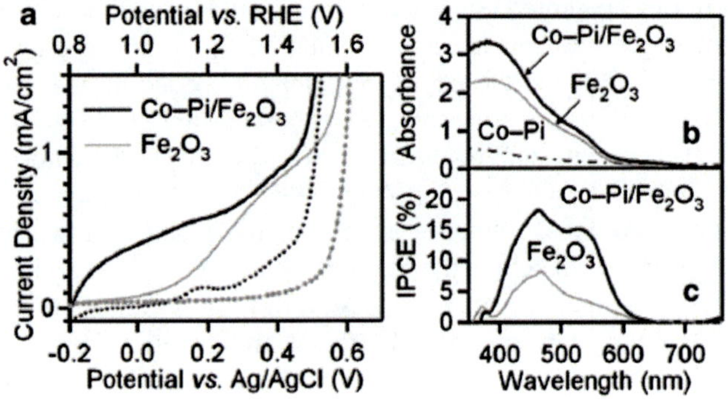

Fig. 4.11 Water splitting with hematite and a cobalt-phosphate-based catalyst: (**a**) Dark (*dashed*) and photocurrent (*solid*) densities for α-Fe_2O_3 (*light*) and Co-Pi/α-Fe_2O_3 (*dark*) photoanodes, collected using simulated AM 1.5 illumination (1 sun, backside illumination) at a scan rate of 50 mV/s. (**b**) Electronic absorption and (**c**) IPCE spectra for α-Fe_2O_3 and Co-Pi/ α-Fe_2O_3 (at 1.23 and 1 V vs. RHE, respectively). The absorption spectrum of Co-Pi on FTO without α-Fe_2O_3 is included in (**b**), but no photocurrent was detected for these anodes. From [116] used with permission

The first surface challenge can be addressed by strong oxidation conditions [46] and careful hematite preparation, but the slow water oxidation kinetics are probably intrinsic to hematite. Nevertheless, methods have recently been found to increase the oxidation rate and thus reduce the overpotential. For example, the water oxidation by cobalt has been extensively studied and is known to be particularly rapid [114]. Indeed the treatment of Fe_2O_3 photoanodes (prepared by APCVD) with a monolayer of Co^{2+} resulted in a ca. 0.1 V reduction of the photocurrent onset potential [105]. Since this treatment also increased the plateau photocurrent it is good evidence that the reaction rate was increased, and the Co^{2+} did not just fill surface traps. Following the report of a remarkably effective cobalt-phosphate (Co-Pi)- based water oxidation catalyst [115], the overpotential was reduced even further on hematite photoanodes by Gamelin and coworkers [116]. Their results are shown in Fig. 4.11.

The application of the Co-Pi catalyst resulting in a shift of the onset potential by 0.25 V. The effect is further supported by the increase in IPCE seen in Fig. 4.11. Applying this treatment to the high performance Fe_2O_3 anodes could possibly result in photocurrents nearing 3 mA/cm^2 at 1.23 V vs. RHE. Further work is necessary to understand the effect of this catalyst over the standard Co^{2+} treatment, and to better characterize the reaction mechanism occurring at the surface of both treated and untreated hematite. Interestingly, recent reports describe a very complex anisotropic reactivity of hematite at the SCLJ due to the anisotropic crystal structure and exposed crystalline planes [117, 118]. Fully exploiting the most reactive crystal face of hematite might further reduce the required overpotential for water oxidation.

4.4.3 Continued Material Considerations

To conclude, it is clear there is a strong dependence of the performance of synthetic hematite on the deposition technique. While methods such as spray pyrolysis and CVD consistently produce electrodes photoactive for water oxidation, solution-based methods such as sol–gel approaches have failed to produce especially photoactive hematite. This is certainly related to the quality of the prepared material in terms of crystallinity and impurity concentrations. Aqueous methods of preparing hematite typically pass through a phase containing iron hydroxide (e.g., akaganeite, lepidocrocite, or goethite) but primarily hematite is detected after at annealing at 500°C. However, it has been shown that at temperatures up to 800°C, a nonstoichiometric composition remains in hematite when prepared in this way [119]. This results in a distortion in the crystal structure due to cation vacancies and remaining hydroxides. A structural temperature dependence has also been shown to affect the color of hematite due to the subtle relaxation of the iron positions in the crystal lattice and the resulting change in the optical reflectivity [120]. This was observed to occur at similar temperatures of 800°C. In addition, nanoparticles of hematite prepared via an aqueous precipitation have shown a size and temperature-dependent concentration of spinel defects at the surface. The formation of these defects on hematite was shown to be governed by the thermodynamics of the α–γ phase transition with decreasing particle size [121]. Raman spectroscopy has shown that similar defects on the surface of highly efficient anodes prepared by APCVD are important to the performance [13].

The above factors can all play a major role in both surface and bulk recombination, and thus care must be taken to prepare and characterize hematite for PEC application. Empirically, these problems can be avoided by simply heating the electrode to sufficiently high temperatures as was done with the sintered polycrystalline work [28] or the recent work by McFarland and coworkers [20, 91]. However, there is a clear tradeoff between phase purity and feature size as at high temperatures rapid sintering of nanostructures occurs. In general, the important material parameters needed to produce highly photoactive hematite must be clarified in order to be able to apply the most appropriate nanostructuring technique and obtain an ideal morphology while still ensuring photoactivity.

In addition to the structural properties, there is an unmistakable need to increase understanding of the hematite–substrate interface. A photoinactive "dead layer" has been observed in hematite near the surface of the supporting electrode [8, 13, 54, 105] which can be ameliorated with the addition of an ultra-thin Si-based insulating layer [105]. While fundamental studies of this "dead layer" have increased understanding and further reduced its effect [96], the ability to create ultra-thin films (~20 nm) of hematite with quantum efficiencies (APCEs) matching the single crystal or Si-doped APCVD electrodes has not yet been demonstrated. A fundamental understanding of hematite/conductor interfaces is needed to overcome this issue which currently limits its application to ETA configurations.

Finally, the low quantum efficiencies (even at short wavelengths in optimized single crystal photoanodes), the ultrafast carrier recombination, and the interbandgap states found in hematite suggest that this material possesses intrinsic limitations that will be difficult, if not impossible, to resolve given our lack of understanding. However, recent combinatorial efforts to optimize both the alloy composition [122, 123] and doping in hematite [124] suggest that some improvement can be made of the standard "single dopant" approach. For example, the proper doping with multiple atoms may increase the hole and electron mobilities by changing the dielectric constant of the crystalline phase. Continued efforts along these lines and further fundamental studies of the PEC properties of hematite are necessary, parallel with the nanostructuring efforts, to completely harness the potential of hematite for solar water splitting. Indeed, since the initial writing of this chapter many important advances already have been reported in the performance and understanding of hematite [125] and many more are sure to follow.

References

1. Fujishima, A., Honda, K.: Electrochemical photolysis of water at a semiconductor electrode. Nature **238**, 37 (1972)
2. Grätzel, M.: Photoelectrochemical cells. Nature **414**, 338 (2001)
3. Khaselev, O., Turner, J.A.: A monolithic photovoltaic-photoelectrochemical device for hydrogen production via water splitting. Science **280**, 425 (1998)
4. Ni, M., Leung, M.K.H., Leung, D.Y.C., Sumathy, K.: A review and recent developments in photocatalytic water-splitting using TiO_2 for hydrogen production. Renew. Sust. Energ. Rev. **11**, 401 (2007). doi:10.1016/j.rser.2005.01.009
5. Santato, C., Ulmann, M., Augustynski, J.: Photoelectrochemical properties of nanostructured tungsten trioxide films. J. Phys. Chem. B **105**, 936 (2001)
6. Murphy, A.B., Barnes, P.R.F., Randeniya, L.K., Plumb, I.C., Grey, I.E., Horne, M.D., Glasscock, J.A.: Efficiency of solar water splitting using semiconductor electrodes. Int. J. Hydrogen Energy **31**, 1999 (2006). doi:DOI 10.1016/j.ijhydene.2006.01.014
7. Cornell, R.M., Schwertmann, U.: The iron oxides: structure, properties, reactions, occurrences, and uses. Wiley-VCH, Weinheim (2003)
8. Sivula, K., Le Formal, F., Grätzel, M.: WO_3–Fe_2O_3 photoanodes for water splitting: a host scaffold. Guest absorber approach. Chem. Mater. **21**, 2862 (2009). doi:10.1021/cm900565a
9. Zboril, R., Mashlan, M., Petridis, D.: Iron(III) oxides from thermal processes-synthesis, structural and magnetic properties, Mossbauer spectroscopy characterization, and applications. Chem. Mater. **14**, 969 (2002). doi:10.1021/cm0111074
10. Iordanova, N., Dupuis, M., Rosso, K.M.: Charge transport in metal oxides: a theoretical study of hematite alpha-Fe_2O_3. J. Chem. Phys. **122**, 144305 (2005)
11. Wang, X.O., Gao, L.S., Zheng, H.G., Ji, M.R., Shen, T., Zhang, Z.: Fabrication and electrochemical properties of alpha-Fe_2O_3 nanoparticles. J. Cryst. Growth **269**, 489 (2004). doi:10.1016/j.jcrysgro.2004.05.081
12. Hermanek, M., Zboril, R., Medrik, N., Pechousek, J., Gregor, C.: Catalytic efficiency of iron (III) oxides in decomposition of hydrogen peroxide: competition between the surface area and crystallinity of nanoparticles. J. Am. Chem. Soc. **129**, 10929 (2007)
13. Cesar, I., Sivula, K., Kay, A., Zboril, R., Grätzel, M.: Influence of feature size, film thickness, and silicon doping on the performance of nanostructured hematite photoanodes for solar water splitting. J. Phys. Chem. C **113**, 772 (2009). doi:10.1021/jp809060p

14. Lindgren, T., Vayssieres, L., Wang, H., Lindquist, S.E.: Photo-oxidation of water at hematite electrodes. In: Kokorin, A.I., Bahnemann, D.W. (eds.) Chemical Physics of Nanostructured Semiconductors, pp. 83–103. VSP International Science Publishers, The Netherlands (2003)
15. Marusak, L.A., Messier, R., White, W.B.: Optical-absorption spectrum of hematite, alpha-Fe_2O_3 near IR to UV. J. Phys. Chem. Solids **41**, 981 (1980)
16. Galuza, A.I., Beznosov, A.B., Eremenko, V.V.: Optical absorption edge in alpha-Fe_2O_3: the exciton-magnon structure. Low Temp. Phys. **24**, 726 (1998)
17. Gardner, R.F.G., Tanner, D.W., Sweett, F.: Electrical properties of alpha ferric oxide. 2. Ferric oxide of high purity. J. Phys. Chem. Solids **24**, 1183 (1963)
18. Kennedy, J.H., Frese, K.W.: Photo-oxidation of water at Alpha-Fe_2O_3 electrodes. J. Electrochem. Soc. **125**, 709 (1978)
19. Beermann, N., Vayssieres, L., Lindquist, S.E., Hagfeldt, A.: Photoelectrochemical studies of oriented nanorod thin films of hematite. J. Electrochem. Soc. **147**, 2456 (2000)
20. Kleiman-Shwarsctein, A., Hu, Y.S., Forman, A.J., Stucky, G.D., McFarland, E.W.: Electrodeposition of alpha-Fe_2O_3 doped with Mo or Cr as photoanodes for photocatalytic water splitting. J. Phys. Chem. C **112**, 15900 (2008)
21. Debnath, N.C., Anderson, A.B.: Optical-spectra of ferrous and ferric oxides and the passive film - a molecular-orbital study. J. Electrochem. Soc. **129**, 2169 (1982)
22. Kennedy, J.H., Frese, K.W.: Flatband potentials and donor densities of polycrystalline alpha-Fe_2O_3 determined from Mott-Schottky plots. J. Electrochem. Soc. **125**, 723 (1978)
23. Dareedwards, M.P., Goodenough, J.B., Hamnett, A., Trevellick, P.R.: Electrochemistry and photoelectrochemistry of iron(III) oxide. J. Chem. Soc. Faraday Trans. **79**, 2027 (1983)
24. Catti, M., Valerio, G., Dovesi, R.: Theoretical-study of electronic, magnetic, and structural-properties of alpha-Fe_2O_3 (hematite). Phys. Rev. B **51**, 7441 (1995)
25. Velev, J., Bandyopadhyay, A., Butler, W.H., Sarker, S.: Electronic and magnetic structure of transition-metal-doped alpha-hematite. Phys. Rev. B **71**, 205208 (2005)
26. Butler, W.H., Bandyopadhyay, A., Srinivasan, R.: Electronic and magnetic structure of a 1000 K magnetic semiconductor: alpha-hematite (Ti). J. Appl. Phys. **93**, 7882 (2003)
27. Ma, Y., Johnson, P.D., Wassdahl, N., Guo, J., Skytt, P., Nordgren, J., Kevan, S.D., Rubensson, J.E., Böske, T., Eberhardt, W.: Electronic structures of alpha-Fe_2O_3 and Fe_3O_4 from O K-edge absorption and emission spectroscopy. Phys. Rev. B **48**, 2109 (1993)
28. Kennedy, J.H., Anderman, M., Shinar, R.: Photoactivity of polycrystalline alpha-Fe_2O_3 electrodes doped with group IVa elements. J. Electrochem. Soc. **128**, 2371 (1981)
29. McGregor, K.G., Calvin, M., Otvos, J.W.: Photoeffects in Fe_2O_3 sintered semiconductors. J. Appl. Phys. **50**, 369 (1979)
30. Turner, J.E., Hendewerk, M., Parmeter, J., Neiman, D., Somorjai, G.A.: The characterization of doped iron-oxide electrodes for the photodissociation of water – stability, optical, and electronic-properties. J. Electrochem. Soc. **131**, 1777 (1984)
31. Launay, J.C., Horowitz, G.: Crystal-growth and photo-electrochemical study of Zr-doped alpha-Fe_2O_3 single-crystal. J. Cryst. Growth **57**, 118 (1982)
32. Cherepy, N.J., Liston, D.B., Lovejoy, J.A., Deng, H.M., Zhang, J.Z.: Ultrafast studies of photoexcited electron dynamics in gamma- and alpha-Fe_2O_3 semiconductor nanoparticles. J. Phys. Chem. B **102**, 770 (1998)
33. Joly, A.G., Williams, J.R., Chambers, S.A., Xiong, G., Hess, W.P., Laman, D.M.: Carrier dynamics in alpha-Fe_2O_3 (0001) thin films and single crystals probed by femtosecond transient absorption and reflectivity. J. Appl. Phys. **99**, 053521 (2006). doi:10.1063/1.2177426
34. Ahmed, S.M., Leduc, J., Haller, S.F.: Photoelectrochemical and impedance characteristics of specular hematite. 1. Photoelectrochemical, parallel conductance, and trap rate studies. J. Phys. Chem. **92**, 6655 (1988)
35. Horowitz, G.: Capacitance voltage measurements and flat-band potential determination on Zr-doped alpha-Fe_2O_3 single-crystal electrodes. J. Electroanal. Chem. **159**, 421 (1983)

36. Morin, F.J.: Electrical properties of alpha-Fe_2O_3 and alpha-Fe_2O_3 containing titanium. Phys. Rev. **83**, 1005 (1951)
37. Morin, F.J.: Electrical properties of alpha-Fe_2O_3. Phys. Rev. **93**, 1195 (1954)
38. Bosman, A.J., Vandaal, H.J.: Small-polaron versus band conduction in some transition-metal oxides. Adv. Phys. **19**, 1 (1970)
39. Chang, R.H., Wagner, J.B.: Direct-current conductivity and iron tracer diffusion in hematite at high-temperatures. J. Am. Ceram. Soc. **55**, 211 (1972)
40. Goodenough, J.B.: Metallic oxides. Prog. Solid State Chem. **5**, 145 (1971)
41. Dimitrijevic, N.M., Savic, D., Micic, O.I., Nozik, A.J.: Interfacial electron-transfer equilibria and flat-band potentials of alpha-Fe_2O_3 and TiO_2 colloids studied by pulse-radiolysis. J. Phys. Chem. **88**, 4278 (1984)
42. Nakau, T.: Electrical Conductivity of α-Fe_2O_3. J. Phys. Soc. Jpn. **15**, 727 (1960). doi:10.1143/JPSJ.15.727
43. Benjelloun, D., Bonnet, J.P., Doumerc, J.P., Launay, J.C., Onillon, M., Hagenmuller, P.: Anisotropy of the electrical-properties of iron-oxide alpha-Fe_2O_3. Mater. Chem. Phys. **10**, 503 (1984)
44. Rosso, K.M., Smith, D.M.A., Dupuis, M.: An ab initio model of electron transport in hematite (alpha-Fe_2O_3) basal planes. J. Chem. Phys. **118**, 6455 (2003). doi:10.1063/1.1558534
45. Shinar, R., Kennedy, J.H.: Photoactivity of doped alpha-Fe_2O_3 electrodes. Solar Energy Mater. **6**, 323 (1982)
46. Leygraf, C., Hendewerk, M., Somorjai, G.: The preparation and selected properties of Mg-doped para-type iron-oxide as a photo-cathode for the photoelectrolysis of water using visible-light. J. Solid State Chem. **48**, 357 (1983)
47. Sastri, M.V.C., Nagasubramanian, G.: Studies on ferric oxide electrodes for the photo-assisted electrolysis of water. Int. J. Hydrogen Energy **7**, 873 (1982)
48. Maruska, H.P., Ghosh, A.K.: Transition-metal dopants for extending the response of titanate photoelectrolysis anodes. Solar Energy Mater. **1**, 237 (1979)
49. Butler, M.A.: Photoelectrolysis and physical-properties of semiconducting electrode WO_3. J. Appl. Phys. **48**, 1914 (1977)
50. Hardee, K.L., Bard, A.J.: Semiconductor electrodes 5. Application of chemically vapor-deposited iron-oxide films to photosensitized electrolysis. J. Electrochem. Soc. **123**, 1024 (1976)
51. Yeh, L.S.R., Hackerman, N.: Iron-oxide semiconductor electrodes in photoassisted electrolysis of water. J. Electrochem. Soc. **124**, 833 (1977)
52. Quinn, R.K., Nasby, R.D., Baughman, R.J.: Photoassisted electrolysis of water using single-crystal alpha-Fe_2O_3 anodes. Mater. Res. Bull. **11**, 1011 (1976)
53. Sanchez, C., Sieber, K.D., Somorjai, G.A.: The photoelectrochemistry of niobium doped alpha-Fe_2O_3. J. Electroanal. Chem. **252**, 269 (1988)
54. Itoh, K., Bockris, J.O.: Stacked thin-film photoelectrode using iron-oxide. J. Appl. Phys. **56**, 874 (1984)
55. Itoh, K., Bockris, J.O.: Thin-film photoelectrochemistry – iron-oxide. J. Electrochem. Soc. **131**, 1266 (1984)
56. Gärtner, W.W.: Depletion-layer photoeffects in semiconductors. Phys. Rev. **116**, 84 (1959)
57. Alexander, B.D., Kulesza, P.J., Rutkowska, L., Solarska, R., Augustynski, J.: Metal oxide photoanodes for solar hydrogen production. J. Mater. Chem. **18**, 2298 (2008). doi:10.1039/b718644d
58. Van de Krol, R., Liang, Y.Q., Schoonman, J.: Solar hydrogen production with nanostructured metal oxides. J. Mater. Chem. **18**, 2311 (2008). doi:10.1039/b718969a
59. Sapieszko, R.S., Matijevic, E.: Preparation of well-defined colloidal particles by thermal-decomposition of metal-chelates. 1. Iron-oxides. J. Colloid Interface Sci. **74**, 405 (1980)
60. Moser, J., Grätzel, M.: Photoelectrochemistry with colloidal semiconductors – laser studies of halide oxidation in colloidal dispersions of TiO_2 and alpha-Fe_2O_3. Helv. Chim. Acta **65**, 1436 (1982)

61. Stramel, R.D., Thomas, J.K.: Photochemistry of iron-oxide colloids. J. Colloid Interface Sci. **110**, 121 (1986)
62. Kiwi, J., Grätzel, M.: Light-induced hydrogen formation and photo-uptake of oxygen in colloidal suspensions of alpha-Fe_2O_3. J. Chem. Soc. Faraday Trans. **83**, 1101 (1987)
63. Chatterjee, S., Sarkar, S., Bhattacharyya, S.N.: Size effect in the photochemical generation of hydrogen from water by colloidal Fe_2O_3 particles. J. Photochem. Photobiol. A **72**, 183 (1993)
64. Zeng, S.Y., Tang, K.B., Li, T.W., Liang, Z.H., Wang, D., Wang, Y.K., Qi, Y.X., Zhou, W.W.: Facile route for the fabrication of porous hematite nanoflowers: Its synthesis, growth mechanism, application in the lithium ion battery, and magnetic and photocatalytic properties. J. Phys. Chem. C **112**, 4836 (2008). doi:10.1021/jp0768773
65. Lian, S.Y., Wang, E.B., Gao, L., Wu, D., Song, Y.L., Xu, L.: Surfactant-assisted solvothermal preparation of submicrometer-sized hollow hematite particles and their photocatalytic activity. Mater. Res. Bull. **41**, 1192 (2006). doi:10.1016/j.materresbull.2005.10.022
66. Hu, X.L., Yu, J.C.: Continuous aspect-ratio tuning and fine shape control of monodisperse alpha-Fe_2O_3 nanocrystals by a programmed microwave-hydrothermal method. Adv. Funct. Mater. **18**, 880 (2008). doi:10.1002/adfm.200700671
67. Björksten, U., Moser, J., Grätzel, M.: Photoelectrochemical studies on nanocrystalline hematite films. Chem. Mater. **6**, 858 (1994)
68. Qian, X., Zhang, X., Bai, Y., Li, T., Tang, X., Wang, E., Dong, S.: Photoelectrochemical characteristics of alpha-Fe_2O_3 nanocrystalline semiconductor thin film. J. Nanopart. Res. **2**, 191 (2000)
69. Gou, X.L., Wang, G.X., Kong, X.Y., Wexler, D., Horvat, J., Yang, J., Park, J.: Flutelike porous hematite nanorods and branched nanostructures: Synthesis, characterisation and application for gas-sensing. Chem. Eur. J. **14**, 5996 (2008). doi:10.1002/chem.200701705
70. Hu, X.L., Yu, J.C., Gong, J.M., Li, Q., Li, G.S.: Alpha-Fe_2O_3 nanorings prepared by a microwave-assisted hydrothermal process and their sensing properties. Adv. Mater. **19**, 2324 (2007). doi:10.1002/adma.200602176
71. Hida, Y., Kozuka, H.: Photoanodic properties of sol-gel-derived iron oxide thin films with embedded gold nanoparticles: effects of polyvinylpyrrolidone in coating solutions. Thin Solid Films **476**, 264 (2005). doi:10.1016/j.tsf.2004.09.063
72. Watanabe, A., Kozuka, H.: Photoanodic properties of sol-gel-derived Fe_2O_3 thin films containing dispersed gold and silver particles. J. Phys. Chem. B **107**, 12713 (2003). doi:10.1021/jp0303568
73. Borse, P.H., Jun, H., Choi, S.H., Hong, S.J., Lee, J.S.: Phase and photoelectrochemical behavior of solution-processed Fe_2O_3 nanocrystals for oxidation of water under solar light. Appl. Phys. Lett. **93** (2008). doi:173103 10.1063/1.3005557
74. Souza, F.L., Lopes, K.P., Nascente, P.A.P., Leite, E.R.: Nanostructured hematite thin films produced by spin-coating deposition solution: application in water splitting. Sol. Energy Mater. Sol. Cells **93**, 362 (2009). doi:10.1016/j.solmat.2008.11.049
75. Yue, W.B., Zhou, W.Z.: Crystalline mesoporous metal oxide. Prog. Nat. Sci. **18**, 1329 (2008). doi:10.1016/j.pnsc.2008.05.010
76. Jiao, F., Harrison, A., Jumas, J.C., Chadwick, A.V., Kockelmann, W., Bruce, P.G.: Ordered mesoporous Fe_2O_3 with crystalline walls. J. Am. Chem. Soc. **128**, 5468 (2006). doi:10.1021/ja0584774
77. Vayssieres, L., Beermann, N., Lindquist, S.E., Hagfeldt, A.: Controlled aqueous chemical growth of oriented three-dimensional crystalline nanorod arrays: application to iron(III) oxides. Chem. Mater. **13**, 233 (2001)
78. Fan, Z.Y., Wen, X.G., Yang, S.H., Lu, J.G.: Controlled p- and n-type doping of Fe_2O_3 nanobelt field effect transistors. Appl. Phys. Lett. **87**, 013113 (2005). doi:10.1063/1.1977203
79. Lindgren, T., Wang, H.L., Beermann, N., Vayssieres, L., Hagfeldt, A., Lindquist, S.E.: Aqueous photoelectrochemistry of hematite nanorod array. Sol. Energy Mater. Sol. Cells **71**, 231 (2002)

80. Peng, L.L., Xie, T.F., Fan, Z.Y., Zhao, Q.D., Wang, D.J., Zheng, D.: Surface photovoltage characterization of an oriented alpha-Fe_2O_3 nanorod array. Chem. Phys. Lett. **459**, 159 (2008). doi:10.1016/j.cplett.2008.05.036
81. Fu, Y.Y., Chen, J., Zhang, H.: Synthesis of Fe_2O_3 nanowires by oxidation of iron. Chem. Phys. Lett. **350**, 491 (2001)
82. Wang, R.M., Chen, Y.F., Fu, Y.Y., Zhang, H., Kisielowski, C.: Bicrystalline hematite nanowires. J. Phys. Chem. B **109**, 12245 (2005). doi:10.1021/jp051197q
83. Wen, X.G., Wang, S.H., Ding, Y., Wang, Z.L., Yang, S.H.: Controlled growth of large-area, uniform, vertically aligned arrays of alpha-Fe_2O_3 nanobelts and nanowires. J. Phys. Chem. B **109**, 215 (2005). doi:10.1021/jp0461448
84. Yu, T., Zhu, Y.W., Xu, X.J., Yeong, K.S., Shen, Z.X., Chen, P., Lim, C.T., Thong, J.T.L., Sow, C.H.: Substrate-friendly synthesis of metal oxide nanostructures using a hotplate. Small **2**, 80 (2006). doi:10.1002/smll.200500234
85. Han, Q., Xu, Y.Y., Fu, Y.Y., Zhang, H., Wang, R.M., Wang, T.M., Chen, Z.Y.: Defects and growing mechanisms of alpha-Fe_2O_3 nanowires. Chem. Phys. Lett. **431**, 100 (2006). doi:10.1016/j.cplett.2006.09.027
86. Mor, G.K., Shankar, K., Paulose, M., Varghese, O.K., Grimes, C.A.: Enhanced photocleavage of water using titania nanotube arrays. Nano Lett. **5**, 191 (2005). doi:10.1021/nl048301k
87. Prakasam, H.E., Varghese, O.K., Paulose, M., Mor, G.K., Grimes, C.A.: Synthesis and photoelectrochemical properties of nanoporous iron (III) oxide by potentiostatic anodization. Nanotechnol. **17**, 4285 (2006). doi:10.1088/0957-4484/17/17/001
88. Rangaraju, R.R., Panday, A., Raja, K.S., Misra, M.: Nanostructured anodic iron oxide film as photoanode for water oxidation. J. Phys. D: Appl. Phys. **42**, 135303 (2009). doi:10.1088/0022-3727/42/13/135303
89. Mor, G.K., Prakasam, H.E., Varghese, O.K., Shankar, K., Grimes, C.A.: Vertically oriented Ti-Fe-O nanotube array films: toward a useful material architecture for solar spectrum water photoelectrolysis. Nano Lett. **7**, 2356 (2007). doi:10.1021/nl0710046
90. Mohapatra, S.K., John, S.E., Banerjee, S., Misra, M.: Water photooxidation by smooth and ultrathin α-Fe_2O_3 nanotube arrays. Chem. Mater. **21**, 3048 (2009)
91. Hu, Y.S., Kleiman-Shwarsctein, A., Forman, A.J., Hazen, D., Park, J.N., McFarland, E.W.: Pt-doped alpha-Fe_2O_3 thin films active for photoelectrochemical water splitting. Chem. Mater. **20**, 3803 (2008). doi:10.1021/cm800144q
92. Spray, R.L., Choi, K.-S.: Photoactivity of transparent nanocrystalline Fe_2O_3 electrodes prepared via anodic electrodeposition. Chem. Mater. **21**, 3701 (2009)
93. Murthy, A.S.N., Reddy, K.S.: Photoelectrochemical behavior of undoped ferric-oxide (alpha-Fe_2O_3) electrodes prepared by spray pyrolysis. Mater. Res. Bull. **19**, 241 (1984)
94. Duret, A., Grätzel, M.: Visible light-induced water oxidation on mesoscopic alpha-Fe_2O_3 films made by ultrasonic spray pyrolysis. J. Phys. Chem. B **109**, 17184 (2005). doi:10.1021/jp044127c
95. Kumari, S., Tripathi, C., Singh, A.P., Chauhan, D., Shrivastav, R., Dass, S., Satsangi, V.R.: Characterization of Zn-doped hematite thin films for photoelectrochemical splitting of water. Curr. Sci. **91**, 1062 (2006)
96. Liang, Y.Q., Enache, C.S., Van de Krol, R.: Photoelectrochemical characterization of sprayed alpha-Fe_2O_3 thin films: influence of Si doping and SnO_2 interfacial layer. Int. J. Photoenergy 739864 (2008). doi:10.1155/2008/739864
97. Satsangi, V.R., Kumari, S., Singh, A.P., Shrivastav, R., Dass, S.: International Workshop on Hydrogen Energy – Production Storage and Application, Jaipur, India, 2006
98. Khan, S.U.M., Akikusa, J.: Photoelectrochemical splitting of water at nanocrystalline n-Fe_2O_3 thin-film electrodes. J. Phys. Chem. B **103**, 7184 (1999)
99. Hagglund, C., Grätzel, M., Kasemo, B.: Comment on "Efficient photochemical water splitting by a chemically modified n-TiO_2" – (II). Science **301**, 1673B (2003)

100. Sartoretti, C.J., Alexander, B.D., Solarska, R., Rutkowska, W.A., Augustynski, J., Cerny, R.: Photoelectrochemical oxidation of water at transparent ferric oxide film electrodes. J. Phys. Chem. B **109**, 13685 (2005)
101. Cesar, I.: Thesis results, Chimie et Génie Chimique Ecole Polytechnique Fédérale de Lausanne, Lausanne (2007)
102. Cesar, I., Kay, A., Martinez, J.A.G., Grätzel, M.: Translucent thin film Fe_2O_3 photoanodes for efficient water splitting by sunlight: Nanostructure-directing effect of Si-doping. J. Am. Chem. Soc. **128**, 4582 (2006). doi:10.1021/ja060292p
103. Orthner, H.R., Roth, P.: Formation of iron oxide powder in a hot-wall flow reactor – effect of process conditions on powder characteristics. Mater. Chem. Phys. **78**, 453 (2002)
104. Wen, J.Z., Goldsmith, C.F., Ashcraft, R.W., Green, W.H.: Detailed kinetic modeling of iron nanoparticle synthesis from the decomposition of $Fe(CO)_5$. J. Phys. Chem. C **111**, 5677 (2007). doi:10.1021/jp066579q
105. Kay, A., Cesar, I., Grätzel, M.: New benchmark for water photooxidation by nanostructured alpha-Fe_2O_3 films. J. Am. Chem. Soc. **128**, 15714 (2006). doi:10.1021/ja064380l
106. Glasscock, J.A., Barnes, P.R.F., Plumb, I.C., Savvides, N.: Enhancement of photoelectrochemical hydrogen production from hematite thin films by the introduction of Ti and Si. J. Phys. Chem. C **111**, 16477 (2007). doi:10.1021/jp0745561
107. Saretni-Yarahmadi, S., Wijayantha, K.G.U., Tahir, A.A., Vaidhyanathan, B.: Nanostructured alpha-Fe_2O_3 electrodes for solar driven water splitting: effect of doping agents on preparation and performance. J. Phys. Chem. C **113**, 4768 (2009). doi:10.1021/jp808453z
108. Saremi-Yarahmadi, S., Tahir, A.A., Vaidhyanathan, B., Wijayantha, K.G.U.: Fabrication of nanostructured alpha-Fe_2O_3 electrodes using ferrocene for solar hydrogen generation. Mater. Lett. **63**, 523 (2009). doi:10.1016/j.matlet.2008.11.011
109. Grätzel, M.: Mesoscopic solar cells for electricity and hydrogen production from sunlight. Chem. Lett. **34**, 8 (2005). doi:10.1246/Cl.2005.8
110. Ernst, K., Belaidi, A., Konenkamp, R.: Solar cell with extremely thin absorber on highly structured substrate. Semicond. Sci. Technol. **18**, 475 (2003)
111. Grätzel, M., Augustynski, J.: Tandem cell for water cleavage by visable light. (2000). PCT Int. Appl., WO 2001002624 (2001)
112. Brillet, J., Cornuz, M., Le Formal, F., Yum, J.H., Grätzel, M., Sivula, K.: Examining architectures of photoanode-photovoltaic tandem cells for solar water splitting. J. Mater. Res. **25**, 17 (2010). doi:10.1557/JMR.2010.0009
113. Hu, Y.S., Kleiman-Shwarsctein, A., Stucky, G.D., McFarland, E.W.: Improved photoelectrochemical performance of Ti-doped alpha-Fe_2O_3 thin films by surface modification with fluoride. Chem. Commun. **19**, 2652 (2009). doi:10.1039/b901135h
114. Brunschwig, B.S., Chou, M.H., Creutz, C., Ghosh, P., Sutin, N.: Mechanisms of water oxidation to oxygen – cobalt(IV) as an intermediate in the aquocobalt(II)-catalyzed reaction. J. Am. Chem. Soc. **105**, 4832 (1983)
115. Kanan, M.W., Nocera, D.G.: In situ formation of an oxygen-evolving catalyst in neutral water containing phosphate and Co^{2+}. Science **321**, 1072 (2008). doi:10.1126/science.1162018
116. Zhong, D.K., Sun, J.W., Inumaru, H., Gamelin, D.R.: Solar water oxidation by composite catalyst/alpha-Fe_2O_3 photoanodes. J. Am. Chem. Soc. **131**, 6086 (2009). doi:10.1021/ja9016478
117. Eggleston, C.M., Shankle, A.J.A., Moyer, A.J., Cesar, I., Grätzel, M.: Anisotropic photocatalytic properties of hematite. Aquat. Sci. **71**, 151 (2009). doi:10.1007/s00027-009-9191-5
118. Yanina, S.V., Rosso, K.M.: Linked reactivity at mineral–water interfaces through bulk crystal conduction. Science **320**, 218 (2008). doi:10.1126/science.1154833
119. Gualtieri, A.F., Venturelli, P.: In situ study of the goethite-hematite phase transformation by real time synchrotron powder diffraction. Am. Mineral. **84**, 895 (1999)

120. Pailhe, N., Wattiaux, A., Gaudon, M., Demourgues, A.: Impact of structural features on pigment properties of alpha-Fe_2O_3 haematite. J. Solid State Chem. **181**, 2697 (2008). doi:10.1016/j.jssc.2008.06.049
121. Chernyshova, I.V., Hochella, M.F., Madden, A.S.: Size-dependent structural transformations of hematite nanoparticles. 1. Phase transition. Phys. Chem. Chem. Phys. **9**, 1736 (2007). doi:10.1039/b618790k
122. Woodhouse, M., Parkinson, B.A.: Combinatorial discovery and optimization of a complex oxide with water photoelectrolysis activity. Chem. Mater. **20**, 2495 (2008)
123. Woodhouse, M., Parkinson, B.A.: Combinatorial approaches for the identification and optimization of oxide semiconductors for efficient solar photoelectrolysis. Chem. Soc. Rev. **38**, 197 (2009). doi:10.1039/b719545c
124. Jang, J.S., Lee, J., Ye, H., Fan, F.R.F., Bard, A.J.: Rapid screening of effective dopants for Fe_2O_3 photocatalysts with scanning electrochemical microscopy and investigation of their photoelectrochemical properties. J. Phys. Chem. C **113**, 6719 (2009). doi:10.1021/jp8109429
125. Sivula, K., Le Formal, F., Grätzel, M.: Solar Water Splitting: Progress Using Hematite (α-Fe_2O_3) Photoelectrodes. Chemsuschem **4**, 432 (2011). doi: 10.1002/cssc.201000416

Chapter 5
Mixed Metal Oxide Photoelectrodes and Photocatalysts

Kazuhiro Sayama

5.1 Introduction

An efficient solar energy conversion system for hydrogen production from water, referred to as a solar–hydrogen system, is most important for genuinely sustainable development. However, for such a system to be practical and economical, we must address several remaining issues related to solar-to-H_2 efficiency, cost, and stability.

Solar–hydrogen systems using semiconductors fall mostly into the following three categories: solar-cell–electrolysis systems, semiconductor–photoelectrode systems, and photocatalyst systems. As Fig. 5.1 shows, each has advantages and disadvantages. The solar-cell–electrolysis system is efficient but expensive; the photocatalyst system is simple and relatively inexpensive, but inefficient; and the semiconductor–photoelectrode system falls somewhere between the two in terms of efficiency and cost. The semiconductor–photoelectrode and photocatalyst systems have been developed in different fields and in different groups. The photoelectrode system has been developed at various places in the world, and the photocatalyst powder system has been developed mainly in Japan [1]. However, as Fig. 5.2 shows, their water-splitting reaction mechanisms are very similar. Therefore, interaction and exchange of information between researchers in both fields are highly desirable. In particular, screening of oxide semiconductor materials with high efficiency under visible light irradiation is a main issue for both fields.

In this chapter, we first discuss mixed metal oxide semiconductors used in both fields. We then discuss photoelectrochemical properties of $BiVO_4$ semiconductors. Finally, we discuss our high-throughput screening system for mixed metal oxide semiconductors.

K. Sayama (✉)
Energy Technology Research Institute, National Institute of Advanced Industrial Science and Technology (AIST), Central 5, 1-1-1 Higashi, Tsukuba, Ibaraki 305-8565, Japan
e-mail: k.sayama@aist.go.jp

R. van de Krol and M. Grätzel (eds.), *Photoelectrochemical Hydrogen Production*, Electronic Materials: Science & Technology 102, DOI 10.1007/978-1-4614-1380-6_5,

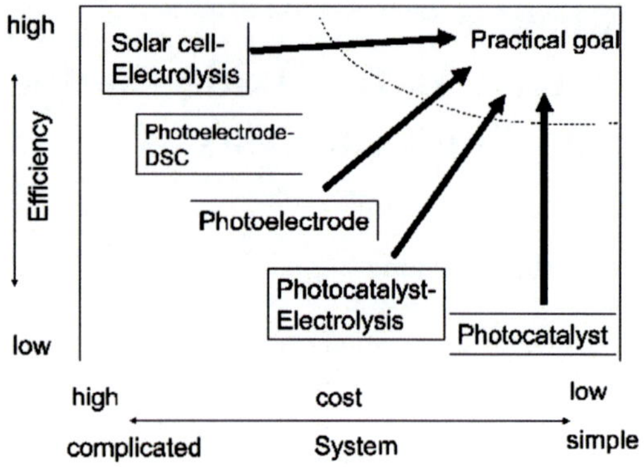

Fig. 5.1 Relationship between efficiency and system cost

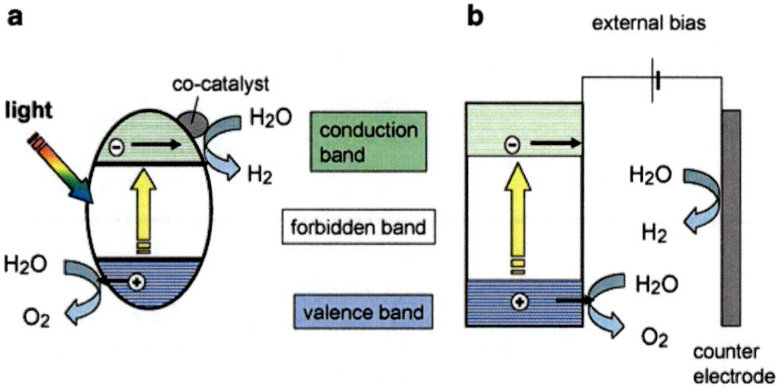

Fig. 5.2 Reaction mechanisms of water splitting using photoelectrode and photocatalyst systems

5.2 Mixed Metal Oxide Semiconductors Used for Photoelectrodes and Photocatalysts

Since the report of the Honda–Fujishima effect using TiO_2 under UV light [2], photoelectrodes using metal oxide semiconductors have been widely investigated for application to solar hydrogen systems. For increased efficiency of solar energy conversion, effective utilization of visible light over a wide wavelength region is necessary. It is noted that metal oxide semiconductors offer many advantages: they are generally stable, suitable for O_2 evolution, and produced easily compared to other types of semiconductor. Moreover, most oxide semiconductors are n-type and H_2 gas collection at the counter electrode is easier than that on the wide-area photoelectrode of p-type semiconductors. However, efficiencies of various conventional-type metal oxide semiconductor electrodes, such as single crystals or sintered pellets, are very low in the visible light region, even with the application of high external bias.

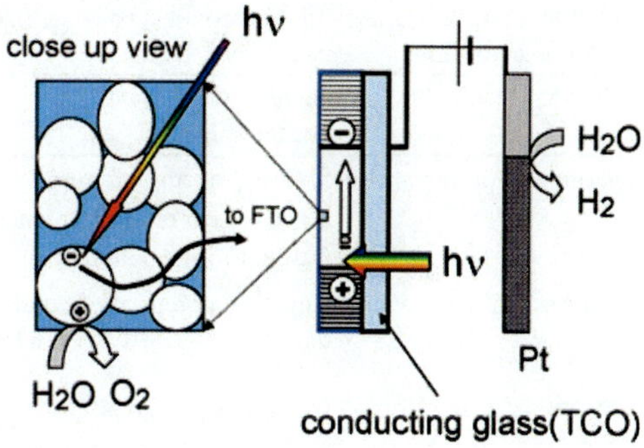

Fig. 5.3 Porous structure of the photoelectrode

An excellent incident-photon-to-current conversion efficiency (IPCE >75%) for water decomposition under visible light irradiation was reported using a nanocrystalline WO_3 thin film on a conducting glass electrode [3, 4]; the film was prepared by spreading a colloidal solution of WO_3 on the conducting glass and firing it. Relatively high IPCEs were also reported for porous Fe_2O_3 thin films on conducting glass electrodes prepared by spraying the glass with a solution of iron salts [5–8]. It is worth mentioning that these high IPCEs were obtained with simple wet-chemical processes, although many defects in the porous film electrodes may be present. Fig. 5.3 shows the porous structure of these nanocrystalline electrodes. Porous thin-film photoelectrodes are advantageous in that the diffusion length of the holes in the valence band formed by band-gap photoexcitation is significantly shorter compared to the length in conventional thin film-based photoelectrodes. This is because the electrolyte solution can penetrate the pores of the entire nanocrystalline semiconductor film, and holes in the nanoparticles can reach the semiconductor–electrolyte interface in a short time. Porous thin-film photoelectrodes have a thickness of a few micrometers or less, whereas conventional photoelectrodes are a few hundred micrometers thick or more. Furthermore, the diffusion path length of the electron is shorter when light is irradiated through the conducting glass. Therefore, this special structure probably accounts for the high IPCE.

Charge separation occurs in the nanoparticles for porous photoelectrodes and in the space charge layer for conventional thick photoelectrodes. The mechanism of charge separation for porous photoelectrodes is similar to that for photocatalysts, because charge separation for powders of the latter also occurs in small particles. Therefore, porous photoelectrodes are sometimes called photocatalyst electrodes.

Table 5.1 lists some essential characteristics of semiconductors for photocatalysts and porous photoelectrodes. Common essential characteristics of the two include stability during illumination, a valence band potential more positive than 1.23 V (vs. RHE), and O_2 evolution capability on the surface. For photocatalytic water splitting, the bottom of the semiconductor's conduction band potential must be <0 V (vs. RHE) for H_2 evolution, and loading with a co-catalyst such as Pt or NiO_x is often needed for H_2 evolution. In contrast, for evolution of O_2 gas at a photoelectrode, a high mobility of electrons to the back contact via the semiconductor particles is essential, so

Table 5.1 Required characteristics of semiconductors for photocatalysts and porous photoelectrodes

	Photocatalysts (for water splitting)	Porous photoelectrode (for O_2 evolution)
Common points	• Stability during photoreaction • Positive potential of valence band top (>1.23 V vs. RHE) • O_2-evolution capability on the surface	
Different points	• Negative potential of conduction band bottom (<0 V vs. RHE)	• n-Type semiconductor with high mobility of e^- to back contact
	• H_2-evolution capability on the surface (with co-catalyst)	

an n-type semiconductor is preferable. Even if the conduction band potential of the photoelectrode is more positive than 0 V vs. RHE, hydrogen evolution at the counter electrode can be readily achieved by applying an external bias potential. Therefore, one can consider the requirements on the band edge positions of photocatalysts to be more restrictive than those of photoelectrodes.

Semiconductor evaluation and screening procedures vary for photocatalysts and photoelectrodes. For photoelectrodes, a semiconductor thin film is prepared on a conducting substrate, and its photocurrent–potential dependence is measured easily by potentiostat, even if the photocurrent is very small. However, quantitative measurement of O_2 gas evolution on photoelectrodes is not generally conducted, therefore, it is important to consider the possibility of oxidation reactions other than O_2 evolution. For photocatalysts, semiconductor powder is suspended in aqueous solution, and evolved H_2 and O_2 gases under light irradiation are measured by on-line gas chromatograph. When the O_2 evolution capability of the semiconductor powder is evaluated, a strong electron scavenger (oxidative sacrificial reagent) such as Ag^+ ion is added to the reaction solution. Preparation of thin films, as is required for photoelectrodes, is not necessary. There are wide variations in the available methods to prepare a single phase of semiconductor powder. Therefore, photocatalysis is an effective method for screening single-phase semiconductor powders of mixed metal oxides. However, the disadvantages include requirements for large glass-made apparatus without air leakage and lengthy times needed for O_2 gas analysis.

There have been some investigations on porous photoelectrodes using visible light active semiconductors which were developed in the photocatalysis field, shown in Table 5.2. WO_3 is a typical simple oxide photocatalyst that evolves O_2 using electron scavengers under visible light irradiation, and the porous WO_3 photoelectrode has the highest IPCE in the visible light region. Bi-based mixed-oxide semiconductors are often used as both photocatalysts and porous photoelectrodes because of their special band structures. For conventional simple oxide semiconductors, the valence band consists mainly of O-2p orbitals if the metal d orbital is unoccupied, and the potential of the O-2p orbital is about +3 V vs. RHE (Fig. 5.4a). Since the required external bias for the photoelectrode increases

Table 5.2 Semiconductors investigated for water decomposition into H_2 or O_2 by photocatalysts and porous photoelectrodes under visible light irradiation

Semiconductor	Bandgap	Photoelectrolysis	Reference	Photocatalysis	Reference
WO_3	2.8	IPCE = 75% (400 nm), V_{op} = +0.5 (RHE) 2.2 mA/cm^2 (1.23 V, RHE, 1 sun)	[3]	$O_2(Ag^+)$: 65 μmol/h (>420 nm) QE = 2% (405 nm)	[1, 16–18]
$BiVO_4$	2.4	IPCE = 44% (420 nm, +1 V, Ag/AgCl, Na_2SO_4) V_{op} = −0.2–0.05 V (ibid)	[9]	$O_2(Ag^+)$: 421 μmol/h (300 W–Xe, >420 nm) QE = 9% (450 nm)	[19]
Bi_2WO_6	2.8	IPCE = 0.1%(400 nm), V_{op} = 0 V (SCE, Na_2SO_4)	[10]	$O_2(Ag^+)$: 3 μmol/h (300 W–Xe, >420 nm)	[20]
Bi_2MoO_6	2.7	V_{op} = 0.15 V (RHE), 3 μA/cm^2 (300 W–Xe)	[11]	$O_2(Ag^+)$: 55 μmol/h (300 W–Xe, >420 nm)	[21]
$BiZn_2VO_6$	2.4	IPCE = 0.4% (420 nm, +1 V, Ag/AgCl, Na_2SO_4)	[12]	$O_2(Ag^+)$: 15 μmol/h (350 W–Xe, >420 nm)	[22, 23]
$BiFeO_3$	2.1	V_{op} = 0.6 V (SCE), 5 μA/cm^2 (1 V, SCE, 500 W–Hg)	[13]	$O_2(FeCl_3)$: 1 μmol/h (500 W–Hg, >420 nm)	[13]
Fe_2TiO_5	2.3	IPCE = 4% (400 nm), V_{op} = +0.3 V (SCE, pH = 7) 2 mA/cm^2 (at +1 V, SCE, 500 W–Xe)	[14]	Acetaldehyde degradation (UV): $FeTiO_3 \gg Fe_2TiO_5$	[24]
TaON	2.5	20 μA/cm^2 (1 V bias, 300 W–Xe, >420 nm)	[15]	$O_2(Ag^+)$: 380 μmol/h (>420 nm), QE = 34%	[25, 26]

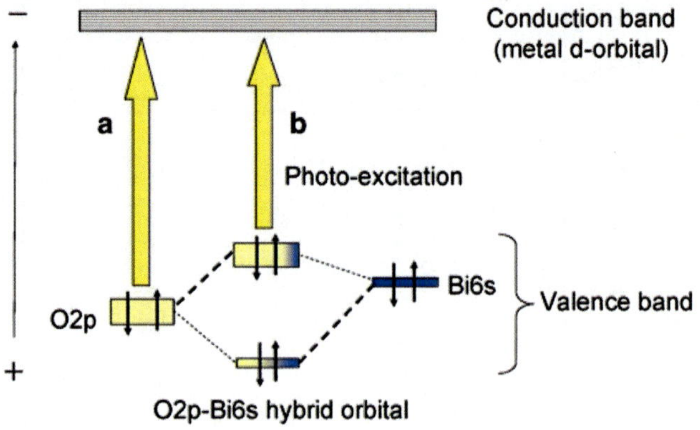

Fig. 5.4 Band structure of Bi-based oxide

with a positive shift of the conduction band, the preferred method to decrease the bandgap is via a negative shift of the valence band potential. For Bi mixed-oxide semiconductors, the valence band potential is shifted negatively when hybrid orbitals of O-2p and Bi-6s are formed due to repulsion of the two orbitals (Fig. 5.4b). Bi mixed oxides are relatively stable, and so Bi-based mixed-oxide semiconductors have been investigated for use as photoelectrodes as well as photocatalysts. Among Bi-based mixed-oxide semiconductors, $BiVO_4$, shows the highest photocatalytic activity for O_2 evolution and the highest IPCE under visible light irradiation [9]. The IPCEs of other Bi-based mixed-oxide semiconductors are very low, but will probably improve if preparation of porous semiconductor film is optimized.

The valence band potential of TaON can also be shifted negatively, due to hybridization of the O-2p and N-2p orbitals. As a result, TaON powders show good photocatalytic activity for not only O_2 evolution but also H_2 evolution [25, 26]. The potential of the conduction band bottom is probably close to the potential of H^+/H_2. Therefore, TaON is a promising semiconductor for photoelectrodes with the application of small external bias, although its photoelectrochemical performance is still poor [15].

There are many reports on Fe-based semiconductor photoelectrodes [5–8], but very few on O_2 evolution over Fe-based mixed-oxide photocatalysts. It is speculated that the reduction capability of electron scavengers such as Ag^+ might be low on the surface of Fe-based semiconductors. On the other hand, it is noted that $BiFeO_3$ photocatalyst has been reported to evolve O_2 gas in the presence of Ag^+ ions under visible light irradiation [13]. Moreover, anion-doped TiO_2 such as N–TiO_2 and S–TiO_2 are widely known to be active photocatalysts under similar irradiation for the degradation of various organic compounds, but their capabilities for O_2 evolution in the presence of Ag^+ ions are negligible, and the photocurrents of anion-doped TiO_2 photoelectrodes are very low [22, 27].

5.3 Development of $BiVO_4$ Photoelectrode for Water Splitting

The fundamental principles by which photoelectrodes and photocatalysts operate are similar; however, quantum and solar energy efficiencies are much higher for the former than for the latter. As mentioned earlier, $BiVO_4$ semiconductors appear to be successful for use as both photoelectrodes and photocatalysts. We have found that $BiVO_4$ photoelectrodes are highly efficient [9], and $BiVO_4$ is also known to be an efficient photocatalyst for O_2 evolution from aqueous $AgNO_3$ [19]. The following discussion introduces the photoelectrochemical and physical properties of $BiVO_4$ semiconductors in detail.

First, the relationship of the photoelectrochemical properties and the morphology of the $BiVO_4$ film prepared by various methods was investigated.

$BiVO_4$ film preparation by the metal organic decomposition (MOD) method: $BiVO_4$ films were prepared by a modified MOD method [9]. Bismuth 2-ethyl-hexanoate in acetylacetone and vanadium (IV) (oxy)acetylacetonate in acetylacetone were mixed in a stoichiometric ratio, PEG 300 was added and the solution was concentrated by evaporation of the solvent. A spin coater was used to coat FTO conducting glass with the solution, and the coated glass was fired at 400–550°C for 30 min.

$BiVO_4$ film preparation by the conventional slurry (CS) method: $BiVO_4$ powder was prepared from $K_3V_5O_{14}$ and $Bi(NO_3)_3$ by the ion-exchange method as described elsewhere [19]. The powder was mixed with water, acetylacetone, and Triton-X in a mortar. The resulting slurry was spread with a glass rod on FTO conducting glass, and the coated glass was annealed at 500°C for 30 min.

$BiVO_4$ film preparation by the chemical-bath deposition (CBD) method: FTO conducting glass was soaked in a solution of $Bi(NO_3)_3$, $NaVO_3$, and EDTA at 100°C for 2 h, and the deposited $BiVO_4$ film was washed thoroughly with distilled water [28].

Three crystalline phases have been reported for synthetic $BiVO_4$: monoclinic scheelite phase, tetragonal scheelite phase, and tetragonal zircon phase [9, 19].

Which phase is present depends on the preparation method. All $BiVO_4$ film electrodes that we prepared were composed of the monoclinic scheelite phase, as confirmed by X-ray diffraction. Table 5.3 shows the dependence of IPCE (at 400 nm) on the preparation method and calcination temperature. The electrode prepared by the MOD method exhibits the best IPCE, and that prepared by the CS method exhibits the worst. The IPCE increases with calcination temperature up to 500°C for the MOD method, but decreases at 550°C.

Figure 5.5a–c shows scanning electron microscopy (SEM) images of $BiVO_4$ thin-film electrodes prepared by the MOD method at different calcination temperatures. The structure is porous, so an electrolyte solution can penetrate the film pores. The electrodes show good mechanical stability and cannot be removed by a hard scratch with a Teflon plate and a nail, suggesting good adhesion of the

Table 5.3 IPCE for various $BiVO_4$ electrodes

Preparation method	Calcination temperature (°C)	IPCE
MOD	400	29
	450	30
	500	33
	550	24
CBD	–	10
	500	15
CS	500	4

film due to strong connections among particles, and between the particles and the conducting glass. The particle size increases with increasing calcination temperature, and the surface area and porosity decrease. Particle sizes at 400 and 500°C are 30–60 and 90–150 nm, respectively; the particle size at 550°C is almost the same as that at 500°C, but the porous structure of the film at 550°C is unclear. Fig. 5.5d shows a SEM image of a $BiVO_4$ thin-film electrode prepared by the CS method. The electrode shows poor mechanical stability and is easily removed by a soft scratch, suggesting poor adhesion among the particles, and between the particles and the conducting glass. Particle sizes range from a few hundred nanometres to 1 μm. The photocatalytic activity for O_2 evolution from aqueous $AgNO_3$ under visible light irradiation is higher than that of the powders used for the MOD and CBD methods. Therefore, we believe that the low IPCE obtained by the CS method is caused mainly by the difficulty of electron transfer between particles, and not by charge separation in the particles. Fig. 5.5e shows a SEM image of a $BiVO_4$ thin-film electrode prepared by the CBD method. The electrode is very strong mechanically and has good crystallinity, both attractive properties for charge separation and charge transfer. The particle size is very large, up to 10 μm. However, the IPCE is smaller than for electrodes prepared by the MOD method. This is attributed to the longer distance that the hole has to travel in order to reach the surface when the particle size is larger and the porosity is smaller.

Considering IPCEs and other factors, then, the MOD method with calcination at <550°C seems to be a good way to achieve porous structure and good adhesion, and hence favorable photoelectrochemical properties.

In the case of water splitting by semiconductor photoelectrodes without any sacrificial reagents in the electrolyte, measurement of O_2 and H_2 evolution is very important; however, there are very few reports on gas measurements in photoelectrochemical cells. We investigated H_2 and O_2 evolution from a $BiVO_4$ cell using a closed gas-circulating system. Evolution of H_2 and O_2 at a stoichiometric ratio ($H_2/O_2 = 2$) was observed on the $BiVO_4$ electrode prepared by the MOD method under visible light irradiation with the application of external bias. There was no gas evolution in the dark under the same conditions. The total amount of evolved gas was almost identical to that calculated by the total passage of photoelectrons. The $BiVO_4$ band structure was investigated by open-circuit potential measurement, density functional theory (DFT) calculations (Fig. 5.6), X-ray photoelectron

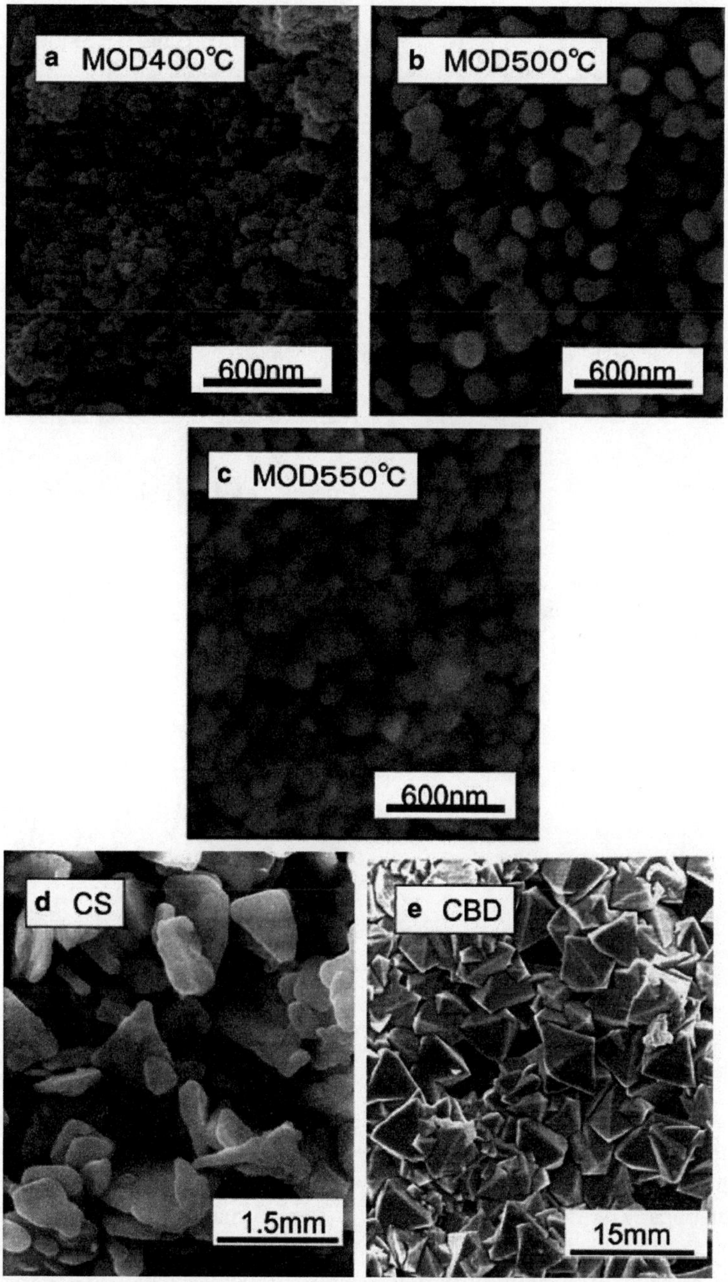

Fig. 5.5 SEM photographs of $BiVO_4$ electrodes prepared by different methods: (**a**–**c**) MOD method at various temperatures; (**d**) CS method; (**e**) CBD method

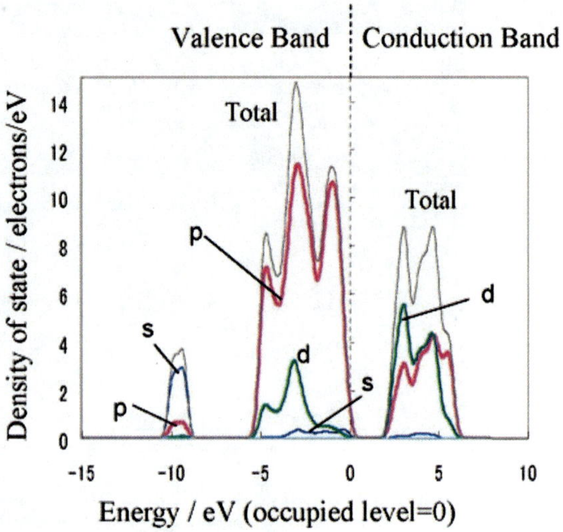

Fig. 5.6 Density of state (DOS) of monoclinic sheelite-type $BiVO_4$ calculated by the DFT method. Note that the *x*-axis is inverted compared to that in Fig. 5.7 and that "zero energy" refers to the occupied level

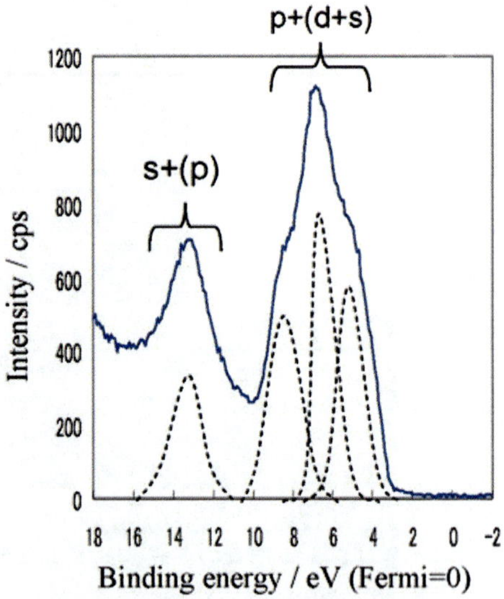

Fig. 5.7 XPS spectrum of $BiVO_4$ on FTO conducting glass. The binding energy was corrected by C-1s (=284.8 eV)

spectroscopy (XPS, Fig. 5.7), and flat-band potential measurement [9]. The shape of the XPS valence band spectrum was similar to that predicted by DFT calculations. The $BiVO_4$ valence band is formed mainly by O-2p, but V-3d and Bi-6s contributions are also observed in the calculated DFT spectrum. The peak at -10 eV is mainly formed by Bi-6s, but O-2p is also observed in the DFT spectrum. It is suggested that these orbitals are hybridized with one another. Valence-band thresholds, determined by XPS for various semiconductors, are in the following order: $Bi_2O_3 < BiVO_4 <$ (TiO_2, WO_3, V_2O_5), suggesting that the top of the valence-band potential for Bi-based oxide semiconductors is shifted negatively

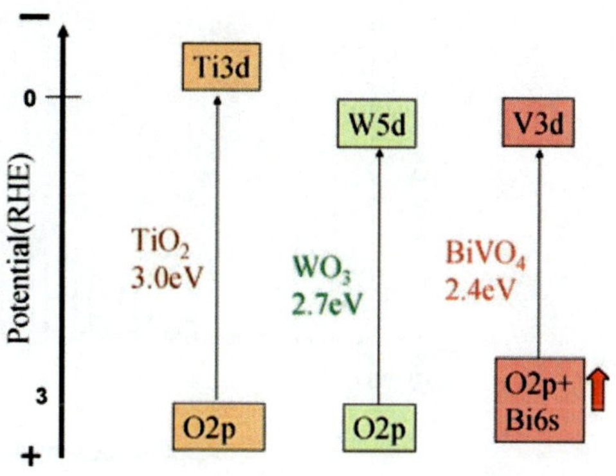

Fig. 5.8 Speculated band structures of TiO_2, WO_3, and $BiVO_4$

compared to the potentials for conventional oxide semiconductors without Bi [9]. The speculated band structures are shown in Fig. 5.8. Hybridization between the O-2p and Bi-6s orbitals and repulsion of two orbitals, as discussed earlier and shown in Fig. 5.4, might contribute to the negative shift of the $BiVO_4$ valence band.

In photocatalyst studies, surface modifications and co-catalyst loading have been investigated for promotion of charge separation and formation of an active reaction site. However, reports on surface modifications of photoelectrodes are limited [6, 29–32]. An improvement in the IPCE of a TiO_2 single-crystal photoelectrode has been reported after adsorption of Fe^{3+} ions [32]. We investigated surface modifications of the $BiVO_4$ electrode by treatment with aqueous solutions (0.01 mol/L, 12 h) containing various metal nitrates and chlorides such as $AgNO_3$, $Pd(NO_3)_2$, $Cr(NO_3)_3$, $Cu(NO_3)_2$, $Ni(NO_3)_2$, $Fe(NO_3)_3$, $RuCl_3$, $RhCl_3$, H_2PtCl_6, and $HAuCl_4$.

Among them, treatment with $AgNO_3$ aqueous solution improves the photocurrent significantly. The maximum IPCE for a Ag^+-treated $BiVO_4$ electrode at 420 nm is 44%. This value is below only that of the nanocrystalline WO_3 electrode among oxide semiconductor electrodes under visible light irradiation and highest among mixed oxide semiconductor electrodes, suggesting great potential for porous thin-film electrodes using various mixed metal oxide semiconductors. In the case of $BiVO_4$ photocatalyst, high O_2 evolution activity was observed using Ag^+ ion as an electron acceptor [19], and the degradation activity of alkylphenols was improved by loading of oxidized Ag particles (AgO_x) as co-catalysts [33]. We investigated $AgNO_3$ treatment for other porous semiconductor electrodes, such as TiO_2 and WO_3, but observed no positive effects on photocurrent and stability. These results suggest a unique interaction between $BiVO_4$ and Ag. $AgNO_3$ treatment also improves photocurrent stability. We surmise that Ag^+ ions in and on $BiVO_4$ catalyze intrinsic O_2 formation with the holes [9]. The observation of this positive interaction between $BiVO_4$ and Ag in both photocatalyst and photoelectrode is interesting.

The photocurrent and IPCE were both improved by the addition of various organic compounds such as alcohols, acids, and aldehydes (Fig. 5.9). Photocurrents were three and six times improved at 0.5 V vs. Ag/AgCl by the addition of methanol and HCOOH. The open circuit potentials were shifted negatively by the addition of

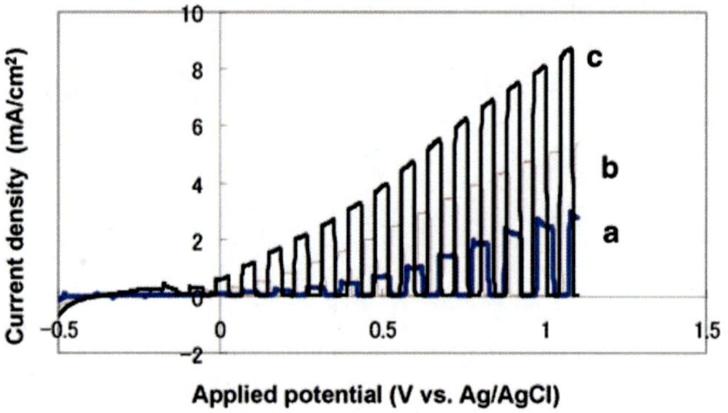

Fig. 5.9 *I–V* curves for the $BiVO_4$ photoelectrode prepared by the MOD method in various electrolyte aqueous solutions: (**a**) Na_2SO_4; (**b**) methanol; (**c**) HCOOH

these organic compounds. The IPCE at 380 nm is >80% when HCOOH is added at 1 V vs. Ag/AgCl. The absorbed-photon-to-current conversion efficiency (APCE), calculated by light absorption loss, is >120%. Therefore, both the hole scavenger effect and the current doubling effect occur when organic compounds are added. We suggest that a photoelectrode system in combination with photodegradation of organic substances can be useful for both environmental purification and energy production.

5.4 Development of A System for High-Throughput Screening of Mixed Metal Oxide Semiconductors for Photoelectrodes and Photocatalysts

In previous study of porous photoelectrodes, we and other groups have mainly adapted mixed metal oxide semiconductors developed for photocatalysts into photoelectrodes. During this process, we have recognized some intimate relationships between photoelectrodes and photocatalysts. For example, for O_2 evolution photocatalysts with sacrificial reagents, most semiconductors are adaptable into porous photoanodes for O_2 evolution, as shown in Table 5.2. Therefore, we conclude that we should continue developing new semiconductor materials for use as both photoelectrodes and photocatalysts. However, it takes many years to manually find efficient semiconductors composed of multiple metals; therefore, we developed a system for high-throughput screening of new visible-light-responsive semiconductors for photoelectrodes and photocatalysts [34].

We selected photoelectrochemical measurements as the method for evaluating the charge-separation ability, and developed an automatic semiconductor synthesis system for preparing porous thin-film photoelectrodes of various materials. The system consists of an automated liquid-handling platform, electric furnace, and

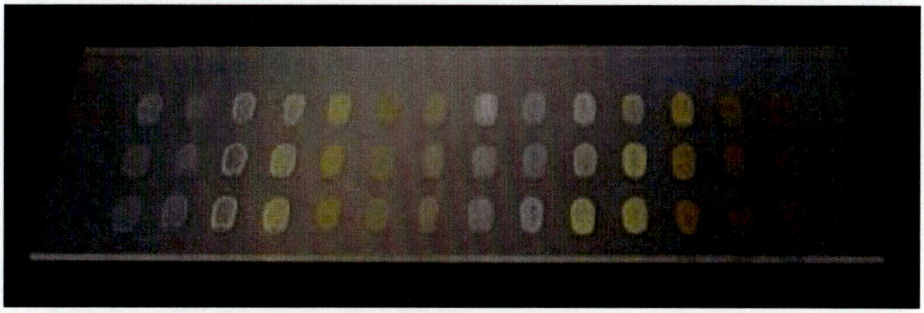

Fig. 5.10 Library of photoelectrode films prepared by the high-throughput screening system

robot arm to transfer samples, and can be used to prepare semiconductors with high element-ratio accuracy. The system uses two types of evaluation programs and two types of libraries. One program is for high-speed evaluation (Fig. 5.10): various samples are printed on the same conductive glass substrate, and their photocurrents are evaluated quickly by scanning light on the substrate. Another program is for detailed evaluation: samples are printed on separate substrates and evaluated individually. The libraries are evaluated photoelectrochemically using focused light and an X–Y stage.

An example of system screening is as follows. An n-type semiconductor, Fe_2O_3, was selected as a target material, and binary iron oxides such as Fe–Ti, Fe–Nb, and Fe–V, with various ratios were synthesized. The Fe–Ti and Fe–Nb binary oxide systems have been reported previously [6], and the result obtained from our system was consistent with those reports, demonstrating the usefulness of our system. For the previously unreported Fe–V system, the highest photocurrent was observed with 50% V, corresponding to $FeVO_4$, a potential new visible-light-responsive material.

Another example of system screening, among iron-based ternary oxides, is shown in Fig. 5.11. We observed excellent photocurrent for the Fe–Ti–Zr system; the highest photocurrent was observed at Fe/Ti/Zr = 60:10:30 with and without UV-cut filter. In yet another example, we screened for a p-type semiconductor material of Bi binary oxides and found $CuBi_2O_4$ to be a new visible-light-responsive p-type semiconductor [34, 35]. Recently, we developed a highly efficient, visible-light-responsive photocatalyst by combining the p-type semiconductor $CuBi_2O_4$ and the n-type semiconductor WO_3. The combined $CuBi_2O_4/WO_3$ shows higher reactivity than do typical TiO_2-based photocatalysts for complete oxidation of acetaldehyde into CO_2 under both solar-simulating light and visible light irradiation. The photoelectrochemical measurements of $CuBi_2O_4$ and WO_3 electrodes suggest that the photocatalytic reaction mechanism can be explained by the model p–n photochemical diode for reductivity/oxidizability improvement (Z-scheme mechanism) [35]. There are some reports on high-throughput screening systems where the semiconductor libraries were prepared by electrochemical deposition or ink-jet printing [36, 37]. We suggest that some new materials developed in some high-throughput screening systems using photoelectrochemical measurements can prove

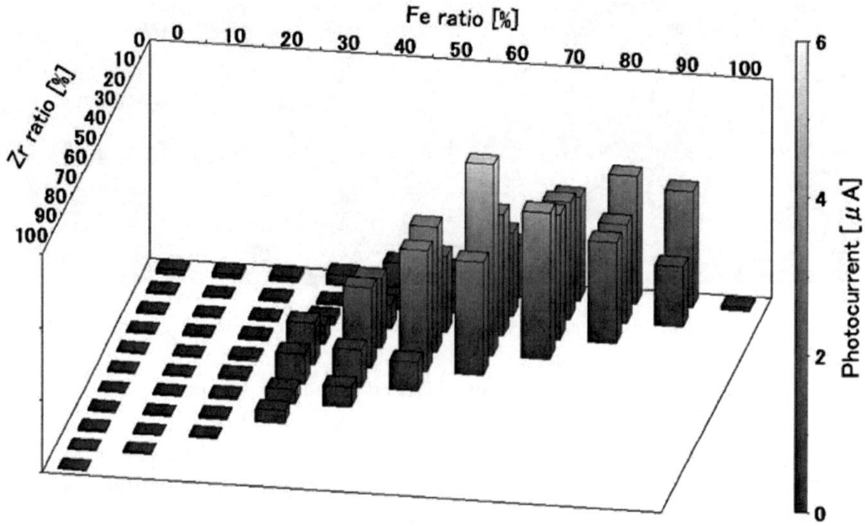

Fig. 5.11 Photocurrent for various metal ratios of Fe–Ti–Zr trinary-oxide photoelectrodes under visible light irradiation

useful in not only photocatalysts for water splitting but also photocatalysts for degradation of harmful compounds or for self-cleaning.

Acknowledgment This study was partially supported by the New Energy and Industrial Technology Development Organization (NEDO) of Japan.

References

1. Kudo, A., Miseki, Y.: Heterogeneous photocatalyst materials for water splitting. Chem. Soc. Rev. **38**, 253 (2009)
2. Fujishima, A., Honda, K.: Electrochemical photolysis of water at a semiconductor electrode. Nature **238**, 37 (1972)
3. Santato, C., Ulmann, M., Augustynski, J.: Photoelectrochemical properties of nanostructured tungsten trioxide films. J. Phys. Chem. B **105**, 936 (2001)
4. Santato, C., Odziemkowski, M., Ulmann, M., Augustynski, J.: Crystallographically oriented Mesoporous WO_3 films: Synthesis, characterization, and applications. J. Am. Chem. Soc. **123**, 10639 (2001)
5. Khan, S., Akikusa, J.: Photoelectrochemical splitting of water at nanocrystalline n-Fe_2O_3 thin-film electrodes. J. Phys. Chem. B **103**, 7184 (1999)
6. Sartoretti, C., Alexander, B., Solarska, R., Rutkowska, W., Augustynski, J., Cerny, R.: Photoelectrochemical oxidation of water at transparent ferric oxide film electrodes. J. Phys. Chem. B **109**, 13685 (2005)
7. Sartoretti, C., Ulmann, M., Alexander, B., Augustynski, J., Weidenkaff, A.: Photoelectrochemical oxidation of water at transparent ferric oxide film electrodes. Chem. Phys. Lett. **376**, 194 (2003)

8. Kay, A., Cesar, I., Grätzel, M.: New benchmark for water photooxidation by nanostructured alpha-Fe_2O_3 films. J. Am. Chem. Soc. **128**, 15724 (2006)
9. Sayama, K., Nomura, A., Arai, T., Sugita, T., Abe, R., Yanagida, M., Oi, T., Iwasaki, Y., Abe, Y., Sugihara, H.: Photoelectrochemical decomposition of water into H_2 and O_2 on porous $BiVO_4$ thin-film electrodes under visible light and significant effect of Ag ion treatment. J. Phys. Chem. B **110**, 11352 (2006)
10. Zhao, X., Wu, Y., Tao, W., Zhu, Y.: Photoelectrochemical properties of thin Bi_2WO_6 films. Thin Solid films **515**, 4753 (2007)
11. Yu, J., Zhang, Y., Gao, X., Kudo, A., Zhao, X.: Enhancement of photo-to-current efficiency over two-dimensional Bi_2MoO_6 nanoplate thin-film photoelectrode. Electrochem. Solid State Lett. **11**, B197 (2008)
12. Murakami, Y., Ikarashi, M., Hashizume, M., Nosaka, A., Nosaka, Y.: Laser ablation of the photocatalytic $BiVO_4$ and $BiZn_2VO_6$ powders in water and their photocurrent properties. Electrochem. Solid State Lett. **11**, H42 (2008)
13. Joshi, U., Jang, J., Borse, P., Lee, J.: Microwave synthesis of single-crystalline perovskite $BiFeO_3$ nanocubes for photoelectrode and photocatalytic applications. Appl. Phys. Lett. **92**, 242106 (2008)
14. Kozuka, H., Kajimura, M.: Sol-gel preparation and photoelectrochemical properties of Fe_2TiO_5 thin films. J. Sol-Gel Sci. Technol. **22**, 125 (2001)
15. Abe, R., Takata, T., Sugihara, H., Domen, K.: The use of $TiCl_4$ treatment to enhance the photocurrent in a TaON photoelectrode under visible light irradiation. Chem. Lett. **34**, 1162 (2005)
16. Krasnovsky, A.A., Brin, G.P., Dokl. Akad. Nauk SSSR **147**, 656 (1962)
17. Darwent, J.R., Mills, A.: Photooxidation of water sensitized by WO_3 powder. J. Chem. Soc. Faraday Trans. 2 **78**, 359 (1982)
18. Erbs, W., Desilvestro, J., Borgarello, E., Grätzel, M.: Visible-light-induced O_2 generation from aqueous dispersions of WO_3. J. Phys. Chem. **88**, 4001 (1984)
19. Kudo, A., Omori, K., Kato, H.: A novel aqueous process for preparation of crystal form-controlled and highly crystalline $BiVO_4$ powder from layered vanadates at room temperature and its photocatalytic and photophysical properties. J. Am. Chem. Soc. **121**, 11459 (1999)
20. Kudo, A., Hijii, S.: H_2 or O_2 evolution from aqueous solutions on layered oxide photocatalysts consisting of Bi^{3+} with 6s configuration and d^0 transition metal ions, Chem. Lett. **28**, 1103 (1999)
21. Shimodaira, Y., Kato, H., Kobayashi, H., Kudo, A.: Photophysical properties and photocatalytic activities of bismuth molybdates under visible light irradiation. J. Phys. Chem. B **110**, 17790 (2006)
22. Murakami, Y., Chatchai, P., Nosaka, Y.: Developments of the Efficient Water-splitting Electrodes under the Visible Light Irradiation. Electrochemistry **77**, 44 (2009)
23. Liu, H., Nakamura, R., Nakato, Y.: A visible-light responsive photocatalyst, $BiZn_2VO_6$, for efficient oxygen photoevolution from aqueous particulate suspensions. Electrochem. Solid-State Lett. **9**, G187 (2006)
24. Ye, F., Tsumura, T., Ohmori, A.: Dependence of photocatalytic activity on the compositions and photo-absorption of functional TiO_2-$Fe3O_4$ coatings deposited by plasma spray. Mater. Sci. Eng. B **148**, 154 (2008)
25. Chun, W.J., Ishikawa, A., Fujisawa, H., Takata, T., Kondo, J.N., Hara, M., Kawai, M., Matsumoto, Y., Domen, K.: Conduction and valence band positions of Ta_2O_5, TaON, and Ta_3N_5 by UPS and electrochemical methods. J. Phys. Chem. B **107**, 1798 (2003)
26. Takata, T., Hitoki, G., Kondo, J.N., Hara, M., Kobayashi, H., Domen, K.: Visible-light-driven photocatalytic behavior of tantalum-oxynitride and nitride. Res. Chem. Intermed. **33**, 13 (2007)
27. Arakawa, H., Shiraishi, C., Takeuchi, A., Yamaguchi, T.: Solar hydrogen production by water splitting using TiO_2 based photoelectrodes. Proc SPIE Int. Soc. Opt. Eng. **6340**, 63400G1 (2006)

28. Neves, M., Trindade, T.: Chemical bath deposition of $BiVO_4$. Thin Solid Films **406**, 93 (2002)
29. Majumder, S., Khan, S.: Photoelectrolysis of water at bare and electrocatalys-covered thinfilm ironoxide electrode. Int. J. Hydrogen Energy **19**, 881 (1994)
30. Watanabe, A., Kozuka, H.: Photoanodic properties of sol-gel-derived Fe_2O_3 thin films containing dispersed gold and silver particles. J. Phys. Chem. B **107**, 12713 (2003)
31. Subramanian, V., Wolf, E., Kamat, P.: Semiconductor-metal composite nanostructures. To what extent do metal nanoparticles improve the photocatalytic activity of TiO_2 films?. J. Phys. Chem. B **105**, 11439 (2001)
32. Ohno, T., Haga, D., Fujihara, K., Kaizaki, K., Matsumura, M.: Unique effects of iron(III) ions on photocatalytic and photoelectrochemical properties of titanium dioxide. J. Phys. Chem. B **101**, 6415 (1997)
33. Kohtani, S., Hiro, J., Yamamoto, N., Kudo, A., Tokumura, K., Nakagaki, R.: Adsorptive and photocatalytic properties of Ag-loaded $BiVO_4$ on the degradation of 4-n-alkylphenols under visible light irradiation. Catal. Commun. **6**, 185 (2005)
34. Arai, T., Konishi, Y., Iwasaki, Y., Sugihara, H., Sayama, K, High-throughput screening using porous photoelectrode for the development of visible-light-responsive semiconductors. J. Comb. Chem. **9**, 574 (2007)
35. Arai, T., Yanagida, M., Konishi, T., Iwasaki, Y., Sugihara, H., Sayama, K.: Efficient complete oxidation of acetaldehyde into CO_2 over $CuBi_2O_4/WO_3$ composite photocatalyst under visible and UV light irradiation. J. Phys. Chem. C **111**, 7574 (2007)
36. Jaramillo, T.R., Baeck, S.-H., Kleiman-Shwarsctein, A., Choi, K.-S., Stucky, G.D., McFarland, E.W.: Automated electrochemical synthesis and photoelectrochemical characterization of Zn1-xCoxO thin films for solar hydrogen production. J. Comb. Chem. **7**, 264 (2005)
37. Woodhouse, M., Herman, G.S., Parkinson, B.A.: Combinatorial approach to identification of catalysts for the photoelectrolysis of water. Chem. Mater. **17**, 4318 (2005)

Chapter 6
Combinatorial Identification and Optimization of New Oxide Semiconductors

Bruce A. Parkinson

6.1 Background

The world must eventually shift from an energy economy primarily based upon fossil fuels to one based on renewable energy sources due to the eventual depletion of fossil energy supplies and concerns about climate change due to the accumulation of carbon dioxide in the atmosphere. Sunlight is the only renewable energy source that has the potential to provide the enormous amounts of energy required by the increasing population and economic development of the world. The production and efficiency of photovoltaic solar cells that produce high value electrical energy directly from sunlight are steadily increasing. However, electrical energy is not easily or economically stored. Since 75% of the world's present energy use is in the form of fuels, methods must be developed to transform solar energy directly into storable fuels. The photoelectrolysis of water directly with sunlight is the most promising method for achieving this result. Hydrogen, the most energy-dense fuel known, can be obtained from a readily available source and, when burned or used in a fuel cell, produces only water in a reversible cycle. Therefore, the development of a cost-effective method for splitting water directly with sunlight would pave the way to a future hydrogen economy, where no pollution or greenhouse gases are produced and that is based on a universally accessible energy source.

Current technology is capable of efficiently producing hydrogen from sunlight and water by connecting conventional solar cells to an electrolyzer [1]. The major disadvantage of this approach is that both of these devices are expensive. In addition, if the electrolyzer is not located near the solar cells, losses will be incurred during transmission of the electrical power, conversion from direct current to alternating current for transmission, and the voltage step-up and step-down at the solar array and electrolyzer,

B.A. Parkinson (✉)
Department of Chemistry and School of Energy Resources
University of Wyoming, Laramie, WY 82071, USA
e-mail: bparkin1@uwyo.edu

R. van de Krol and M. Grätzel (eds.), *Photoelectrochemical Hydrogen Production*, Electronic Materials: Science & Technology 102, DOI 10.1007/978-1-4614-1380-6_6,

respectively. A potentially more efficient approach is to combine these two expensive components into an inexpensive single device that both absorbs the solar energy and performs the oxidation and reduction of water. In this approach, the generated hydrogen can be transported through pipelines (with significantly less loss than the present transport of electrical power on transmission lines) or in high-pressure tanks.

Although oxygen evolution was observed as early as 1968 when illuminating a rutile electrode in solution [2], application of this concept to water photoelectrolysis was first pointed out by Fujishima and Honda in a series of experiments that used the n-type semiconductor rutile form of TiO_2 [3, 4]. Although TiO_2 is stable under illumination in aqueous electrolytes, its large bandgap (3.0 eV) limits its utilization to the UV portion of the solar spectrum and thus limits its ultimate efficiency. It should also be pointed out that the conduction band of rutile is not negative enough to reduce water and so in the original publications a "pH bias" was present where the oxygen-producing side of the cell was basic with respect to the hydrogen-producing electrolyte. Later work using heterojunctions of III–V materials as photoelectrodes considerably increased the visible light conversion efficiency but also increased the cost and decreased the effective lifetime of the device due to corrosion in aqueous electrolytes [5]. Stability of a photoelectrolysis electrode is crucial for a viable system since the significant capital investment needed for the support structures, electrolyte handling and gas handling systems requires that the system last for many years. Since the photoelectrode will be continually immersed in an electrolyte and illuminated with direct solar radiation, a long lifetime is a daunting challenge that must be adequately addressed even in the basic research stages. It is primarily for their stability that metal oxide semiconductors arguably hold the most promise for constructing a stable photoelectrolysis system.

In addition to low cost and stability, any efficient photoelectrolysis material must also effectively utilize the solar spectrum and generate sufficient photovoltage to drive the water-splitting reaction. Either a single or tandem semiconductor electrode system may be used [6]. The obtained photovoltage must be greater than the thermodynamic value for the difference between the water oxidation and water reduction potentials (1.23 eV). The material(s) should have some catalytic activity for hydrogen or oxygen production so as to minimize the additional photopotential above 1.23 V that is required to overcome the electrochemical overpotentials needed to drive the water oxidation and reduction reactions at the desired rate. Current densities (rates) on the order of 15–25 mA/cm^2 at illumination intensities around 1 sun are also needed for a single photoelectrode system and half these values for a tandem system. In addition, a single semiconductor photoelectrode must have a valence band edge that is more positive than the water oxidation potential and a conduction band edge that is more negative than the water reduction potential so that the photogenerated holes and electrons have the electrochemical driving force for the water-splitting reaction. If a dual photoelectrode system is used, the conduction band of the p-type material must be negative of the water reduction potential and the valence band of the n-type material must be positive of the water oxidation potential.

As stated above, either a single semiconductor electrode or a two-semiconductor electrode photoelectrolysis system can be envisioned. The single illuminated electrode system will need a bandgap greater than about 1.7 eV in order to supply the photovoltage needed for water photoelectrolysis given that photovoltage of about 2/3 the band gap is often the maximum obtainable from a good semiconductor material operating at its maximum power point. A single photoelectrode system would need to be attached to a catalytic metal (perhaps Pt or Pt modified) electrode that can accomplish the complementary water-splitting half reaction. A two-semiconductor system, with both a p-type and an n-type semiconductor photoelectrode, can be configured in several ways. In one configuration the electrodes could be placed with the larger gap material absorbing the higher energy portion of the solar spectrum in front of the smaller bandgap material of the opposite carrier type in which the wavelengths transmitted through the large bandgap material would be absorbed in the smaller gap material. The two materials must absorb nearly equal numbers of solar photons since the current through each semiconductor must be matched due to the recombination at the ohmic contacts that provides the summation of the two photovoltages. Therefore, in this configuration the quantum yield for water photoelectrolysis will be half that of a single photoelectrode system; however, since there is considerable photon flux in the red and near IR portion of the solar spectrum, the lower quantum yield may be offset by utilizing this portion of the solar spectrum. This configuration has the advantage that one glass substrate, with a transparent conducting layer on both sides, could be used for both materials reducing the total system costs. The ideal bandgaps for the two materials have not yet been identified but they will most likely be somewhat different than for a two-bandgap heterojunction solar cell where bandgaps of 1.9 and 1.2 eV give the best theoretical light to electricity conversion efficiency but yield a photovoltage larger than necessary for water photoelectrolysis. A larger photovoltage would provide some extra driving force for a photoelectrode that does have optimal catalytic properties for water oxidation or reduction. A schematic of a cross section of such a device is shown in Fig. 6.1. Systems that convert the extra photovoltage to electrical power can also be designed but the electrical power produced would most likely be more expensive than if produced from a conventional photovoltaic device.

Another configuration would be to place the n and p type materials side by side rather than putting a larger bandgap material in front resulting in full spectrum sunlight illumination for both electrodes. In this case, an n-type and p-type electrode of the same material or two different oppositely doped smaller bandgap materials can be paired as long as the sum of their photovoltages at maximum power was greater than the ~1.7 eV needed for water photoelectrolysis and the photocurrents in both electrodes were matched. If two materials were used the bandgaps would have to be close to each other since they need to absorb nearly the same flux of solar photons in order for the photocurrents to be matched. Two stable materials would be needed for this configuration to work, a p-type material with a conduction band at least 0.8 eV negative of both the n-type material conduction band and the hydrogen potential and valence band offsets of about 0.8 V. Two substrates would also be required, but they need not be transparent (as in the case of the stacked system discussed above and shown in Fig. 6.1) and so

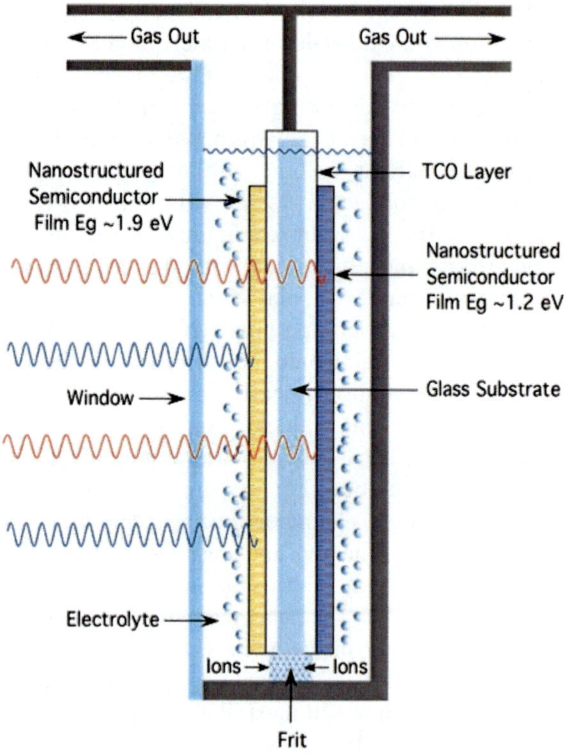

Fig. 6.1 A design for a tandem photoelectrolysis system that uses nanostructured semiconductor films of opposite carrier type that are deposited on opposite sides of a single substrate that is covered on both sides with a continuous transparent conducting oxide film. The larger bandgap material is transparent to the lower energy radiation that is absorbed by the smaller gap film on the backside of the substrate

cheaper and more abundant materials may be utilized. Table 6.1 summarizes the advantages and disadvantages of the different photoelectrolysis configurations discussed above. Systems that contain buried junctions, where charge separation is remote from the electrolyte, are really solid state solar cells in series with metal electrodes, are not considered since they are essentially equivalent to hooking up photovoltaic cells to an electrolysis cell.

Any photoelectrolysis system will have to be designed with an ion conducting separator or electrolyte path between the two electrodes and with separate gas collection plumbing for hydrogen and perhaps oxygen. Oxidation reactions, other than water oxidation, have been discussed since there is much less demand for pure oxygen than there is for hydrogen. The commercial demand for pure oxygen would be rapidly saturated once any water-splitting photoelectrolysis system with the capacity to produce meaningful amounts of hydrogen is deployed. The only other commercial chemical currently produced on a very large scale by aqueous electrolysis is chlorine, but again the demand for chlorine would be quickly met by any system producing enough hydrogen fuel to make an impact on world energy demand. However, photoelectrolysis technology could get an initial boost from the additional revenue obtained from the photoelectrolysis of hydrochloric acid or brine.

Table 6.1 A comparison of the advantages and disadvantages of various configurations for a photoelectrolysis device

Configuration	Advantages	Disadvantages
Single photoelectrode	Only one substrate and one p or n type semiconductor needed. Dark electrode can be conventional.	Poor utilization of the solar spectrum since large bandgap (>~1.7 eV) is needed.
Tandem photoelectrodes (Fig. 6.1)	Needs only one substrate. More efficient utilization of the solar spectrum.	Need to identify one p and one n type semiconductor and needs a transparent conducting substrate. Two photons and current balancing needed.
Separately illuminated photoelectrodes	Cheap nontransparent substrates can be used. More efficient utilization of the solar spectrum.	Need two semiconductors or ability to dope one both n and p type. Two photons and current balancing needed. Twice the collector area of tandem.

Often materials are synthesized and tested for photochemical water-splitting activity as either colloidal solutions or as powdered slurries. While this method may be useful for screening for water-splitting activity, it is unlikely that homogeneous colloidal solutions or slurries will be useful for a practical water-splitting system. Several reasons for this include (1) The products are not produced in separate compartments, resulting in highly explosive mixtures of hydrogen and oxygen. (2) Energy is required to separate hydrogen and oxygen reducing the overall efficiency of the water-splitting process. (3) Since catalysts for hydrogen and oxygen production from water are also good catalysts for the recombination of hydrogen and oxygen, illumination of such systems will eventually result in a photostationary state where forward and back reactions have equal rates and no more water splitting can occur. The exchange current at the photostationary state will be related to the solubility of the gases in the electrolyte and the rate at which they are removed from the reactor.

To avoid undesirable back reactions, and the poor kinetics for hydrogen or oxygen production, sacrificial reagents are often used in the photoreactors to scavenge either holes or electrons. The most common examples of hole scavenging agents introduced into solution are alcohols (usually methanol), amines (usually triethanolamine or EDTA), or sulfite salts. Electron scavengers such as the easily reduced Ag^+ can also be added. Electron or hole scavengers can be useful to study *one* of the water-splitting half-reactions without complications associated with the kinetics of the other half-reaction (although using a three-electrode potentiostat to study a single photoelectrode would achieve the same end) but these additives are not viable for any practical system for sustainable energy production.

As a comprehensive review of photoelectrochemical hydrogen generation [7] and a recent book [8] have been published on this subject, this chapter will concentrate on the combinatorial aspects of the search for appropriate materials for a practical device.

6.2 Survey of Known Materials and Their Photoactivity

As discussed above, the positions of the conduction band and valence bands with respect to water oxidation and reduction potentials are one requirement for a water-splitting photoelectrode. Among the first materials to be exhaustively studied were the perovskite structure oxides $SrTiO_3$ [9–13] and $BaTiO_3$ [9, 14, 15]. Both of these materials are better suited than TiO_2 for water photoelectrolysis because their conduction band edges are more negative than the water reduction potential. Unfortunately, as is also true for TiO_2, their bandgaps are so large (~3.2 eV) that they also required UV light to produce photocurrent resulting in very poor solar efficiencies.

A practical solar photoelectrolysis material must also be stable under illumination for many years while immersed in an aqueous electrolyte. Photocorrosion is the main problem with nonoxide semiconductors that have near optimum band gaps [16] and band positions for efficient photocatalytic water splitting such as transition metal chalcogenides [17–19], cadmium and zinc chalcogenides, GaAs [20], GaP [21], and InP [5, 22–24]. Thermodynamic stability is assured only if the oxidation decomposition potential of the semiconductor is more positive than the valence band edge and the reduction decomposition potential lies more negative than the conduction band edge [25]. Because the redox potentials for the decomposition reactions of these materials are generally within the band gap, and additionally because the decomposition reactions are kinetically favored, these experiments generally lead to electrode photocorrosion. If even only a tiny fraction of the photoexcited carriers are involved in photocorrosion reactions the electrode will not survive for many years under solar illumination intensities.

The most stable rocks and minerals that survive many years of weathering are all oxides, suggesting that a semiconducting oxide material will have the best chance to fulfill the stability criteria for a water-splitting photocatalyst. Smaller bandgap metal oxide semiconductors, notably WO_3 [26–29], Fe_2O_3 [30–35], NiO [36, 37], and Cu_2O [38], have also been extensively studied. Although these materials harvest a larger percentage of the solar spectrum, their conduction band edge energies are often not optimally positioned for the production of hydrogen [39]. They can also be difficult to electrically dope, resulting in being either poor conductors of electrons and/or holes and some of them (e.g., Cu_2O) are still prone to corrosion [16].

The valence band positions of oxide materials are usually similar and quite positive due to the fact that the band is derived from oxygen 2p states. Holes at such a positive potential can easily oxidize water but can waste considerable energy in doing so either due to the large difference in potential between the holes and the water oxidation potential or poor electrochemical kinetics for water oxidation. One idea to both extend the spectral response of large gap oxide semiconductors and to raise the valence band energy is to incorporate sulfur, carbon, or nitrogen into the lattice. A series of these compounds including $Sm_2Ti_2S_2O_5$ [40], TaON [41], $Y_2Ta_2O_5N_2$ [42], Ta_3N_5 [43], and $LaTiO_2N$ [44, 45] as well as incorporating S, N, or C into TiO_2 [46–49] have recently been examined by several research groups. These groups often use density functional theory to calculate that the bottom of the conduction bands consists of empty d-orbitals and, depending upon the material, the tops of the valence bands consist mainly of N 2p or S 3p orbitals resulting in smaller band gaps for improved solar energy utilization. The smaller band gaps, and fortuitous band positions relative to the redox potentials of water, imply that these materials could function as water photoelectrolysis materials; and indeed H_2 and/or O_2 is produced at quantum efficiencies up to 30% [41]. Unfortunately, however, most of these systems are typically not reported to function without the presence of the "sacrificial reagents" in solution mentioned earlier. It is likely that the sacrificial reagents are necessary to camouflage inherent oxidation or reduction decomposition reactions and that these materials will ultimately have the same stability problems as the earlier materials that incorporated nonoxide anions since nitrogen, sulfur and carbon are easily oxidized in solution.

In principle, if one introduces an additional metal or metals into the structure that have electronegativities greater than that of the parent metal and less than that of oxygen, then it is possible that a valence band created from mixed metal d states will be more negative in energy (closer to the oxidation potential of water) than that of a pure oxygen 2p band, thereby reducing the bandgap to 1.5–2.0 eV [50]. As an example, a mixed metal oxide semiconductor could be constructed with conduction band edge derived mainly from titanium. In principle, it should then be possible to create mixed bands, lowering the band gap of the parent titanium oxide, by mixing the transition metal orbitals with the oxygen 2p orbitals. For example, the ions Mn^{2+}, Ni^{2+}, Cr^{3+}, Rh^{3+}, V^{4+}, Os^{3+}, Sn^{2+}, Sb^{3+}, Pd^{2+}, and Pt^{2+} all have d orbital energies that are more positive (further down an absolute scale) than the oxidation potential of water but more cathodic than that of oxygen's 2p orbital [50]. And so, in theory at least, by combining these metals with titanium it should be possible to create a material with a band gap can absorb a large fraction of the solar spectrum and is optimally positioned to photoelectrolyze water. Other compositions such as $LuRhO_3$ [51], $SrZr_{0.25}Ti_{0.75}O_3$ [50], $LaVO_3$ [52], $K_2La_2Ti_3O_{10}$ [53], and $RbPb_2Nb_3O_{10}$ [54], have also been reported to be viable photoelectrolytic systems and provide useful starting points in a combinatorial search.

6.3 Combinatorial Production and Screening of Metal Oxide Photocatalysts

Multicomponent materials are often required to optimize a special property. High temperature superconductivity is a good example of a special property requiring multicomponent materials. The oxide with the highest known transition temperature contains four or even five metals ($HgBa_2CaCu_2O_{6+d}$, $Hg_{0.8}Tl_{0.2}Ba_2Ca_2Cu_3O_{8.33}$, $T_c \sim 125$ K). Since our present theoretical knowledge is insufficient to a priori calculate the behavior of such complex systems, or estimate the identity and atomic ratios of the specific metals needed to perform the efficient photoelectrolysis of water, empirical methods will be necessary to identify the best material. Thousands, or perhaps hundreds of thousands, of mixed metal oxide photocatalyst compositions may need to be produced and tested for efficiency and corrosion resistance until an effective material is discovered. Fortunately, the recent development of combinatorial methods has provided tools for the production of large libraries of compounds and the methods for screening these libraries for the presence or optimization of a property of interest. In this section, we will review the combinatorial approaches that have been applied to the discovery and optimization of water-splitting photocatalysts.

We can first do an analysis of the number of possible combinations of mixed metal oxides that could be produced and screened. If we prepare all possible ternary oxides, mixing N materials three at a time there are $N!/3!(N-3)!$ combinations if we only consider 1:1:1 stoichiometries. Since there are ~60 possible candidate metals in the periodic table this results in >34,000 combinations to be produced and tested. If a quaternary material is needed then there are $N!/4!(N-4)! > 480{,}000$ possible combinations. Again these numbers do not consider producing and screening the many possible stoichiometries. It is also possible that a ternary or quaternary material will require small amounts of an additional element that acts as an electrical dopant to optimize its conductivity. These small amounts can often exist as impurities in the metal oxide precursors or may have to be discovered by additional combinatorial screening (discussed later in this chapter) greatly multiplying the number of experiments that must be done. However, some combinations could be excluded because it would not make sense to, for example, mix only large bandgap oxide materials (i.e., Ti–Zr–Si–Y oxides) and expect visible light absorption. It is clear that given the large number of possible mixtures that need to be produced, a "high-throughput" approach is needed, as well as multidimensional experimental design, to explore the parameter space more rapidly. And while it may not be reasonable to produce and test ALL of the possible combinations, it is certainly imperative that the throughput of materials screening be rapid relative to the serial approaches that have historically been used.

We can also use some chemical knowledge to reduce the number of combinations that need to be tested. We can very loosely classify metals into one of four categories according to their potential role in a photoelectrode material; structural, since they form very stable high heat of formation oxides (i.e., Ti, W, Zr, Ta, Si, Mo, Nb, Hf, In, Sn, Y, Sc whose oxides have very high bandgaps and so are not useful by themselves), light absorbing, where multiple elements may be needed to absorb light across the solar spectrum [55, 56] (i.e., Fe, Cr, V, Co, Mn, Ni, Cu, and some rare earths such as Ce), catalytic, to lower overpotential losses in the electrochemical reactions (i.e., Ru, Rh, Pd, Ni, Pt, Ir, Os, Re) [55, 57, 58] and ionic charge compensators (i.e., Ca, Sr, Ba, Mg, Zn, Cd, Li, Na, K, Rb, Cs)[1] that are often present in stable structures such as spinels and perovskites to balance the charge in the compound since oxygen is almost always present in stable oxides as O^{2-}.

[1] We purposely do not include toxic metals such as Pb, Tl, Cd, and Hg since we envision eventual large-scale implementation of any discovered photocatalysts and we want to avoid the environmental consequences of these elements.

This chemical guidance greatly reduces the number of possible combinations (but it is still in the tens of thousands). Finally, it may be necessary to deposit catalytic materials such as Pt, Pd, Ir, Ru, or Ni [59–61] or small molecule catalysts onto the surface of the electrode if the surface of the semiconductor itself has little catalytic activity.

With the appropriate substrate a combinatorial library can be tested for photoactivity by a number of different methods. While the ideal method would be to *directly* measure hydrogen production with, for example, gas chromatography, the amount of hydrogen expected to be produced from a thin film photoelectrode that does not absorb a large fraction of the incident light (the reason for very thin films is discussed below) is typically too small to generate hydrogen at a rate needed to overcome the solubility of the gas in aqueous solution. In light of this limitation, screening for photocatalytic activity has largely been carried out by measuring spatially resolved photocurrents or by observation of pH changes in the electrolyte near the array.

The use of pH-dependent fluorescent indicators [62] or molecular sensitizers [63] is based upon the principle that the local concentration of protons increases at sites catalytically active for water oxidation (oxygen production) and decreases at sites active for water reduction (hydrogen production). The reverse processes of hydrogen oxidation or oxygen reduction also cause a change in pH making the method useful for screening fuel cell catalysts as well as photoelectrolysis catalysts. A visual or CCD camera inspection of an irradiated catalyst library can be sufficient to ascertain activity if the amount of redox chemistry and/or the fluorescence of the indicator is sufficient. Otherwise, a spatial analysis of absorbance or emission changes must be carried out. The advantage of this approach is that it is a rapid detection method and avoids some of the spurious signals that can result from electrochemical detection methods. On the other hand, it is not high resolution and one must assume that the indicator molecules do not interfere with the overall redox process occurring in solution (such as competitive charge transfer to the indicator in lieu of photocatalysis) and the appropriate indicator must be found for the chosen electrolyte. Alternative screening techniques with higher resolution will be described in the following section.

6.4 Methods for Combinatorial Production and Screening

6.4.1 Inkjet Printing and Electrochemical Scanning Light Beam Analysis

Inkjet printing has been utilized for many combinatorial chemistry applications including the production of luminescent metal oxide libraries [64]. The hallmarks of inkjet printing for materials deposition are its speed, ease of use, low cost (depending upon the printing platform chosen) and its versatility. The combinatorial search strategy within our group [58], and later adopted by the Lewis group [65],

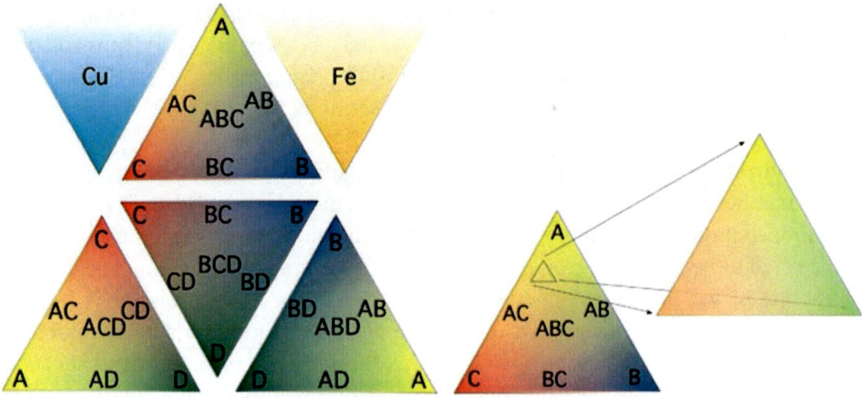

Fig. 6.2 The faces of a four-component tetrahedral phase diagram where the four metals are mixed three at a time in triangles that correspond to ternary phase diagrams. When a promising composition was found, for example, the small triangle inset of the ABC ternary mixture on the right, the gradients could be adjusted to expand that region of the ternary phase diagram and the necessary number of iterations could be performed until the best mixture is found. Copper oxide and iron oxide are also printed in a gradient triangle as p and n type simple oxide internal standards, respectively. Mallouk et al. used this scheme for screening of precious metal fuel cell catalysts [62]

utilizes inkjet printing of overlapping patterns of soluble metal oxide precursors onto a conductive substrate. Metal nitrate salts have been the most widely printed precursors since they are highly soluble and subsequent pyrolysis at temperatures below 550°C yields metal oxides. Temperatures must be kept below about 550°C since above these temperatures the glass begins to soften and the conducting FTO layer becomes unstable.

Our work has used a research grade printer manufactured by FujiFilm Dimatix [58] and off-the-shelf consumer grade printers made by Hewlett Packard [66]. Originally, films were deposited in overlapping triangular gradient patterns as shown in Fig. 6.2.

While inkjet printing such a pattern ideally produces smooth gradients in composition, in reality the extent of uneven mixing of compositions within the triangles can be unknown. Assuming that the actual deposition amounts of the material precursors is linear from the vertices to the edges of the triangle, other factors such as miscibility, phase segregation, annealing effects, etc., might also play a role and make it difficult to know the exact composition in a particular spot. One particular complication that has emerged during our studies is that the metal nitrate salts generally melt before they decompose and coalesce into droplets, resulting in less uniform films of the oxide after pyrolysis. The use of metal oxide nanoparticles, polyoxometalates, and various alkoxide compounds as precursors that decompose directly into the oxide will be useful to avoid this complication.

To analyze for photoelectrolysis activity several groups use custom-built scanning laser screening systems where a laser is rastered over the printed and pyrolyzed substrates that are immersed in an aqueous electrolyte. Scanning the laser over the sample produces a false color image of photoactivity by measuring the photocurrent at each "pixel." To detect small signals the laser can be chopped or modulated and a lock-in amplifier can be used to isolate the photocurrent signal even if there is substantial dark current present. A potentiostat can also be used to apply a positive or negative external bias to the sample to aid the oxidation or reduction of water, respectively. The external bias will not be useful for a practical system but is used so that a good semiconducting material that might not be electrocatalytically active may still be identified and can have a surface catalytic layer added later.

Figure 6.3 shows a template, a photograph, a false color photocurrent image and a false color photocurrent image of a compositional zoom of a promising compositional region. In the false color photocurrent images in Fig. 6.3, the yellow to white pixels corresponds to higher relative photocurrent where a Fe–Cs–Nd composition is most photoactive. The dark blue regions represent areas where there is virtually no photoactivity either from the FTO substrate or the material. Also printed and screened using this template are n- and p-type "internal standards" of a-Fe_2O_3 and CuO, respectively, printed because it not only affords a direct comparison to these well-known semiconducting oxide materials [20], but also between all printed sample plates.

Hundreds of materials have been produced and screened using the unfolded tetrahedron template, with most of the material combinations showing no or very little photoresponse but some showing significantly better performance than the internal reference materials. One of the materials discovered in our search, composed of Fe, Co, and Al, has reproducibly shown p-type photoactivity and was not previously known to function as a photoelectrolysis material. The regions of high activity were found in the Co-rich regions of the Co–Al–Fe triangles. While this original high-throughput method was useful for identifying promising materials, the exact stoichiometry of the metal components in the active mixture was uncertain since the exact amounts of the various "inks" delivered by the printer driver software is unknown.

In order to determine the optimum stoichiometry of the promising mixture, a series of samples with an array of areas with known compositions was prepared using a research grade inkjet printer manufactured by FujiFilm Dimatix. This printer has a camera that focuses on the printhead nozzles and software control of parameters such as the waveform applied to the piezoelectric jetting nozzle, drop spacing, drop volume, and jetting frequency. By controlling these parameters, production of a broad or narrow compositional range with known stoichiometries within each printed area is possible. The thickness of each sample can also be controlled with these factors or, if desired, by depositing multiple layers. An example of a sample plate prepared in this way is shown in Fig. 6.4. In this pattern, four different densities of Co and Fe were printed as double horizontal and double vertical rows, respectively, while different amounts of Al were deposited in a repeating pattern within groups of four squares containing identical Fe and Co amounts as shown in the dashed inset in the upper left-hand corner. The precise values of the drop spacing and density printed within each square in

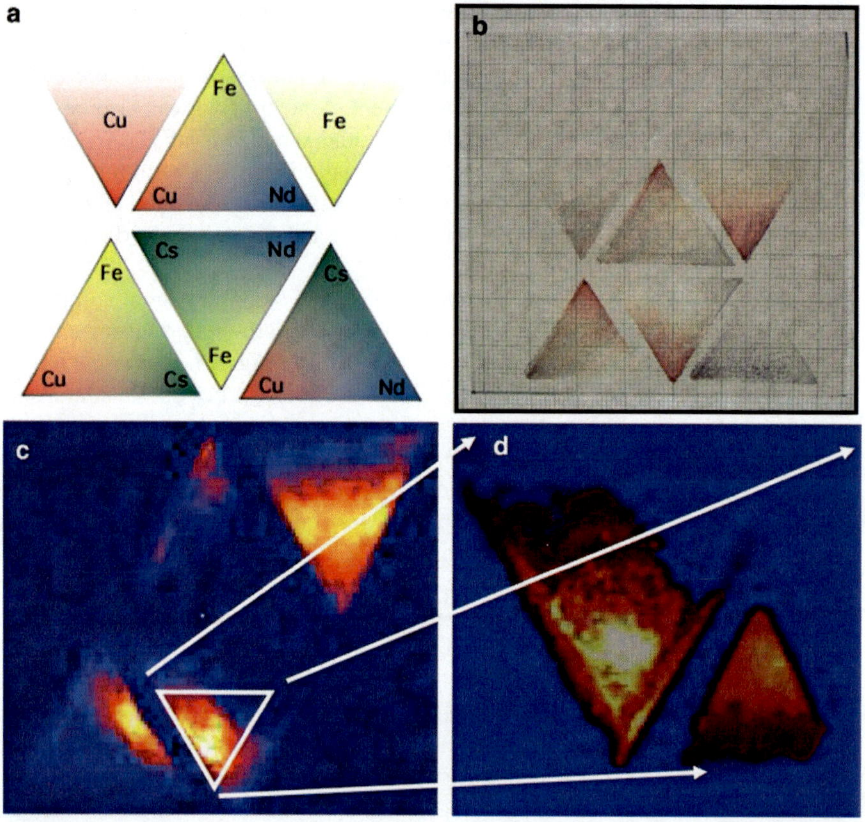

Fig. 6.3 Printing and screening a four-metals-three-at-a-time pattern and a compositional zoom for a Fe–Cs–Nd–Cu system. (**a**) False color template showing the positions and gradients used for printing the four metal precursor solutions. (**b**) Photograph of the printed and fired film. Note the triangular internal standards of a-Fe_2O_3 and CuO (*upper right* and *left*, respectively) with thickness gradients (*bottom* to *top*) that are used as internal standards. (**c**) False color photocurrent image of the film shown in B using 514.5 nm illumination under 0.5 V bias in a 0.5 M NaOH solution. The photocurrent produced at a particular "pixel," relative to the others in the two-dimensional array, is represented by its relative brightness with the most photocurrent in a particular direction (water oxidation in this case) being the brightest. (**d**) Photocurrent scan at 514.5-nm of a triangular composition zoom in on the brightest area of the Fe–Cs–Nd triangle shown in (**c**) that has a maximum IPCE value approximately twice that of the a-Fe_2O_3 internal standard (*smaller triangle* to the *lower right*). Expanding the printing gradients within the brightest region of the n-type material created the "zoom in." Reproduced with permission from [66], copyright 2005 American Chemical Society

Fig. 6.4 for the Fe and Cu internal standards, as well as the Fe–Co–Al Oxide precursors, can be found in reference [66]. As an example, the brightest square in Fig. 6.4, third square from the right, third row down, is given the label $Fe_BCo_DAl_A$ which correlates to $Fe_{1.7}Co_{14.4}Al_1$ (In atomic percentages 10% Fe, 84% Co, and 6% Al) and this composition was used in future photoelectrochemical and structural

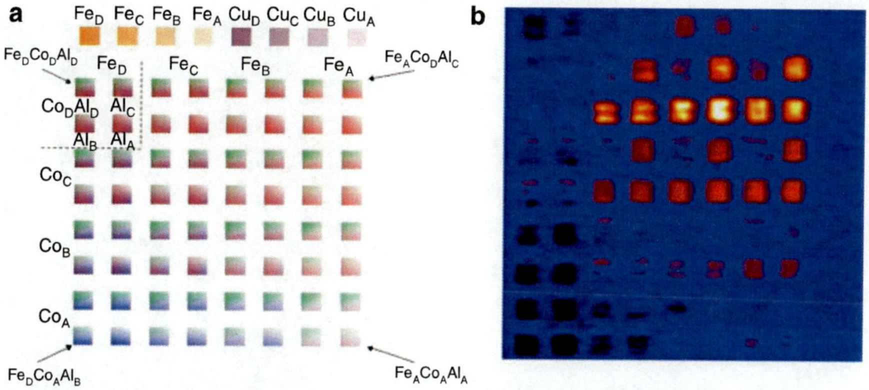

Fig. 6.4 (**a**) Printing template used for quantifying the optimal stoichiometry in the Co–Al–Fe system where known amounts of the components are printed into individual 3 × 3 mm^2. The density of precursor drops (and the resulting stoichiometry) corresponding to each stoichiometry for different thickness is described in reference [58]. (**b**) False color photocurrent map with no applied bias with 532 nm laser excitation in a 0.1-M NaOH solution. The cathodic photocurrent measured by laser scanning the brightest square of the film is nearly three times that generated by the best pure CuO internal standard labeled Cu_D. The compositions within the solid white box were expanded in Fig. 6.5. Reproduced with permission from [58], copyright 2008 American Chemical Society

experiments where bulk samples were prepared as it consistently appeared to be a near optimal composition.

In contrast to the "zoom in" shown in Fig. 6.3, created by adjusting the printed gradients, the most promising stoichiometries within array-based samples can be investigated much more quantitatively. To further refine the optimal composition of the photocatalyst described above, another template was created that encompasses stoichiometries outlined by the white triangle as well as some similar stoichiometries. The template, and the result of the photocurrent screening, is shown in Fig. 6.5. In this pattern, the amount of cobalt was kept constant (labeled B in the figure with a drop spacing equal to that of all mixtures within the white box of Fig. 6.4) while the Fe and Al contents (labeled w, x, y, and z) were varied and calculated using the rationale described previously. The internal standards were printed with the same drop densities as in Fig. 6.4 and the overall amount of printed internal standard material relative to the working mixtures is approximately the same between Figs. 6.4 and 6.5 in order to not overestimate the contribution of sample thickness to performance. The resulting photocurrent for all working Co–Al–Fe oxides is within ±20% for all squares indicating flexibility in the stoichiometries of the photoactive compositions. Indeed, the Fe and Alcontent can vary by as much as a factor of two (between $Co_{17.8}Fe_1Al_{2.1}$ and $Co_{17.8}Fe_{2.1}Al_1$) and still produce a material with comparable photoactivity.

Currently inkjet printing allows one to rapidly produce and screen many materials and quantify the most promising stoichiometries. Lewis, Katz and Gingrich [65] have devised a variation of the technique described above that allows

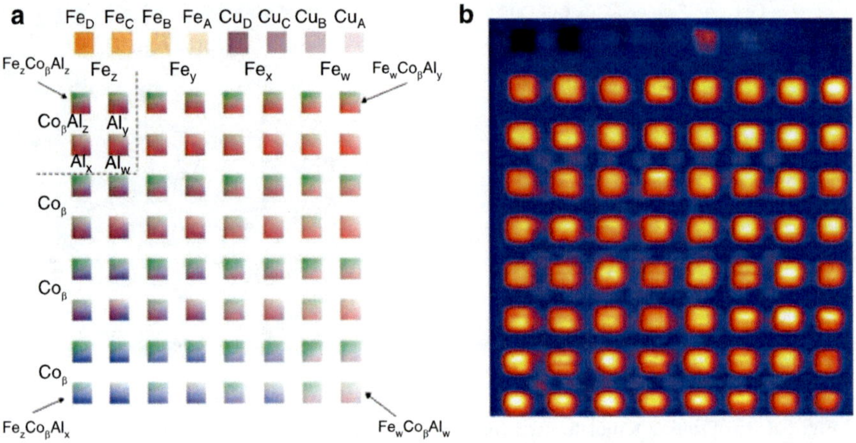

Fig. 6.5 Compositional library produced by expansion of the compositional region outlined by the white box in Fig. 6.4. The template shown in (**a**) was created by including all of the mixtures created within the inset of Fig. 6.4 as well as some intermediate stoichiometries. The pure Fe and Cu internal standards were printed as in Fig. 6.4. (**b**) Photocurrent map of the array of 3 × 3 mm^2 measured in 0.1-M NaOH with a 532-nm laser showing that a rather large range of composition space appears to work better than the CuO internal standards. Reproduced with permission from [58], copyright 2008 American Chemical Society

for the measurement of the open-circuit photovoltage as well as the photocurrent response of each composition. The photovoltage provides a valuable estimate of the position of the quasi-Fermi level of the majority carriers under illumination, which indicates whether the photogenerated carriers are thermodynamically capable of a particular photoelectrolysis reaction. Since the photovoltage measurement is for the entire electrically connected substrate, to measure it for a specific area or composition requires electronic isolation of each metal oxide mixture, as well as passivation of the exposed conductive substrate. To achieve this up to 260 regions on the FTO substrate are laser etched to remove the FTO from the regions between the areas to be printed and electrically isolate them. Up to eight precursors are then printed in the isolated regions and pyrolyzed. An insulating epoxy is then screen printed over any exposed FTO. A multiple contact connector is then clamped to the substrate and connected to a custom-designed printed circuit board where each contact on the circuit board is associated with a particular spot in the materials library and connected through ribbon cables to a multiplexer for the electrical measurements.

The Sayama group also currently utilizes a scanning light beam analysis system for high-throughput screening of photoelectrolysis materials [67]. In their system the libraries are produced by an automated robotic pipetting of precursors onto a substrate with a subsequent firing. The photocurrent of each sample in an array is then measured at a constant potential in a three-electrode cell by scanning the photoelectrode with a focused beam from a xenon lamp guided by an optical fiber with a UV cutoff filter. Their high-throughput production and screening system has been used to optimize the stoichiometry of some porous iron oxide binary compositions.

6.4.2 Combinatorial Dopant Optimization Using Inkjet Printing

Low levels of impurities are well-known to control the electrical properties of semiconductors via electrical doping or to produce potent deep trap states that can promote carrier recombination and destroy the photoresponse of the material. It is often difficult to control or determine these low levels of active impurities; however, in some photoelectrode materials purposely adding impurities can dramatically improve the photoresponse. One such material, hematite (α-Fe_2O_3), has received much attention because of its bandgap of 2.2 eV, abundance, nontoxicity, and electrochemical stability. Many methods have been investigated to improve the photoelectrochemical performance of α-Fe_2O_3 including the incorporation of various other elements into the hematite films. Grätzel found that the introduction of a small amount of Si into the film dramatically increased the photooxidation current. Many other "dopant" species have also been added including Ca^{2+}, Mg^{2+}, Cu^{2+}, Zn^{2+}, Si^{4+}, Ge^{4+}, Ti^{4+}, Pt^{4+}, Cr^{4+}, V^{5+}, Nb^{5+}, and Mo^{6+} [68–82]. A number of these elements have improved the photoresponse of α-Fe_2O_3 but the mechanism for the enhancement has not often been explained. Some reasons for improvements of the photoactivity from adding impurities include lowering the resistivity of the film due to an increase in the majority carrier density, improvements in the electrocatalytic activity for water (OH^-) oxidation, directing the morphology of the film (nanostructuring) to allow more carriers to reach the semiconductor/electrolyte interface, passivating grain boundaries or producing different crystallographic orientations of the α-Fe_2O_3 crystallites to favor the growth of faces with a lower surface recombination velocity. The use of the term "dopants," when referring to the addition of more than parts per thousand levels of impurities, does not conform to the usual semiconductor physics definition of electrically active shallow dopants that can control the electrical conductivity at the level of parts per million. However, since this term has come into wide use, even when adding amounts of elements at levels more properly called solid solutions (if in fact all the additive is incorporated into the lattice), we will continue to use the term "doping" with the understanding that the reason these added elements influence the photocurrent and/or photovoltage are not usually known.

Given the multitude of possible enhancement effects, adding multiple elemental impurities might result in synergistic effects and produce films with much higher photoactivity than with just one impurity. However, given now the large number of possible additives at multiple concentration levels the number of samples that must be prepared becomes very large. Therefore, we adapted our inkjetting technique to quickly produce many "doped" α-Fe_2O_3 compositions and screen them for their water photooxidation activity. Since we will be printing arrays of α-Fe_2O_3 we first optimized the thickness of printed films of this material by varying the drop density and the number of times we overprint the pattern. An 8×8 array of squares was printed onto SnO_2:F-coated glass with 0.5 M $FeCl_3$ with drop spacing of 80 μm. This pattern, shown in Fig. 6.6a, was divided into four double rows printed once, twice, three times, and four times from the bottom to the top, to determine the

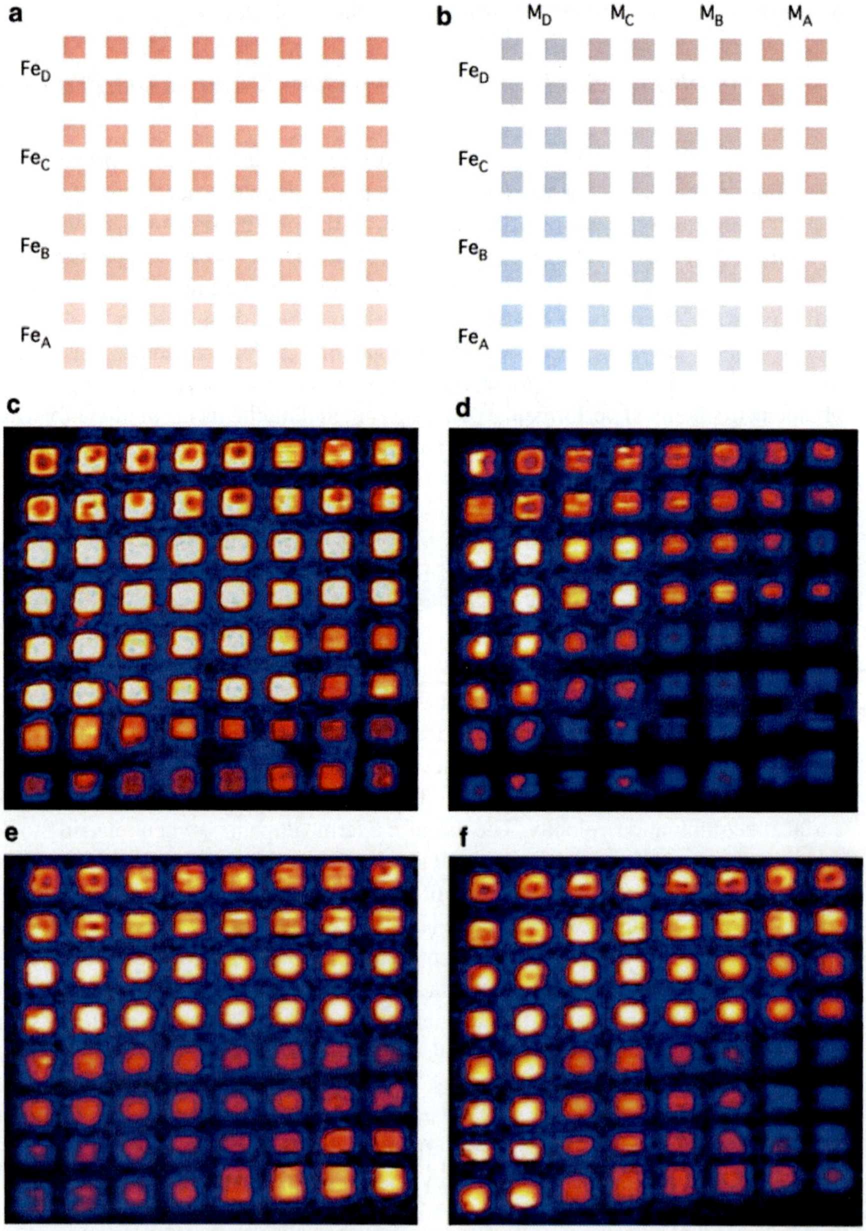

Fig. 6.6 (**a**) Printing template used for undoped α-Fe_2O_3 with ~80 μm drop spacing and double rows printed from the bottom up: once, twice, three times, and four times corresponding to Fe_A to Fe_D. (**b**) Printing template used for the Fe–M (M = Ti, Si, or Al) series where known amounts of the components are printed on top of the Fe rows from (**a**) into individual 3 × 3 mm^2 resulting in groups of four squares with identical compositions. The density of precursor drops and resulting stoichiometries for (**d–f**) are listed in Table 6.2. False color photocurrent maps of libraries printed using template (**b**): (**d**) Ti-doping; (**e**) Si-doping; (**f**) Al-doping. All libraries were scanned at 532 nm with +0.6 bias in a 0.1-M NaOH solution. Note that the false colors are relative within a scan and not absolute values therefore comparisons of color between scans are not quantitative. Refer to Table 6.3 for more quantitative comparisons

Table 6.2 Density of printed precursor drops used to produce libraries shown in Fig. 6.6d–f

			Relative stoichiometry (FeM_x)			
Row label	1D DS (μm)	Drops per mm^2	Column D	Column C	Column B	Column A
D	40	642	0.0067	0.0017	0.00078	0.00047
C	79	169	0.025	0.0067	0.003	0.0018
B	120	75	0.057	0.015	0.0067	0.004
A	160	45	0.095	0.025	0.011	0.0067

DS drop spacing

optimum α-Fe_2O_3 thickness for the maximum photocurrent signal. It was found that printing this concentration of $FeCl_3$ three times gave the best photocurrent signal (Fig. 6.6c). The top double row (Fe_D) was too thick resulting in darker areas in the center of the squares and bright edges due to the film being slightly thinner near the edges. The short diffusion length of carriers in this material results in decreases in photocurrent when the film is too thick. Note that in all cases most of the light still passes through the very thin film of α-Fe_2O_3 and so incident photon current efficiency (IPCE) values are low.

According to the literature [68, 72, 83], the addition of small amounts of Ti, Si or Al improve the photoelectrochemical performance of α-Fe_2O_3. Therefore, we selected Ti, Si, and Al to investigate the effects of the incorporation of small amounts of these elements and combinations of them on thin film α-Fe_2O_3 photoanodes. Combinatorial printing templates such as those shown in Fig. 6.4 were created where the total amount of precursor deposited into each square is determined from the concentration of precursor salts in the printing solution, the drop volume of a printed drop, the number of layers printed and the number of drops inkjetted per unit area. This amount is determined by the areal density of dots and the number of ink-jetting nozzles actually firing on each printer cartridge. In the pattern, in Fig. 6.6b, four different densities of Fe and other metal (M = Ti, Si, or Al) were printed as double horizontal and double vertical rows, respectively, resulting in groups of four squares of nominally identical composition. The result of the photocurrent screening is shown in Fig. 6.6d–f.

Controlling the density of the printed drops allows the production of a quantifiable broad or narrow compositional range of dopant additions on each substrate. Table 6.2 contains the values of the drop spacings and densities used in Fig. 6.6d–f for the Fe and M (M = Ti, Si, or Al) oxide precursors taken individually. For example, the bottom four squares on the far left, corresponding to Fe_AM_D, were printed with a drop spacing of 160 μm for Fe (three layers) and 40 μm for M (M = Ti, Si, or Al for Fig. 6.6d, e and f, respectively). This set of squares corresponds to a nominal stoichiometry of $Fe_1M_{0.095}$ or 8.7% dopant, whereas the smallest percentage square corresponds to Fe_DM_A with a nominal stoichiometry of $Fe_1M_{0.00047}$ or 0.047% dopant. The corresponding numerical maximum IPCE data are summarized in Table 6.3. These values can be used to evaluate the relative responses when comparing the false color photocurrent scans since the brightest yellow shade was used to represent the maximum photocurrent response in each image. We found that in general only Ti-incorporation enhanced the photoelectrochemical performance

Table 6.3 Maximum IPCE data for Fig. 6.6c–f measured at two different biases

	Undoped Fig. 6.6c	Ti-doped Fig. 6.6d	Si-doped Fig. 6.6e	Al-doped Fig. 6.6f
0 V bias	2.60×10^{-4}	4.06×10^{-4}	2.58×10^{-4}	1.85×10^{-4}
+0.6 V bias	11.47×10^{-4}	15.54×10^{-4}	7.77×10^{-4}	6.02×10^{-4}

in comparison with an undoped iron oxide thin film, while the incorporation of Si or Al shows somewhat detrimental effects although there seems to be an increase in response for the thinner iron oxide films with the smallest amount of Si incorporation (Fe_ASi_A or $Fe_1Si_{0.0067}$ - bottom right corner of Fig. 6.6e) suggesting that perhaps even smaller additions of Si might improve the relative photoresponse. In previous work for films prepared by ultrasonic spray pyrolysis the corresponding nominal stoichiometry showing highly improved photoresponse is $Fe_1Si_{0.01}$ or ~1% dopant [37]. As for the Ti-doped iron oxide thin films, the best photocurrent response was observed at the relative stoichiometry of $Fe_1Ti_{0.025}$ while it ranges from $Fe_1Si_{0.0018}$ (or 0.18% dopant) to $Fe_1Si_{0.025}$ (or 2.4% dopant) for Si-doping, and $Fe_1Al_{0.0067}$ for Al-doping.

There have been numerous reports about the effects on photoelectrochemical response of incorporation of small amounts of the other elements into the binary α-Fe_2O_3 system. However, it is challenging to investigate the effects of adding multiple elements to α-Fe_2O_3 thin films especially if individual films have to be prepared at each stoichiometry. Our high-throughput combinatorial method can be used to quickly measure the effects of additions of multiple elements to find an optimal composition. Therefore we printed films with combinations of Ti, Si, and Al simultaneously added to α-Fe_2O_3. A printing template shown in Fig. 6.7 was created where four different densities of Ti and Si were printed as double horizontal and double vertical rows, respectively, and different amounts of Al were deposited in a repeating pattern within groups of four squares containing identical Fe, Ti, and Si amounts. The template was printed four times with different iron precursor drop spacings. The result is that the total thickness of the α-Fe_2O_3 decreases from Fig. 6.8a–d, whereas the relative concentration of the added dopants increases as we make the α-Fe_2O_3 film thinner. The top row of squares in the template are equal thickness squares of α-Fe_2O_3 where the first three pairs of squares have only one added metal dopant at the A and D levels (Al, Si, Ti left to right) with the far right pair of squares being just α-Fe_2O_3 printed at 40 and 160 μm spacings (Fe_D and Fe_A). Table 6.4 contains the values of drop spacings and drop densities used to print the template in Fig. 6.7 for the Fe–Ti–Si–Al oxide precursors. For illustrative purpose, the bottom square on the far left of the template, labeled $Si_DTi_AAl_B$, was printed with a drop spacing of 40, 160, and 120 μm for Si, Ti, and Al, respectively, while the Fe precursor was printed in three layers with different drop spacings of 40, 80, 120, 160 μm for Fig. 6.8a–d, respectively. So this lower left square in Fig. 6.8a–d will correspond to stoichiometries of $Fe_1Si_{0.0067}$-$Ti_{0.00047}Al_{0.00078}$, $Fe_1Si_{0.025}Ti_{0.0018}Al_{0.003}$, $Fe_1Si_{0.057}Ti_{0.0049}Al_{0.0067}$ and $Fe_1Si_{0.095}Ti_{0.0067}Al_{0.011}$, respectively. This printing pattern results in libraries of decreasing α-Fe_2O_3 thickness but increasing doping levels from Fig. 6.8a–d. The results of the photocurrent screening of these libraries are shown in Fig. 6.8a–d.

Fig. 6.7 Template used for the printed pattern in the Fe-Ti-Si-Al system where known amounts of the precursors are printed into individual 3 × 3 mm^2. The density of precursor drops and resulting stoichiometries for the various produced libraries are listed in Table 6.4

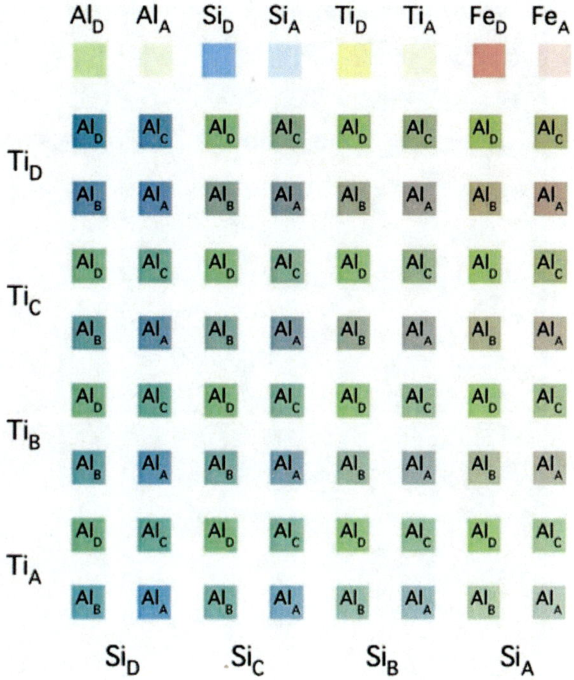

A series of pairs of internal standards consisting of undoped Fe and the Fe–Ti, Fe–Si, and Fe–Al binary systems with different drop densities (thickness) are also printed in the top row of the pattern as seen in Fig. 6.7.

IPCE values from the false color photocurrent data in Fig. 6.8a–d are summarized in Table 6.5. Again these values can be used to evaluate the relative responses when comparing the false color photocurrent scans since the brightest yellow shade was used to represent the maximum photocurrent response in each image. The highest overall IPCE data was obtained when the $FeCl_3$ precursor was printed three times with a drop spacing of 80 μm followed by drop spacings of 120, 40, and 160 μm (Figures 6.8b, c, a, and d, respectively). In Fig. 6.8a ($FeCl_3$ DS = 40 μm), the highest IPCE was observed at the Ti-rich area in this pattern ($Fe_1Ti_{0.0067}Si_xAl_x$ top two rows) but the overall thickness of this library was too large for efficient carrier harvesting revealed by the dark squares in the centers of the squares. Fig. 6.8b (DS = 80 μm) has the brightest spots when the relative stoichiometry is near $Fe_1Ti_{0.0067}Si_{0.0067}Al_x$ that occurs in the center section of the library. In Fig. 6.8c (DS = 120 μm) the relative stoichiometry of $Fe_1Ti_{0.0017}Si_xAl_x$ in the second set of double rows appears to have the highest photocurrent. It is quite interesting to observe there is tendency in Fig. 6.8d for the highest photocurrents appearing in the Al-rich areas with stoichiometry of $Fe_1Ti_xSi_xAl_{0.095}$. Another interesting aspect is that in all of the libraries the pure α-Fe_2O_3 (top right two squares) and the binary internal standards (top row pairs 1–3) almost always have a lower photocurrent response than the ternary additions in their respective libraries.

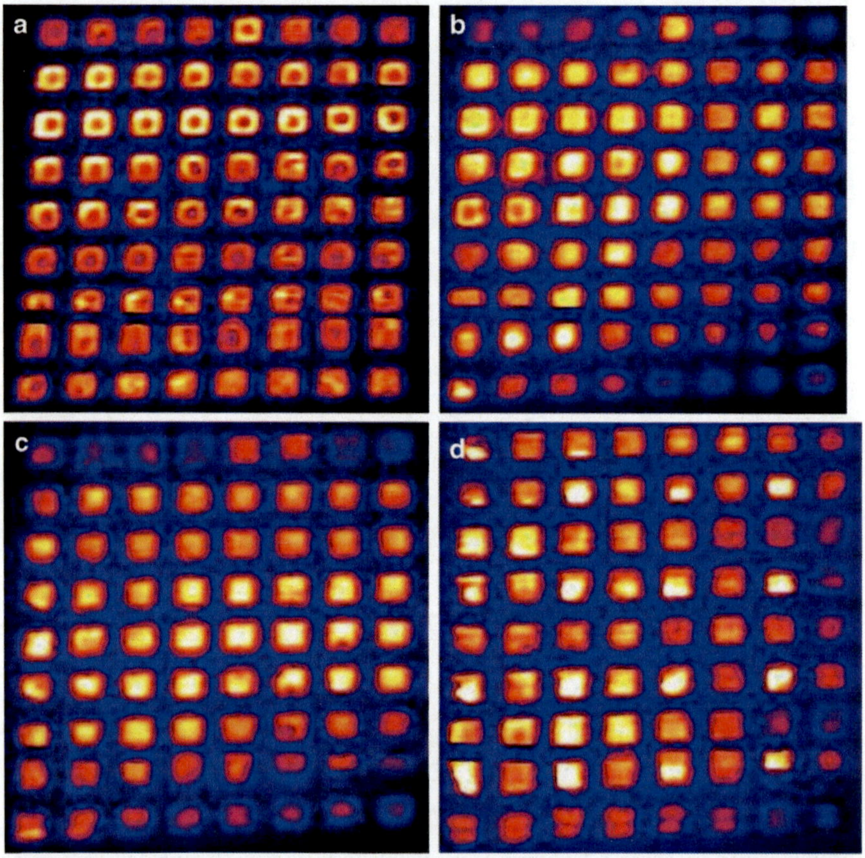

Fig. 6.8 **(a–d)** False color photocurrent maps with 532 nm laser excitation of the Fe–Ti–Si–Al system with no applied bias in a 0.1-M NaOH solution (Top Left $FeCl_3$ precursor drop spacings: $a = 40$ μm, $b = 80$ μm, $c = 120$ μm, $d = 160$ μm). Note that the false colors are relative within a scan and not absolute values therefore comparisons of color between scans are not quantitative. Refer to Table 6.5 for quantitative comparisons and the top row of standard squares with only one dopant or no dopant (see text)

Table 6.4 Density of printed precursor drops used to produce the mixtures shown in Fig. 6.8

Label	1D DS (μm)	Drops per mm^2	Relative stoichiometries (%) $FeCl_3$, 3 × DS = 40 μm	$FeCl_3$, 3 × DS = 79 μm	$FeCl_3$, 3 × DS = 120 μm	$FeCl_3$, 3 × DS = 160 μm
D	40	642	0.67	2.5	5.7	9.5
C	79	169	0.17	0.67	1.5	2.5
B	120	75	0.078	0.3	0.67	1.1
A	160	45	0.047	0.18	0.49	0.67

The iron precursor ($FeCl_3$) was over-printed three times
DS drop spacing

Table 6.5 Summary of maximum IPCE data from Fig. 6.8

	Fig. 6.8a DS = 40	Fig. 6.8b DS = 80	Fig. 6.8c DS = 120	Fig. 6.8d DS = 160
0 V bias	2.36×10^{-4}	4.52×10^{-4}	4.15×10^{-4}	0.98×10^{-4}
+0.6 V bias	10.16×10^{-4}	15.54×10^{-4}	13.32×10^{-4}	6.26×10^{-4}

This indicates that there are synergistic effects of adding multiple dopants to the binary oxide semiconductor α-Fe_2O_3. As stated in the introduction this empirical study does not provide information about which of the many possible mechanisms for photocurrent improvement provided by the impurity is actually occurring in the α-Fe_2O_3 film. However, it does provide improved stoichiometries that should prove fruitful with more detailed studies of the actual electrical doping, morphological changes, carrier diffusion lengths and grain boundary and orientation effects attributable to the various impurities.

Due to the simplicity and low cost of the inkjet printing and screening protocols for discovering metal oxides capable of the photoelectrolysis of water, we embarked on a project to enlist multiple universities and even high schools in the search. We have produced the first beta printing and screening kits based on inexpensive components such as Lego Midstorms® kits, laser pointers and off-the-shelf inkjet printers that were distributed in the summer of 2008 to several universities with undergraduate summer research programs. The Solar Hydrogen Activity research Kit project (SHArK Project - www.thesharkproject.org) has recently expanded to over 20 sites world-wide and will be expanding to perhaps hundreds of sites in the near future. The SHArK Web site acts as a database of material combinations that have been investigated and as a communication tool for the various sites to exchange information. This "distributed research" project, initially funded by the Camille and Henry Dreyfus Foundation and now by the National Science Foundation, will lay the groundwork for a new generation of young scientists dedicated to helping solve the world energy problem.

6.4.3 Electrochemical Synthesis and Screening

To our knowledge McFarland et al. were the first to publish work on the combinatorial production of materials for water splitting. They demonstrated the electrochemical deposition of metal oxide compositions using robotics to electroplate and screen for water photooxidation activity individually created oxide materials using multiplexed photoelectrochemical cells [84]. Their work focused on variations in the electrochemical synthesis conditions, surface redox catalysts, control over structure [85] (by diverse surfactant and templating agents) and/or variable doping [86–88]. As illustrated in Fig. 6.9, the electrochemical deposition system utilizes a three-electrode electrochemical cell with coiled Pt and Ag wires serving as the

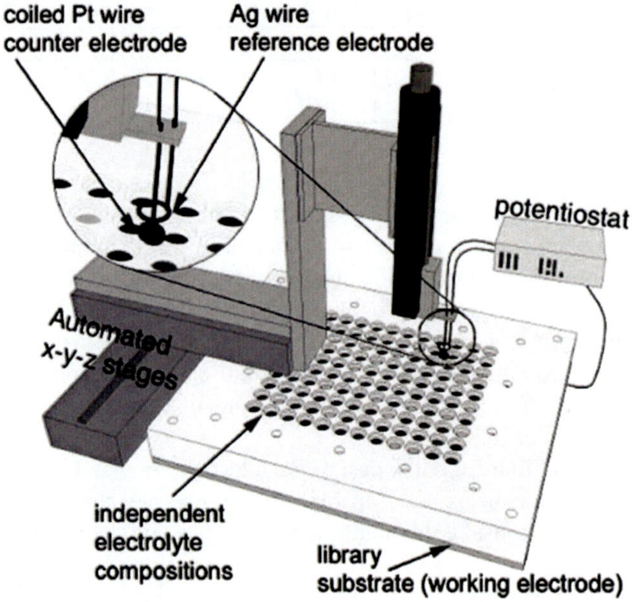

Fig. 6.9 An automated electrochemical deposition system used for the production of metal oxides. The custom-made well plate shown has been designed for a 120-sample array, each sealed by "O-rings" underneath a patterned Teflon block, and each containing a unique electrolyte composition. Reprinted with permission from [86], copyright 2005 American Chemical Society

counter and reference electrodes, respectively, that can be moved with a computer-controlled x–y–z stage. Metal ligand systems (citrates, peroxo complexes, etc.) or electrolyte conditions (pH and relative concentrations) are carefully designed to produce similar deposition potentials for the species in solution as the precursors are electrodeposited onto the conductive substrate (FTO or Ti) surface and then connected as the working electrode for evaluation in the sample well. The desired voltage, with a necessary overpotential to ensure deposition of both species, is applied at each well until a specified amount of charge has passed before moving the scanning electrodes to the next well.

In this high-throughput configuration, each cell is manually filled using a robotic pipetter before electrodeposition. Subsequent calcination produces mixtures of known stoichiometries that were verified by XPS measurements coupled to Ar^{+} ion etching to examine the bulk and surface composition. Structural information was obtained with XRD. The screening system utilized for the electrochemically deposited samples incorporates measurements of flat-band potential, bias dependent photocurrent, and action spectra with a scanning photoelectrochemical probe [86]. Briefly, the probe steps from sample-to-sample across the library and an automated pump fills the probe with solution from a reservoir of stock electrolyte. An illuminated optical fiber targets each individual sample well and a three-electrode configuration is used to obtain photoelectrochemical data. A monochromator is placed between a chopped Xe lamp and the optical waveguide for the IPCE measurements. Using the multiplexed photoelectrochemical screening system, the measured photoactivity of large bandgap materials for oxygen production, including ZnO [86] and WO_3 [84], and hydrogen production, including Al_2O_3 [72], has shown improvement after doping with certain elements around particular stoichiometries and complete photoelectrochemical information is obtained for each sample.

6.4.4 Sol Gel Routes

The recent work of Maier et al. [88, 89] has also added to the arsenal of combinatorial production and screening strategies to the search for novel hydrogen-producing photocatalysts, particularly for producing doped mixtures of compounds with sol gel precursors such as titanium. The reported technique incorporates computer-controlled pipetting robots to quantitatively dispense stock solutions of precursors into an array of small vials. When the pipetting process is finished, the samples are placed in an orbital shaker for mixing and a consistent aging, drying, and calcination protocol is followed for each sample to produce catalyst powders. The powders are then manually ground and suspended in an aqueous methanol solution in small gas chromatography (GC) vials. After flushing the headspace within the vials with an inert carrier gas, the GC sample holder is illuminated with a visible light source and the direct production of hydrogen gas can be monitored.

Using this method, after several generations of optimization, a newly discovered material, $Al_{40}Bi_{40}Pb_{20}O_x$, was identified for future studies. There are several problems associated with this technique, including the use of powders and sacrificial agents such as methanol, both of which were previously discussed. Nevertheless, it would be interesting to deploy the materials discovered by this technique as thin film photoelectrodes to test their photoelectrolysis activity.

6.4.5 Other Thin Film Deposition Techniques

Although they have not yet been applied to the production and screening of photoelectrolysis materials, we will briefly mention some other approaches to produce metal oxide libraries using well-established thin film deposition techniques (meant here to include physical vapor deposition, sputtering, pulsed laser deposition, and molecular beam epitaxy). These techniques have been used for the production of combinatorial metal oxide libraries in the search for more effective luminescent materials [90, 91], transparent conducting oxides [92, 93], and dielectrics [94, 95]. It would presumably be straightforward to apply the same techniques to the production of material libraries to be screened for photoelectrolysis activity.

6.5 From Materials Search to Bulk Photoelectrodes

Identification of stable semiconducting oxide materials with appropriate band gaps and band positions is only the first step toward a viable photoelectrolysis system. In general, two factors conspire to limit the photoconversion efficiency of oxide semiconductors. First, the optical transitions in these materials are usually forbidden d–d transitions and so they have much lower absorption coefficients when compared to direct bandgap semiconductors. The result is that light penetrates deeply into the material before being absorbed to create an electron hole pair. Other indirect photovoltaic materials such as Si have a similar absorption problem but it is overcome by having very long carrier lifetimes, resulting in long carrier diffusion lengths, so that photogenerated electrons and holes are eventually separated. Carriers in metal oxide semiconductor d-bands usually have very low mobilities due to the limited overlap between the metal d orbitals, resulting in shorter charge carrier diffusion lengths and increased bulk recombination of photogenerated carriers. Therefore, even if we could use costly single crystals of a metal oxide semiconductor, the carrier collection would be restricted to the narrow space charge layer where only a small fraction of the solar photons will be absorbed. Luckily, these same two problems have been solved by the architecture of the nanocrystalline dye-sensitized solar cell, that has reached laboratory efficiencies of over 10% compared to sensitized single crystals of the same oxide semiconductor that show a quantum yield of at most 1–2% due to the low light absorption of the monolayer of dye. Simply making the dye layer thicker does not solve the problem as excitons produced away from the dye/semiconductor interface recombine before they are separated due to the short exciton diffusion lengths in the organic dye layer. The high surface area morphology works to increase the effective optical absorption for these systems as the multitude of dye monolayers absorb all the light and inject their electrons into the TiO_2 conduction band and the holes (or oxidized dye molecules) are quickly scavenged by a regenerator (reducing agent) in solution. A similar nanostructured morphology is probably necessary for photoelectrolysis with metal oxides in order to compensate for the low absorption and carrier mobilities characteristic of metal oxide semiconductors with forbidden d–d optical transitions.

Another illustration of the importance of morphology in a metal oxide semiconductor uses α-Fe_2O_3 (hematite) as a photoelectrolysis material. Although iron (III) oxide is primarily used as a photoanode, it has been extensively studied to optimize its performance because it is inexpensive, easy to fabricate, has a bandgap (2.2 eV) that can absorb a good portion of the solar spectrum, and it is stable in aqueous solutions under illumination. High surface area nanostructured morphologies [96–98] of this material have been shown to more effectively harvest visible light relative to smooth films. For water cleavage, the best reported efficiencies (a solar-to-chemical conversion efficiency of 2.1%, based on the heat of hydrogen combustion), are achieved when used in a tandem configuration with a dye-sensitized solar cell.

The thickness of perpendicularly aligned α-Fe_2O_3 sheets is between 15 and 25 nm allowing for carrier diffusion to the surface. This morphology appears to be a critical parameter, since samples produced by different methods with similar thicknesses, crystal structure, and identical optical characteristics do not perform nearly as well. The authors argue that this particular grain size, orientation and morphology (leaf-shaped sheets) are required due to the short hole diffusion length of only a few nanometers for α-Fe_2O_3. This particular morphology affords facile transfer of holes to the electrolyte, before recombination occurs, while still shuttling electrons to the back contact. This morphology was produced using ultrasonic spray pyrolysis of their iron precursor solutions [86] to prepare their thin films. Addition of silicon produced further improvements in photoresponse again attributed to morphology changes rather than to electronic effects [84].

A nanostructured morphology also reduces the microscopic current density for water oxidation and reduction reactions resulting in lower overpotential losses than at a planar electrode. This will lower the electrocatalytic activity requirements for the semiconductor. In addition, many of the cell designs discussed herein or envisioned allow for thin nanostructured films deposited on a glass substrate with a transparent conducting layer. Such a configuration allows for back illumination, much like in the nanocrystalline solar cell. Back illumination of a photoelectrolysis electrode will reduce or eliminate the efficiency losses due to light scattering by the gas bubbles being evolved from the electrode surface.

6.6 Conclusions

We have surveyed the state of water photoelectrolysis using semiconductor electrodes and can conclude that there are currently no known suitable materials to efficiently and economically carry out this process. Due to the stability requirements we surmise that semiconducting oxide photoelectrodes are required. There are many potential oxide semiconductors that need to be prepared and tested for photoelectrolysis activity and therefore high-throughput combinatorial techniques are needed to discover promising materials. The first steps have been taken in the development of the combinatorial methods to solve this problem and these methods have been briefly reviewed. Optimization of the morphology of the semiconducting oxide thin film, probably with some kind of nanostructure, is required to overcome the limitations of light absorption and carrier transport in most oxide semiconductors. Mobilization of both experts and nonexpert researchers to discover and optimize new semiconductor materials is needed to produce the hydrogen needed to help move toward a clean energy economy based on renewable resources.

References

1. Ogden, J.M., Williams, R.H.: Electrolytic hydrogen from thin film solar cells. Int. J. Hydrogen Energy **15**, 155 (1990)
2. Boddy, P.J.: Oxygen evolution on potassium tantalate anodes. J. Electrochem. Soc. **115**, 199 (1968)
3. Fujishima, A., Honda, K.: Electrochemical evidence for the mechanism of the primary stage of photosynthesis. Bull. Chem. Soc. Jpn. **44**, 1148 (1971)
4. Fujishima, A., Honda, K.: Electrochemical photolysis of water at a semiconductor electrode. Nature **238**, 37 (1972)
5. Khaselev, O., Turner, J.A.: A monolithic photovoltaic-photoelectrochemical device for hydrogen production via water splitting. Science **280**, 425 (1998)
6. Rajeshwar, K.: Hydrogen generation at irradiated oxide semiconductor–solution interfaces. J. Appl. Electrochem. **37**, 765 (2007)
7. Bak, T., Nowotny, J., Rekas, M., Sorrell, C.C.: Photo-electrochemical hydrogen generation from water using solar energy. Materials-related aspects. Int. J. Hydrogen Energy **27**, 991 (2002)
8. Grimes, C.A., Varghese, O.K., Ranjan, S.: Light, water hydrogen. Springer, New York (2008)
9. Kung, H., Jarrett, H., Sleight, A., Ferretti, A.: Semiconducting oxide anodes in photoassisted electrolysis of water. J. Appl. Phys. **48**, 2463 (1977)
10. Wrighton, M.S., Ellis, A.B., Wolczanski, P.T., Morse, D.L., Abrahamson, H.B., Ginley, D.S.: Strontium titantate photoelectrodes: efficient photoassisted electrolysis of water at zero applied potential. J. Am. Chem. Soc. **98**, 2774 (1976)
11. Mavroides, J.G.: In: Heller, A. (ed.) Semiconductor liquid-junction solar cells. The Electrochemical Society, Princeton, NJ (1977)
12. Watanabe, I., Matsumoto, Y., Sato, E.: Photoelectrochemical properties of the single-crystal $SrTiO_3$ doped in the surface region. J. Electroanal. Chem. **133**, 359 (1982)
13. Mavroides, J.G., Kafalas, J.A., Kolesar, D.F.: Photoelectrolysis of water in cells with $SrTiO_3$ anodes. Appl. Phys. Lett. **28**, 241 (1978)
14. Nasby, R.D., Quinn, R.K.: Photoassisted electrolysis of water using a $BaTiO_3$ electrode. Mater. Res. Bull. **11**, 985 (1976)
15. Kennedy, J.H., Karl, J., Freese, W.: Photooxidation of water at barium titanate electrodes. J. Electrochem. Soc. **123**, 1683 (1976)
16. Harris, L.A., Wilson, R.H.: Semiconductors for photoelectrolysis. Annu. Rev. Mater. Sci. **8**, 99 (1978)
17. Hodes, G., Fonash, S.T., Heller, A., Miller, B.: In: Gerischer, H. (ed.) Advances in electrochemistry and electrochemical engineering. Wiley, New York (1985)
18. Rajeshwar, K., Singh, P., Dubow, J.: Energy conversion in photoelectrochemical systems – a review. Electrochim. Acta **23**, 1117 (1978)
19. Fan, F.R.F., White, H.S., Wheeler, B.L., Bard, A.J.: Semiconductor electrodes XXIX. High efficiency photoelectrochemical solar cells with n-WSe_2 electrodes in an aqueous iodide medium. J. Electrochem. Soc. **127**, 518 (1980)
20. Heller, A.: Conversion of sunlight into electrical power and photoassisted electrolysis of water in photoelectrochemical cells. Acc. Chem. Res. **14**, 154 (1981)
21. Strehlow, W.H., Cook, E.L.: J. Phys. Chem. Ref. Data **2**, 163 (1973)
22. Levy-Clement, C., Heller, A., Bonner, W.A., Parkinson, B.A.: Spontaneous photoelectrolysis of HBr and HI in two photoelectrode semiconductor liquid junction cells. J. Electrochem. Soc. **129**, 1701 (1982)
23. De, G.C., Roy, A.M., Bhattacharya, S.S.: Effect of n-Si on the photocatalytic production of hydrgen by Pt-loaded CdS and CdS/ZnS catalyst. Int. J. Hydrogen Energy **21**, 19 (1996)
24. Rauh, R.D., Buzby, J.M., Reise, T.F., Alkaitis, S.A.: Design and evolution of new oxide photoanodes for the photoelectroysis of water with solar energy. J. Phys. Chem. **83**, 2221 (1979)

25. Gerischer, H.: Photodecompostion of semiconductors, thermodynamics, kinetics, and applications to solar cells. Faraday Disc. **70**, 1A (1980)
26. Butler, M., Nasby, R., Quinn, R.: Tungsten trioxide as an electrode for photoelectrolysis of water. Solid State Commun. **19**, 1011 (1976)
27. Butler, M.A.: Photoelectrolysis and physical properties of the semiconducting electrode WO_3. J. Appl. Phys. **48**, 1914 (1977)
28. Berak, J.M., Sienko, M.J.: Effect of oxygen-deficiency on electrical transport properties of tungsten trioxide crystals. J. Solid State Chem. **2**, 109 (1970)
29. Wang, H., Lindgren, T., He, J., Hagfeldt, A., Lindquist, S.-E.: Photolelectrochemistry of nanostructured WO_3 thin film electrodes for water oxidation: mechanism of electron transport. J. Phys. Chem. B **104**, 5686 (2000)
30. Hardee, K., Bard, A.: Photoelectrochemical behaviour of several polycrystalline metal oxide electrodes in aqueous solutions. J. Electrochem. Soc. **123**, 1024 (1976)
31. Nasby, R., Quinn, R.: Photoassisted electrolysis of water using a $BaTiO_3$ electrode. Mater. Res. Bull. **11**, 985 (1976)
32. Kennedy, J.H., Frese, J.K.W.: Photooxidation of water at α-Fe_2O_3 electrodes. J. Electrochem. Soc. **125**, 709 (1978)
33. McGregor, K.G., Calvin, M., Otvos, J.W.: Photoeffects in Fe_2O_3 sintered semiconductors. J. Appl. Phys. **50**, 369 (1979)
34. Frelein, R.A., Bard, A.J.: Semiconductor electrodes XXI. The characterization and behavior of n-type Fe_2O_3 electrodes in acetonitrile solutions. J. Electrochem. Soc. **126**, 1892 (1979)
35. Wilhelm, S.M., Yun, K.S., Ballenger, L.W., Hackerman, N.: Semiconductor properties of iron oxide electrodes. J. Electrochem. Soc. **126**, 419 (1979)
36. Dare-Edwards, M.P., Goodenough, J.B., Hammett, A., Nicholson, N.D.: Photoelectrochemistry of NiO. J. Chem. Soc. Faraday Trans. **77**, 643 (1981)
37. Koffyberg, F.P., Benko, F.A.: A photoelectrochemical determination of the position of the conduction and valence band edges of p-type CuO. J. Electrochem. Soc. **128**, 2476 (1981)
38. Nikitine, S., Zielinger, J.P., Coret, A., Zouaghi, M.: Phys. Lett. **18**, 105 (1965)
39. Goodenough, J.B., Hamnett, A., Dare-Edwards, M.P., Campet, G., Wright, R.D.: Inorganic materials for photoelectrolysis. Surface Science **101**, 531 (1980)
40. Ishikawa, A., Takata, T., Kondo, J., Hara, M., Kobayashi, H., Domen, K.: Oxysulfide $Sm_2Ti_2S_2O_5$ as a stable photocatalyst for water oxidation and reduction under visible light irradiation. J. Am. Chem. Soc. **124**, 13547 (2002)
41. Hitoki, G., Takata, T., Kondo, J.N., Hara, M., Kobayashi, H., Domen, K.: Ta_3N_5 as a novel visible light-driven photocatalyst. Chem. Commun. 1698 (2002).
42. Liu, M., You, W. Lei, Z., Zhou, G., Yang, J., Wu, G., Ma, G., Luan, G., Takata, T., Hara, M., Domen, K., Li, C.: Water reduction and oxidation on Pt-Ru/$Y_2Ta_2O_5N_2$ catalyst under visible light irradiation. Chem. Commun. 2192 (2004).
43. Hitoki, G., Ishikawa, A., Takata, T., Kondo, J., Hara, M., Domen, K.: Ta_3N_5 as a novel visible light-driven photocatalyst. Chem. Lett. **1**, 736 (2002)
44. Kasahara, A., Nukumizu, K., Hitoki, G., Takata, T., Kondo, J., Hara, M., Kobayashi, H., Domen, K.: Photoreactions on $LaTiO_2N$ under visible light irradiation. J. Phys. Chem. A **106**, 6750 (2002)
45. Kasahara, A., Nukumizu, K., Takata, T., Kondo, J., Hara, M., Kobayashi, H., Domen, K.: $LaTiO_2N$ as a visible light driven photocatlayst. J. Phys. Chem. B **107**, 791 (2003)
46. Yu, J.C., Zhang, L., Zheng, Z., Zhao, J.: Synthesis and characterization of phosphated mesoporous titanium dioxide with high photocatalytic activity. Chem. Mater. **15**, 2280 (2003)
47. Asahi, R., Morikawa, T., Ohwaki, T., Aoki, K., Taga, Y.: Visible light photocatalysts in nitrogen-doped titanium oxides. Science **293**, 269 (2001)
48. Serpone, N.: Is the band Gap of pristine TiO_2 narrowed by anion and cation doping of titanium dioxide in second generation photocatalysts. J. Phys. Chem. B **110**, 24287 (2006)
49. Livraghi, S., Paganini, M.C., Giamello, E., Selloni, A., Valentin, C.D., Pacchioni, G.: Origin of photoactivity of nitrogen doped titanium dioxide under visible light. J. Am. Chem. Soc. **128**, 15666 (2006)

50. Bin-Daar, G., Dare-Edwards, M.P., Goodenough, J.B., Hamnett, A.: New anode materials for photolectrolysis. J. Chem. Soc. Faraday Trans. **79**, 1199 (1983)
51. Jarrett, H.S., Sleight, A.W., Kung, H.H., Gillson, J.L.: Photoelectrochemical and solid-state properties of $LuRhO_3$. J. Appl. Phys. **51**, 3916–3925 (1980)
52. Guruswamy, V., Keillor, P., Campbell, G.L., Bockris, J.O.M.: The photoelectrochemical response of the lanthanides of chromium, rhodium, vandium, and gold on a titanium base. Sol. Energ. Mater. Sol. Cell. **4**, 11 (1980)
53. Takata, T., Shinohara, K., Tanaka, A., Hara, M., Kondo, J., Domen, K.: Photocatalytic decomposition of water on spontaneously hydrated layered perovskites. J. Photochem. Photobiol. **106**, 45 (1997)
54. Yoshimura, J., Ebina, Y., Kondo, J., Domen, K.: Visible light induced photocatalytic behavior of a layered perovskite type niobate, $RbPb_2Nb_3O_{10}$. J. Phys. Chem. **97**, 1970 (1993)
55. Ghosh, A.K., Maruska, P.: Photoelectrolysis of water in sunlight with sensitized semi-conductor electrodes. J. Electrochem. Soc. **124**, 1516 (1977)
56. Kato, H., Kudo, A.: Visible light response and photocatalyltic activities of TiO2 and $SrTiO_3$ photocatalyts Co doped with antimony and chromium. J. Phys. Chem. B **106**, 5029 (2002)
57. Ishii, T., Kato, H., Kudo, A.: H_2 evolution from an aqueous methanol solution on $SrTiO_3$ photocatalyts codoped with chromium and tantalum ions under visible light irradiation. J. Photochem. Photobiol. A **163**, 181 (2004)
58. Woodhouse, M., Parkinson, B.A.: Combinatorial discovery and optimization of a complex oxide with water photoelectrolysis activity. Chem. Mater. **20**, 2495 (2008)
59. Tsuji, I., Kato, H., Kobayashi, H., Kudo, A.: Photocatalytic H_2 evolution reaction from aqueous solutions over band controlled $AgIn_xZn_{2-x}S_2$ solid solution photocatalyts with visible light response and their surface nanostructures *J*. Am. Chem. Soc. **126**, 13406 (2004)
60. Zou, Z., Ye, J., Sayama, K., Arakawa, H.: Direct splitting of water under visible light irradiation with an oxide semiconductor photocatalyst. Nature **414**, 625 (2001)
61. Kudo, A., Kato, H.: Effect of lanthanide doping into $NaTaO_3$ photocatalyts for efficient water splitting. Chem. Phys. Lett. **331**, 373 (2000)
62. Reddington, E., Sapienza, A., Gurau, B., Viswanathan, R., Sarangapeni, S., Smotkin, E.S., Mallouk, T.E.: Combinatorial electrochemistry: a highly parallel, optical screening method for discovery of better electrocatalysts. Science **280**, 1735 (1998)
63. Morris, N.D., Mallouk, T.E.: A high-throughput optical screening method for the optimization of colloidal water oxidation catalysts. J. Am. Chem. Soc. **124**, 11114 (2002)
64. Chen, L., Bao, J., Gao, C.: Combinatorial synthesis of insoluble oxide library from ultrafine/ nanoparticle suspension using a drop on demand inkjet delivery system. J. Comb. Chem. **6**, 699 (2004)
65. Woodhouse, M., Herman, G., Parkinson, B.A.: Combinatorial approach to identification of catalysts for the photoelectrolysis of water. Chem. Mater. **17**, 4318 (2005)
66. Katz, J.: Metal oxide based photoelectrochemical cells for solar energy conversion. PhD Thesis, California Institute of Technology, Pasadena, CA (2007)
67. Arai, T., Konishi, Y., Iwasaki, Y., Sugihara, H., Sayama, K.: High throughput screening using porous photoelectrodes for the development of visible-light responsive semiconductors. J. Comb. Chem. **9**, 574 (2007)
68. Sartoretti, C.J., Alexander, B.D., Solarska, R., Rutkowska, W.A., Augustynski, J., Cerny, R.: Photoelectrochemical oxidation of water at transparent ferric oxide film electrodes. J. Phys. Chem. B **109**, 13685 (2005)
69. Kay, A., Cesar, I., Gratzel, M.: A new benchmark for water photooxidation by nanostructured Fe_2O_3 films. J. Am. Chem. Soc. **128**, 15714 (2006)
70. Murphy, A.B., Barnes, P.R.F., Randeniya, L.K., Plumb, I.C., Grey, I.E., Horne, M.D., Glasscock, J.A.: Efficiency of solar water splitting using semiconductor electrodes. Int. J. Hydrogen Energy **2006**, 31 (1999)
71. Mohanty, S., Ghose, J.: Studies on some α-Fe_2O_3 photoelectrodes. J. Phys. Chem. Solids **53**, 81 (1992)

72. Glasscock, J.A., Barnes, P.R.F., Plumb, I.C., Savvides, N.: Enhancement of photoelectrochemical hydrogen production from hematite thin films by the introduction of Ti and Si. J. Phys. Chem. C **111**, 16477 (2007)
73. Arai, T., Konishi, Y., Iwasaki, Y., Sugihara, H., Sayama, K.: High throughput screening using porous photoelectrodes for the development of visible-light responsive semiconductors. J. Comb. Chem. **9**, 574 (2007)
74. Shinar, R., Kennedy, J.H.: Iron oxide photoanaodes. Sol. Energy Mater. **6**, 323 (1982)
75. Leygraf, C., Hendewerk, M., Somorjai, G.A.: Photodissociation of water by p- and n-type polycrystalline iron oxides by using visible light ($\leq$2.7 eV) in the absence of external potential. Proc. Natl. Acad. Sci. U.S.A. **79**, 5739 (1982)
76. Turner, J.E., Hendewerk, M., Parmeter, J., Neiman, D., Somorjai, G.A.: The characterization of doped iron oxide electrodes for the photodissociation of water. J. Electrochem. Soc. **131**, 1777 (1984)
77. Khader, M.M., Vurens, G.H., Kim, I.K., Salmeron, M., Somorjai, G.A.: Photoassisted catalytic dissociation of water to produce hydrogen on partially reduced alpha.-iron(III) oxide. J. Am. Chem. Soc. **109**, 3581 (1987)
78. Bjorksten, U., Moser, J., Gratzel, M.: Photoelectrochemical studies on nanocrystalline hematite films. Chem. Mater. **6**, 858 (1994)
79. Duret, A., Gratzel, M.: Visible light-induced water oxidation on mesoscopic α-Fe_2O_3 films made by ultrasonic spray pyrolysis. J. Phys. Chem. B **109**, 17184 (2005)
80. Aroutiounian, V.M., Arakelyan, V.M., Shahnazaryan, G.E., Stepanyan, G.M., Khachaturyan, E.A., Wang, H., Turner, J.A.: Investigations of the metal-oxide semiconductors promising for photoelectrochemical conversion of solar energy. Sol. Energy **80**, 1098 (2006)
81. Hu, Y.S., Kleiman-Shwarcstein, A., Forman, A.J., Hazen, D., Park, J.N., McFarland, E.W.: Pt-doped α-Fe_2O_3 thin films active for photoelectrochemical water splitting. Chem. Mater. **20**, 3803 (2008)
82. Kleiman-Shwarsctein, A., Hu, Y.S., Forman, A.J., Stucky, G.D., McFarland, E.W.: Electrodeposition of α-Fe_2O_3 doped with Mo or Cr as photoanodes for photocatalytic water splitting. J. Phys. Chem. C **112**, 15900 (2008)
83. Cesar, I., Kay, A., Cesar, I., Martinez, J.A.G., Grätzel, M.: Translucent thin film Fe_2O_3 photoanodes for efficient water splitting by sunlight: nanostructure-directing effect of Si doping. J. Am. Chem. Soc. **128**, 4582 (2006)
84. Baeck, S.H., Jaramillo, T.F., Brandli, C., McFarland, E.W.: Combinatorial electrochemical synthesis and characterization of tungsten-based mixed metal oxides. J. Comb. Chem. **4**, 563 (2002)
85. Jaramillo, T.F., Baeck, S.-H., Kleiman-Shwarsctein, A., McFarland, E.W.: Combinatorial electrochemical synthesis and screening of mesoporous ZnO for photocatalysis. Macromol. Rapid Commun. **25**, 297 (2004)
86. Jaramillo, T.F., Baeck, S.-H., Kleiman-Shwarsctein, A., Choi, K.-S., Stucky, G.D., McFarland, E.W.: Automated electrochemical synthesis and photoelectrochemical characterization of Zn-Co-oxide thin films for solar hydrogen production. J. Comb. Chem. **7**, 264 (2005)
87. Jaramillo, T.J., Ivanovskaya, A., McFarland, E.W.: High throughput screening system for catalytic hydrogen producing materials. J. Comb. Chem. **4**, 17 (2002)
88. Seyler, M., Stoewe, K., Maier, W.F.: New hydrogen producing photocatalysts – a combinatorial search. Appl Cat. B **76**, 146 (2007)
89. Lettmann, C., Hinrichs, H., Maier, W.F.: Combinatorial discovery of new photocatalysts for water purification with visible light. Angew. Chem. Int. Ed. **40**, 3160 (2001)
90. Takeuchi, I., Lauterbach, J., Fasolka, M.J.: Combinatorial synthesis and evaluation of functional inorganic materials using thin-film techniques. Mater. Today **8**, 18 (2005)
91. Wang, J., Yoo, Y., Gao, C., Takeuchi, I., Sun, X., Chang, H., Xiang, X.-D., Schultz, P.G.: Identification of a blue photoluminiscent composite material from a combinatorial library. Science **279**, 1712 (1998)

92. Hest, M.F.A.M.v., Dabney, M.S., Perkins, J.D., Ginley, D.S., Taylor, M.P.: Titanium-doped indium oxide: a high-mobility transparent conductor. Appl. Phys. Lett. **87**, 032111 (2005).
93. Hest, M.F.A.M.v., Dabney, M.S., Perkins, J.D., Ginley, D.S.: High-mobility molybdenum doped indium oxide. Thin Solid Films **496**, 70 (2006).
94. Dover, R.B.v., Schneemeyer, L.F., Fleming, R.M.: Discovery of a useful thin-film dielectric using a composition spread approach. Nature **392**, 162 (1998).
95. Dover, R.B.v., Schneemeyer, L.F.: The codeposited composition spread approach to high-throughput discovery/exploration of inorganic materials. Macromol. Rapid Commun. **25**, 150 (2004).
96. Cesar, I., Kay, A., Martinez, J.A.G., Gratzel, M.: Translucent thin film Fe_2O_3 photoanodes for efficient water splitting by sunlight: nanostructure-directing effect of Si doping. J. Am. Chem. Soc. **128**, 4582 (2006)
97. Sartoretti, C.J., Alexander, B.D., Solarska, R., Rutkowska, I.A., Augustynski, J., Cerny, R.: Photoelectrochemical oxidation of water at transparent ferric oxide film electrodes. J. Phys. Chem. B **109**, 13685 (2005)
98. Duret, A., Gratzel, M.: Visible light-induced water oxidation on mesoscopic α-Fe_2O_3 films made by ultrasonic spray pyrolysis. J. Phys. Chem. B **109**, 17184 (2005)

Part II
Materials Properties and Synthesis

Chapter 7
Multijunction Approaches to Photoelectrochemical Water Splitting

Eric L. Miller, Alex DeAngelis, and Stewart Mallory

7.1 Introduction

Hydrogen is an extremely valuable chemical commodity, not only in today's industrial marketplace, but even more so in the emerging *Green Economies* so vital to the future of our planet and its people. *Green* futures will need to rely less-and-less on fossil fuels, and more-and-more on solar, wind, biomass, geothermal, and other renewable energy resources. Hydrogen is envisioned as one of the primary media for the storage and distribution of energy derived from this renewable portfolio. To achieve the vision, much work is needed in the development of new more practical technologies and infrastructures for hydrogen production, storage, delivery, and utilization. Nonpolluting technologies for large-scale hydrogen-production utilizing renewable energy are of particular importance.

Among the viable renewable hydrogen-production approaches, solar options, particularly solar-powered water-splitting, offer the best hope for large-scale renewable production with low carbon emissions. Several solar-to-hydrogen (STH) pathways are possible, including solar electrolysis, where electricity generated by photovoltaic (PV) or concentrated solar thermal (CST) power plants is used to drive electrolyzer systems, for example, commercial "alkaline" or "PEM" electrolyzers. Among a number of alternative, more direct conversion pathways, photoelectrochemical (PEC) water-splitting using semiconductor photoelectrodes or photocatalysts remains one of the most promising, though still elusive based on challenges in developing stable and efficient PEC materials, devices, and systems. Broadly speaking, a PEC system combines the harnessing of solar energy and the electrolysis of water into a single conversion system. As illustrated in Fig. 7.1,

E.L. Miller (✉)
U.S. Department of Energy, Washington, DC, USA
e-mail: eric.miller@hq.doe.gov

A. DeAngelis • S. Mallory
University of Hawaii at Manoa, Honoulu, HI, USA

R. van de Krol and M. Grätzel (eds.), *Photoelectrochemical Hydrogen Production*, Electronic Materials: Science & Technology 102, DOI 10.1007/978-1-4614-1380-6_7,

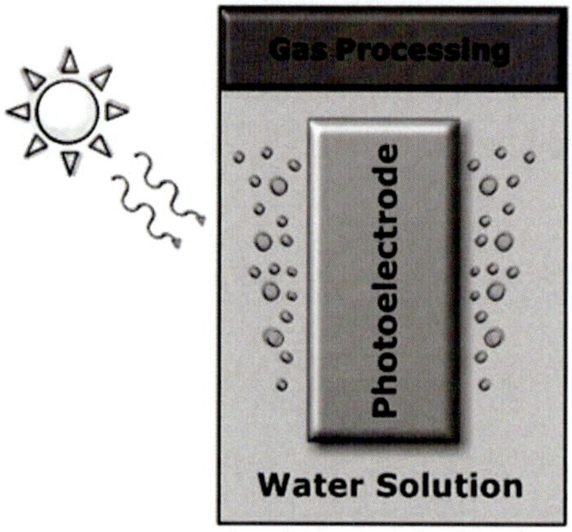

Fig. 7.1 A PEC photosensitive electrode uses sunlight to split water, evolving hydrogen and oxygen gases from solution

sunlight shining on a suitable PEC semiconductor device immersed in a water-based solution can simultaneously drive water oxidation and hydrogen reduction electrochemical reactions to split water. This process evolves hydrogen and oxygen gases, effectively converting the solar energy in to stored chemical energy.

The PEC approach is simple and elegant in concept, but is challenging to implement in practice since it relies on complex interactions involving sunlight, semiconductors, and liquid solutions. In a practical system, the semiconductor material must efficiently absorb sunlight and generate sufficient photovoltage to split water, while the semiconductor interface must be favorable to sustaining the hydrogen- and oxygen-gas evolution reactions. In addition, the PEC system needs to remain stable in solution, and must be inexpensive for compatibility with large-scale deployment. To date, no known semiconductor system achieves *all of the above criteria* – though some have come close. PEC devices based on multijunction III–V semiconductor technology have been demonstrated at the National Renewable Energy Laboratory (NREL) with impressive STH conversion efficiencies (i.e., the ratio of the chemical energy available in the generated hydrogen to the incident solar energy) exceeding 12% [1–4]. Unfortunately, these III–V devices have lacked long-term stability, and moreover, they utilize a number of prohibitively expensive materials as seen in the multilayered device structure illustrated in Fig. 7.2. Lower cost PEC devices based on thin-film semiconductors, also in multijunction configurations, have demonstrated stable hydrogen conversion efficiencies ranging from 3 to 8% [5–8], but progress toward higher performance in such devices has been slow. Before practical PEC hydrogen production can become a reality, key scientific advancements in the development of new PEC materials and devices are still needed.

Toward this end, the most advanced scientific techniques in materials theory, synthesis, and characterization are being brought to the table on an international scale, and powerful synergies among researchers in the PEC, photovoltaics, and nanotechnology fields are emerging in collaborative pursuit of the PEC *quest* [9].

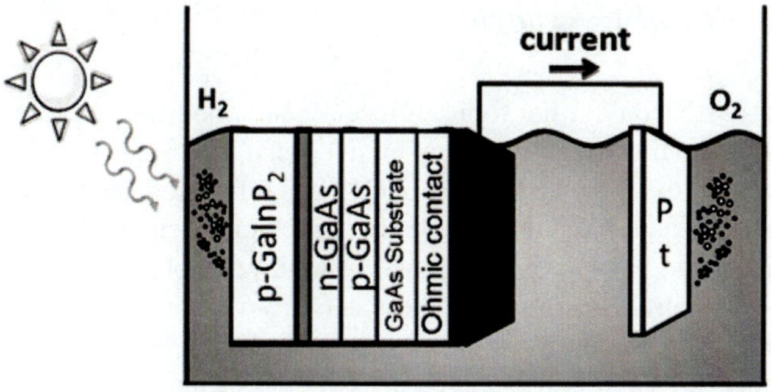

Fig. 7.2 Illustration of multilayered III–V semiconductor device structure that has demonstrated high solar-to-hydrogen conversion efficiencies for limited durations

In fact, PEC researchers often stake claim to PEC as the *Holy Grail of hydrogen production* [10, 11]. This may seem somewhat overly dramatic, though PEC is unique in offering an efficient low-cost approach for harnessing solar energy to produce high-purity hydrogen from water, at low operating temperatures, with no carbon emissions. The key to successful deployment will be in the identification and development of innovative materials systems, likely involving multijunction semiconductor configurations similar to that seen in Fig. 7.2.

Multijunction approaches offer some of the best hope for achieving practical PEC hydrogen production in the near term, but complicated materials and interface issues need to be addressed by the scientific community. This chapter explores the challenges and benefits of large-scale solar water splitting for renewable hydrogen production, with specific focus on the multijunction PEC production pathways. The technical motivation and approach in the R&D of multijunction PEC devices are considered, and examples of progress in laboratory scale prototypes are presented.

7.2 Solar-to-Hydrogen Conversion

Hydrogen is a valuable chemical commodity. Today, it is a vital component in the industrial processing of important chemical commodities, such as fuels and fertilizers. In the near future, it could take on new significance as a viable alternative to gasoline or electricity as an energy carrier for stationary power and transportation applications. On a commercial scale, STH conversion, specifically in the form of solar water-splitting, offers the best hope for *renewable* production with low carbon emissions. This approach for large-scale industrial deployment is truly *renewable* and *sustainable* considering the enormity of the solar resource and the abundance of water as a low-cost feedstock for hydrogen production. The "re-cycling" of water as a by-product of hydrogen fuel cell and combustion utilization schemes also factors into the *sustainability* of solar water splitting.

7.2.1 Value of Hydrogen

Hydrogen has become a hot topic in recent years, both in political and scientific circles. A national spotlight was cast on hydrogen in Former President George W. Bush's State of the Union Speech in 2004, which featured such memorable quotes as: "America can lead the world in developing clean, hydrogen powered automobiles" and "With a new national commitment, our scientists and engineers will overcome obstacles to taking these cars from laboratory to showroom" [12]. Similar optimism has been expressed across the board by politicians, economists, and scientists, as exemplified in Jeremy Rifkin's canonical work "The Hydrogen Economy" [13]. Still, there is by no means universal acceptance. From the opposing perspective, energy expert Joseph J. Romm in his book *The Hype About Hydrogen* asserts: "Neither government policy nor business investment should be based on the belief that hydrogen cars will have meaningful commercial success in the near- or medium-term" [14].

Whatever view you take on the future importance of hydrogen in our society, the world's most abundant element is in fact an extremely valuable chemical commodity today. In the contemporary industrial marketplace, hydrogen is a high volume chemical with US production exceeding 20 billion kg annually [15]. Important industry uses include the production of chemicals, processing of materials, semiconductor manufacturing, generator cooling, and fertilizer production. In fact, the demand for hydrogen has been expanding significantly in recent years, especially in the fuels processing sector. As petroleum crude becomes dirtier, and as biomass feedstocks enter the fuels mix, growing amounts of hydrogen are being required for "hydrocracking" in the refining processes [16].

Hydrogen's low density, high thermal conductivity, and strong chemical reducing properties make it ideal for the broad ranging industrial applications discussed above. To satisfy the expanding demand, current hydrogen production continues to rely primarily on established fossil-fuel technologies. Worldwide, over 95% of hydrogen is produced from natural gas, oil, or coal, with production relying largely on steam-methane reforming (SMR) with downstream purification processes such as pressure swing absorption (PSA) [17]. Of course, fossil fuels are vulnerable to dwindling availability and rising cost, and result in carbon emissions and other forms of environmental contamination. The current fossil-based industries for hydrogen production and utilization still suffer the risks and disadvantages of our current *Fossil Fuel Economy*. New, cleaner, and long-term approaches to hydrogen production and utilization are essential, not only to sustain current levels of demand, but also to accommodate future expanded *Hydrogen Economy* markets [13, 18–20].

In visions of a future *Hydrogen Economy*, hydrogen is an ideal energy carrier for the storage and distribution of renewable energy resources such as solar, wind, geothermal, and hydroelectric. Using fuel-cell or combustion engine technologies, hydrogen can be converted simply and cleanly to power or heat with no carbon emissions, and with water as the primary by-product. Unfortunately, although hydrogen is nature's most abundant element, it exists primarily in strongly bonded chemical compounds, and extracting it involves difficult and energy-intensive processes. With current technologies, production, storage, and utilization of this ideal energy carrier are all challenging. An enormous amount of technology and infrastructure development would be needed to attain a pure *Hydrogen Economy*. More realistically, a *Green Economy* will emerge comprising a broad portfolio of alternative energy sources and carriers, including hydrogen. As our present reliance on petroleum-based fossil fuels become increasingly difficult to sustain, both economically and environmentally, this will become inevitable. As a result, new, more distributed approaches to national and world energy management will take hold. Different locations rich in their own renewable resources will manufacture energy currencies such as electricity or hydrogen for large-scale distribution to the broader energy marketplace. Electricity is a key energy carrier today, and will remain so long into the future. Hydrogen, however, will also emerge in an important complementary role, providing important benefits in large-scale energy storage and long-distance distribution.

Hydrogen is indisputably valuable today, and will become increasingly valuable as an energy carrier with the future development of new renewable energy production and distribution infrastructures. In the process, new and improved technologies will emerge for the economical and environmentally friendly production, storage, delivery, and utilization of hydrogen. Production technologies using solar energy to split water are enormously attractive, motivating accelerated PEC research and development. Current research in multijunction PEC devices will be particularly important to near-term deployment of large-scale solar hydrogen production, as discussed in the following sections.

7.2.2 The Solar Resource

The true value of the solar resource as a renewable energy commodity becomes painfully apparent considering the current situation of oil. According to the US Energy Information Administration (EIA), the current worldwide oil in reserves totals approximately 213 trillion liters [21]. The amount of oil consumed globally as of 2008 was determined to be 5 trillion liters [22]. The projected oil consumption growth rate between 2004 and 2030 is expected to be approximately 1.6% [23]. Figure 7.3 shows projected oil consumption in relation to known oil reserves, both at current consumption levels and for a projected growth of 1.6% per year. As can be seen, even if the current consumption does not increase, the current oil reserves will become significantly depleted over the next 50 years.

New oil reserves will be found, yet the fact remains that this is a shell game of diminishing returns. Oil is a finite commodity, and before it runs out entirely, it will become increasingly energy-intensive and costly to extract from the earth and process into fuels. Creating a renewable alternative to satisfy our world's energy appetite is likely to be the challenge of our Age. As Dr. Carl Safina aptly noted: "Whoever owns the new sources of energy will own the future economy" [24].

Our best hope for the future is the sun, the ultimate renewable energy resource, continuously bombarding Earth with about 180,000 TW of radiant power. This is enough to power about 10 quadrillion compact fluorescent light bulbs; over

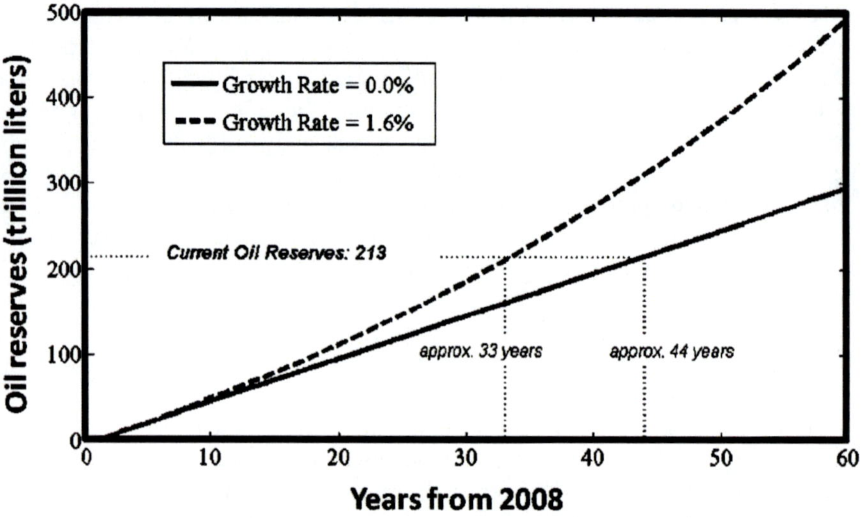

Fig. 7.3 Projected exponential growth of oil consumption plotted for 0.0 and 1.6% growth rates given current global oil reserves

10 million bulbs for every person on the planet! [25, 26]. Approximately 36,000 TW of the incident solar power strikes earth's land masses where terrestrial-based solar-energy conversion plants could be constructed. To put this in perspective, our society on average consumes 13–15 TW, with projected increases up to 30 TW over the next 50 years, consistent with trend shown in Fig. 7.3 [27]. These energy consumption numbers are enormous, but still represent a small fraction of the sun's influx of radiant power. Of course, the planet relies on the sun for many things, including sustaining plant-life and driving its weather patterns, so not all of the incident solar energy is available for mankind's energy hunger. Still there is significant solar energy to spare. Using 10% solar energy conversion systems, 30 TW could be produced using only 0.25% of the land.

Unfortunately, it is not that easy. The large-scale conversion of solar energy to mechanical work or electricity is currently quite challenging and costly. At peak times of daylight, the solar intensity available for terrestrial conversion scales to approximately 1,000 W/m^2. Large collection areas and significant landmass would therefore be needed for commercial scale power production. Such expansive commercial deployment requires an enormous capital investment. For example, commercial photovoltaic technologies today can convert sunlight to electricity at efficiencies between 10 and 20%, at $2–5 per installed Watt [28]. A single Gigawatt plant would cost billions of dollars and span over 2,500 acres. As an added disadvantage, these expensive behemoths would be inoperative at nighttime and ineffectual under any significant cloud coverage.

There are practical difficulties and challenges, but the sun is still the most promising of renewable resources, and the most underutilized in modern society. Currently, less than 0.1% of the world energy production is from solar energy plants, though this number is starting to increase [29]. Encouragingly, improved technologies for solar energy conversion, storage, and utilization are emerging to make their impact on the world energy scene. New and improved solar-to-electric and STH conversion technologies are all poised to be part of the new energy mix. Developing the most practical and efficient technologies to convert sunlight to useable energy carriers such as electricity or hydrogen is a critical priority. With its potential benefits, PEC hydrogen production remains an attractive conversion route for large-scale solar hydrogen production.

7.2.3 Solar Conversion Pathways

In converting sunlight, whether to electricity or to hydrogen, fundamental thermodynamic principles govern the energy conversion process. The sun radiates energy at a *black body temperature* of 5,780 K, while the Earth's black body temperature is 300 K. The Carnot limit between these source and sink temperatures is readily calculated as 95%, representing the amount of radiant energy can be converted into other more useable energy forms [30]. Though this is encouraging, actual conversion of sunlight to practical end-uses is always further limited by unavoidable losses associated with the available energy conversion routes. Every added conversion step in the process adds losses and reduces overall conversion efficiency.

A primary mechanism for converting solar energy on earth is the photoexcitation of electrons in terrestrial matter. With this electronic excitation as an initial step, the two basic routes for further energy conversion of the photoexcited electrons are *solar–thermal* and *solar-potential*. In the *solar–thermal routes*, the energized electrons thermalize to their surroundings, converting the energy to heat, which can be converted further, for example, using heat engines to produce work. Such routes are further restricted by lower Carnot limits based on the intermediate source temperatures driving the heat engines. In the alternative *solar-potential* routes, the elevated electrochemical potential of the energized electrons can directly drive further conversion processes, for example, producing electricity or chemical products. Thermal energy is not being converted to heat, so additional Carnot limits are not imposed.

Based on these photoexcitation models for energy exchange, solar water-splitting for hydrogen production can follow several different conversion routes, as shown in Fig. 7.4. The solar–thermal route shown as process (A) in the figure is a two-step process, with a photon-to-thermal energy conversion step followed by a thermal-to-chemical (TC) conversion step. The other two-step process shown as process (B) represents PV-electrolysis, where a photon-to-electric conversion step is followed by an electric-to-chemical conversion process. The three-step process (C) represents a concentrated-solar–thermal–electricity/electrolysis route, involving photon-to-heat, heat-to-electricity, and electricity-to-chemical conversion steps. The final pathway depicted as process (D), representing a single-step direct conversion from photon-to-chemical energy, is the PEC water-splitting process.

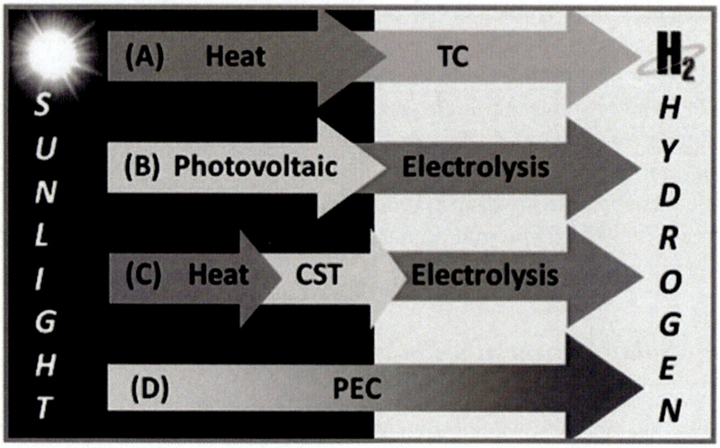

Fig. 7.4 Four viable single and multistep solar-to-hydrogen conversion pathways

Although a direct route in theory, practical implementation, for example, with multijunction semiconductor devices, will entail thermal losses which need to be minimized for high-efficiency conversion. Other STH pathways are also possible, including photobiological routes [31, 32] and the ultrahigh temperature thermolysis route [33], all with their own sets of challenges.

A number of different pathways can contribute to the renewable production of hydrogen for future "Green Economies," but economics will determine which ones will play the most important roles. In general, the most direct conversion processes, such as PEC water-splitting, could have some inherent economic advantages. In fact, PEC hydrogen production as a low-temperature single-stage process could be one of the front-running alternatives, contingent on development of sufficiently efficient, stable, and inexpensive material systems. Under the general umbrella of PEC solar hydrogen production, there are several possible plant strategies and configurations that could be practically feasible. These are described below.

7.2.4 *Photoelectrochemical Hydrogen Plants*

There are numerous hydrogen-production approaches utilizing solar energy for the PEC water splitting. With these different approaches, there are several possible commercial scale plant configurations [34, 35]. Ultimately, the practicality of any plant will be measured by the cost of the produced hydrogen. This will be determined by the efficiency, cost, and lifetime of the active PEC materials used, in addition to associated plant capital costs and operating expenses. As discussed in the following section, initial techno-economic analyses have been performed to compare and contrast economic advantages and disadvantages of different PEC plant concepts [35], but more comprehensive follow-on analyses will be needed,

Fig. 7.5 (a) Interconnected PV–electrolyzer system with power conditioning unit; (b) area-matched integrated PV–electrolyzer system

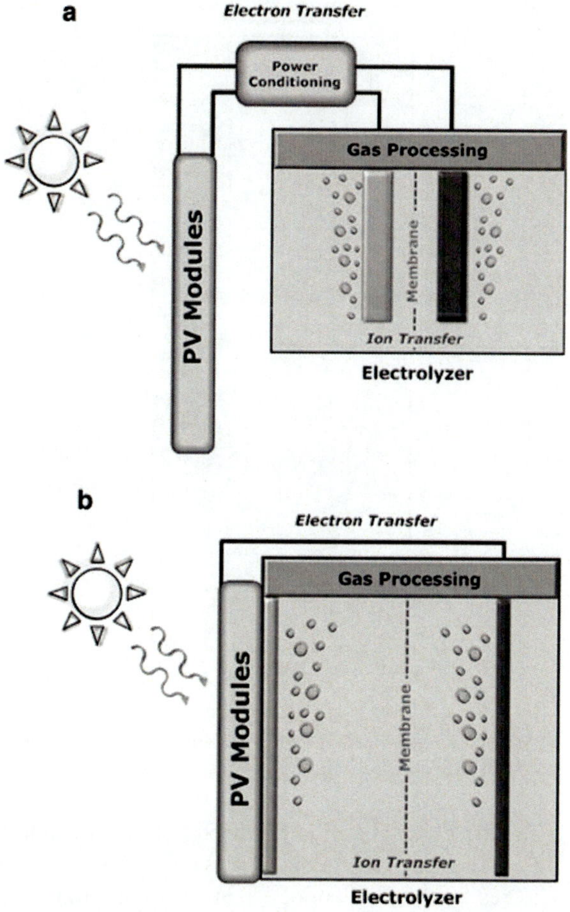

especially as PEC conversion processes evolve and become more refined. Broadly speaking, PEC plant configurations fall into three categories: (1) combined PV-electrolysis systems; (2) photoelectrode-based systems; and (3) photocatalyst-based slurry systems. Many of the different single- and multijunction PEC materials and devices discussed in the following sections can be integrated into one or more of these plant types. Hydrogen production scales ranging from small-scale 1,500 kg/day for distributed production up to large-scale 50,000 kg/day central-plant production are being envisioned for all three categories. Some of the likely reactor schemes are illustrated in the following figures.

Figure 7.5a, for example, illustrates the most direct implementation of the "Combined PV-electrolysis" configuration. Commercially available photovoltaic (PV) panels are coupled with separate commercial electrolyzer units, such as alkaline or PEM electrolyzers; and appropriate power-conditioning equipment is utilized to load-match the processes. This is the clear path to near-term renewable solar hydrogen, but it is by no means inexpensive. Based on recent cost studies from the NREL, hydrogen production cost would exceed \$10/kg for PV electricity cost at

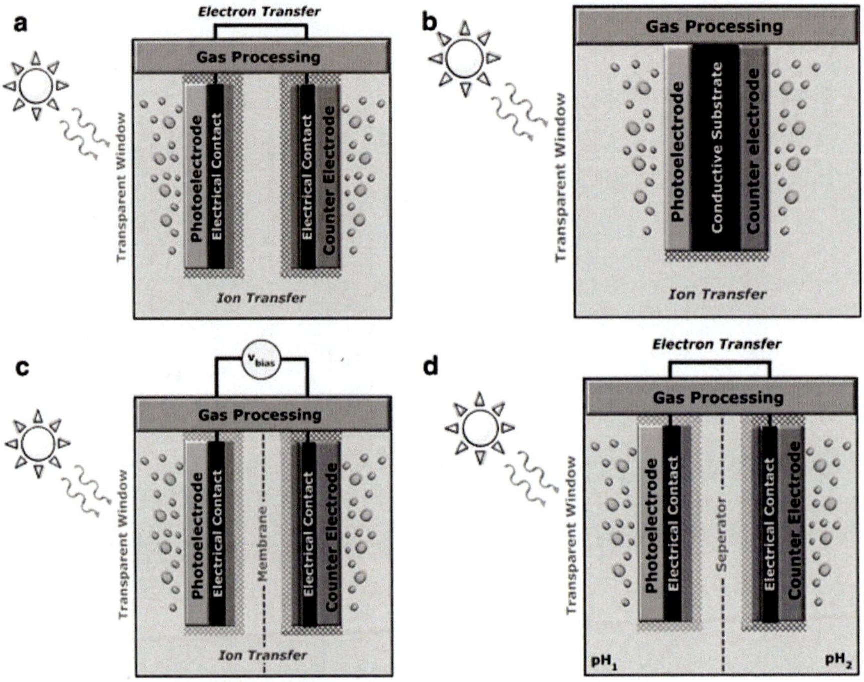

Fig. 7.6 (**a**) Two-electrode PEC system; (**b**) monolithically integrated single-electrode PEC system; (**c**) two-electrode PEC system with external voltage biasing; (**d**) two-electrode PEC system with chemical biasing

15–25¢/kW h [36, 37]. The need for power conditioning equipment contributes to this high cost. Additional factors include the electrolyzer capital costs as well as the losses from lateral collection and transmission of electricity.

A variation of the PV-electrolysis approach is the integrated PV-electrolyzer unit shown in Fig. 7.5b. In such systems, the solid-state PV panels and large area electrolysis components are designed to be better matched in terms of both operational voltage and current density. Power conditioning is eliminated and transmission losses would be reduced. Based on the direct coupling to the solar flux rage, the electrolysis process would run at low current densities, for example, below 50 mA/cm^2, compared with rates over 1,000 mA/cm^2 in commercial electrolyzers. This relaxes the need for precious metal catalysts in the electrolysis process, and therefore reduces capital cost. Added complexity and cost of the dual-function panels for large-scale deployment is a significant tradeoff.

Unlike the systems in the "PV-electrolysis" category, the photoactive components in the "photoelectrode based systems" configuration are immersed in solution, and typically include one or more PEC solid/liquid rectifying junctions.

Figure 7.6a shows a classic two-electrode configuration with a photoelectrode immersed in aqueous solution and electrically interconnected to a metallic counter electrode. Sunlight incident on the photoelectrode energizes the water-splitting process, promoting hydrogen evolution at the "cathode" surface and oxygen evolution at the "anode" surface. The PEC process entails both electron exchange in the external wire and ion exchange in solution. To minimize electronic and ionic conductivity losses, the counter electrode is placed in close proximity to the photoelectrode surface. In some cases, the counter electrode is a small catalyst wire placed right in front of the photoelectrode. This, however, introduces some optical shading loss; and it also requires a front-surface gas-separation membrane, adding further loss.

Figure 7.6b illustrates a single-electrode variation of the photoelectrode system comprising anode and cathode surfaces at opposite faces of a single monolithically integrated device. In this configuration, electron transfer is readily achieved through the body of the electrode, but pathways for ionic transfer in the electrolyte need to be provided around the electrode body to minimize solution conductivity losses. Electron transmission loss is reduced, as is system part count. This offers potential cost reduction.

The photoelectrode schemes in Fig. 7.6a, b require photoactive electrode materials and devices capable of spontaneously splitting water with solar illumination. As discussed at length in the following sections, this will most likely require the use of multijunction structures to generate sufficient voltage to sustain the process. Alternative external-voltage and chemical biasing strategies are possible in photoelectrode-based systems as illustrated in Fig. 7.6c, d. External biasing is specifically useful for single-junction photoelectrodes incapable of efficient unassisted solar water splitting. The added costs associated with the supplemental electricity needed in Fig. 7.6c or chemical feedstocks needed to maintain the pH gradient in Fig. 7.6d generally makes these approaches unattractive from a practical perspective.

In the photoelectrode-based systems, active semiconductor layers or thin-films are coated on fixed electrode structures which are immersed in solution. In contrast, the "photocatalyst"-based reactors utilize photoactive semiconductor particles free-floating as slurry in a solution bed. Figure 7.7a represents a conceptual single-bed photocatalyst reactor, comprising concentrations of one or more functionalized photocatalyst particles suspended in solution. In the example shown in Fig. 7.7a, the semiconductor particle would need to be capable of internally generating enough photovoltage to split water. In addition, specific surface sites on the particle would need to be functionalized, for example, with catalyst nanoparticles, to promote hydrogen and oxygen evolution and facilitate charge separation. An interesting variation of the single-bed photocatalyst system is the "Z-scheme," which is somewhat analogous to the multijunction approach in the photoelectrode case. As described in more detail in later sections, the Z-scheme utilizes different photocatalysts for hydrogen and oxygen evolution, but the two particle types are mixed together in the same reactor bed with an ion mediator in solution to couple the reactions. Single-bed photocatalyst reactors are extremely simple, with perhaps the best potential for low cost at large scales. The obvious drawback is that potentially explosive mixtures of hydrogen and oxygen gas are generated in the same solution, necessitating safe and reliable gas separation mechanisms.

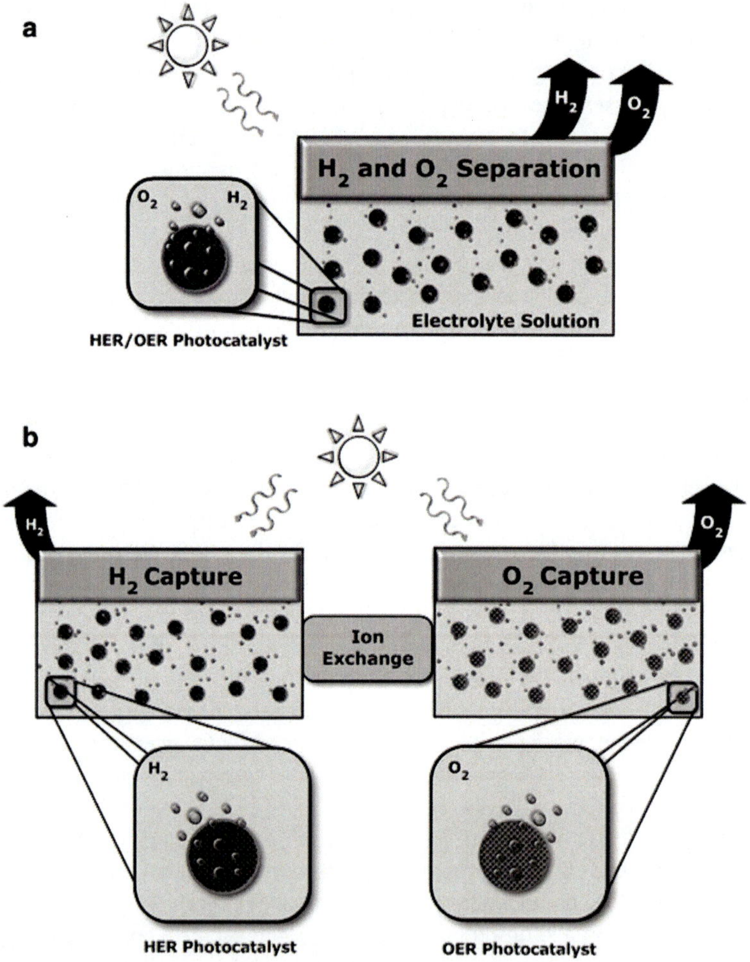

Fig. 7.7 (**a**) Single-bed photocatalyst particle suspension system, co-generating oxygen and hydrogen in the same reactor; (**b**) two-bed photocatalyst particle suspension system, with separated oxygen-evolution and hydrogen-evolution beds

Separation technologies, such as pressure swing absorption, are commercially available, but would add complexity and cost.

Figure 7.7b illustrates a two-bed photocatalyst reactor. Similar to the Z-scheme, two separate photocatalyst particles are used, one for hydrogen evolution and the other for oxygen evolution. In the two-bed case, however, the different particle types are suspended in separate reactors which are connected by an ionic bridge-way to allow for mediator coupling. In the two-bed reactors, the hydrogen and oxygen gases are generated separately and safely, eliminating the cost of separations equipment. The drawback is the reduced hydrogen production efficiency inherent in this configuration. The ion bridge introduces conductivity loss; but more importantly, a larger solar collection area is necessary for water splitting when hydrogen and oxygen reactions are stacked side-by-side, reducing conversion efficiency and increasing overall cost.

7.2.5 Techno-economic Analyses

The US Department of Energy (DOE) recently commissioned *Directed Technologies Inc.* (DTI) of Arlington, Virginia, to conduct a techno-economic evaluation of conceptual PEC hydrogen production systems [35] incorporating performance and processing cost feedback from the broader materials R&D efforts at DOE. The objective was to provide a basis for evaluating the long-term feasibility of large-scale PEC production technologies in comparison with other renewable approaches. The DTI study, which was completed in 2009, comparatively evaluated two photocatalyst particle bed configurations (compatible with the reactor configurations illustrated in Fig. 7.7) as well as two photoelectrode-based systems (compatible with the reactor configurations illustrated in Fig. 7.6). The four conceptual systems in the study were designated "Type-1" through "Type-4" as follows:

- *Type-1: Single-Bed Photocatalyst System*: A single electrolyte-filled reactor bed containing a colloidal suspension of PEC photocatalyst particles which produce a mixture of H_2 and O_2 product gases.
- *Type-2: Dual-Bed Photocatalyst System*: Two electrolyte-filled reactor beds containing colloidal suspensions of PEC photocatalyst particles, with one bed carrying out the oxygen evolution half-reaction, the other bed carrying out the hydrogen evolution half-reaction, and including a mechanism for circulating the ions between beds.
- *Type-3: Photoelectrode System*: A fixed array of planar PEC photoelectrodes immersed in an electrolyte reservoir, tilted toward the sun at local latitude angle, producing hydrogen at atmospheric pressure.
- *Type-4: Photoelectrode System with Moderate Concentration*: A PEC solar concentrator system, using reflectors to focus the solar flux at a 10:1 intensity ratio onto planar PEC photoelectrodes immersed in an electrolyte reservoir and pressurized to 300 psi.

In the DTI analyses, each of the modeled system types included a PEC reactor generating H_2 and O_2: photocatalyst-based on Types 1 and 2, and photoelectrode-based on Types 3 and 4. Each also included a gas processing system for compressing and purifying the output gas stream, in addition to ancillary balance of plant equipment. For each system type, the total cost of hydrogen production was projected for a 10-metric ton per day plant assuming technology-readiness of the PEC reactor. The baseline for readiness included targeted values of 10% for STH conversion efficiency and 5 years for PEC cell lifetimes. Sensitivity analyses around the baseline assumptions were performed to assess the effects on production

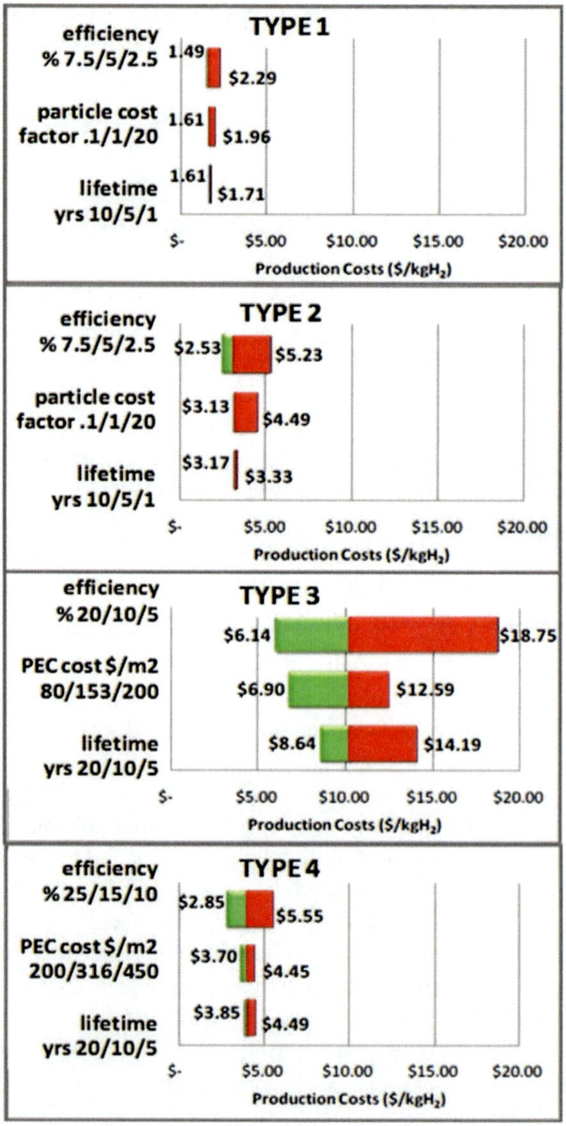

Fig. 7.8 Hydrogen cost sensitivity analyses for the four reactor types evaluated in the DTI techno-economics report on *Photoelectrochemical Hydrogen Production*. Projected hydrogen costs as a function of efficiency, lifetime, and materials cost are shown

costs of variations in the PEC performance and cost parameters. A summary of the sensitivity analyses for the four system types is shown in Fig. 7.8.

As concluded in the DTI report, the hydrogen production costs are lowest for the Type 1 and Type 2 photocatalyst systems, ranging from \$1.50 to \$5.25/kg H_2. These are encouraging results, consistent with the US DOE cost targets of \$2–4/kg H_2 [38] for future central hydrogen production pathways. The report recognized, though, that significant research and development work is still needed to demonstrate particle-based PEC reactors operating at the baseline performance levels assumed in the study. In contrast, the Type 3 photoelectrode system is currently the most mature of the concepts, with numerous prototypes demonstrated to date on the laboratory scale. The report concluded, however, that hydrogen production costs were the highest using this route, ranging from \$6.14 to 18.75/kg H_2. The high costs in this case were dominated by substantial material and capital costs, similar to those plaguing the PV industry. The Type 4 photoelectrode system is a modification of Type 3 system with the addition of modest solar concentration (up to 10×) and moderate internal hydrogen compression (to 300 psi). Interestingly, these added features resulted in substantial cost reductions, specifically due to the reduction in PEC-electrode area, reduced amount of PEC materials needed, and fewer postprocessing compression stages. The resulting hydrogen production costs were reduced to \$2.85–5.55/kg H_2, falling within the DOE targeted range.

Although preliminary, this study has shown that hydrogen production by PEC systems can be economically viable in both photocatalyst and photoelectrode configurations. However, key research challenge remains for attaining PEC material systems compatible with the baseline levels of 10% STH conversion efficiency and 5-year lifetime. As discussed in the following sections, multijunction approaches are likely to be critical for near-term realization of such performance levels. As research and development of PEC materials and devices progresses, more techno-economic analyses will be needed to further evaluate the tradeoffs between different large-scale PEC reactor approaches. Two of the most promising avenues appear to be photoelectrode systems using low-cost thin-film materials [39] and large-scale slurry bed reactors utilizing low-cost functionalized semiconductor particles [40]. The path forward for practical PEC solar hydrogen production needs to navigate all the economic as well as the scientific obstacles that remain.

7.3 Photoelectrochemical Water Splitting

In general, standard chemical processes involve interactions between chemical and ionic species. Electrochemical processes also involve interfacial interactions between ionic conductors, such as electrolytes, and solid-state electronic conductors, such as semiconductors. Involving even more complexity, PEC processes are exposed to light, so that optical photons can also interact with the electrochemical reactions. In semiconductor photoelectrochemistry, energetic photons of light typically create electron–hole pairs within the semiconductor that can participate in redox reactions at the semiconductor/electrolyte interfaces. The complicated set of fundamental electrochemical and solid-state optoelectronic principles that govern the behavior of such systems are well-documented in the literature [41–46]. Within the realm of photoelectrochemistry, the PEC water-splitting process itself presents a unique set of energetic and kinetic challenges [47–50]. Some useful simplifications to the problem, presented in the following sections, can provide a broad overview of PEC water-splitting, and help to highlight the loss mechanisms and conversion efficiency limitations inherent in PEC hydrogen production.

7.3.1 PEC Water-Splitting Reactions

Any light-driven water splitting process can be expressed as a generalized chemical reaction representing the addition of photonic energy to liquid water to split the H_2O molecule, evolving gaseous hydrogen and oxygen:

$$H_2O + 2\gamma \rightarrow H_2(g) + 1/2O_2(g) \quad \text{Net water-splitting reaction} \tag{7.1a}$$

$$\Delta G^0 = +\ 237.18\ \text{kJ/mol} \quad \text{Standard Gibbs free energy.} \tag{7.1b}$$

Here, γ is the photon energy and ΔG° is the standard Gibbs free energy. As indicated in (7.1), energetic photons, such as those in sunlight, can supply the energy to drive the uphill reaction converting the feedstock H_2O into hydrogen and oxygen gases. Specifically, the standard Gibbs free energy change of +237.18 kJ/mol is a quantification of the thermodynamic minimum energy needed for splitting water into the constituent gases at standard conditions of 25°C and 1 bar.

PEC implementations of light-driven water splitting utilize one or more photoactive semiconductor electrodes to convert the photon energy. In a PEC system, absorbed photons induce electron–hole pairs in the semiconductor bulk which can be separated and extracted to promote two "half reactions" in solution. Specifically, photogenerated electrons drive the hydrogen evolution half-reaction (HER) at the cathode interface, while photogenerated holes drive the oxygen evolution half-reaction (OER) at the anode interface. Both half reactions must be sustained simultaneously, coupled by their exchange of electrons in the solid state, and ions in solution. In real systems, multiple reaction pathways and reaction steps are possible at both anode and cathode, and physical conditions shape if and how the process will proceed. As one example, the half-reactions for water splitting can be written explicitly in terms of the H^+ ionic exchange between anode and cathode:

$$2\gamma \rightarrow 2e^- + 2h^+ \quad \text{Photon-induced electron-hole pair generation} \tag{7.2a}$$

$$H_2O + 2h^+ \rightarrow 2H^+ + 1/2O_2(g) \quad \text{OER: anodic water-oxidation half-reaction} \tag{7.2b}$$

$$2H^+ + 2e^- \rightarrow 2H_2(g) \quad \text{HER: cathodic } H^+ \text{reduction half-reaction} \tag{7.2c}$$

$$H_2O\ + 2\gamma \rightarrow H_2(g) + 1/2O_2(g) \quad \text{Net water-splitting reaction} \tag{7.2d}$$

$$\Delta G^0 = +\ 237.18\ \text{kJ/mol} \quad \text{Standard Gibbs free energy} \tag{7.2e}$$

$$\Delta V^0_{\text{rev}} = \Delta G^0/nF = 1.23\ \text{V} \quad \text{Standard reversible potential} \tag{7.2f}$$

$$V_{\text{op}} = \Delta V^0_{\text{rev}} + \eta_a + \eta_c + \eta_\Omega + \eta_{\text{sys}} \quad \text{Operating voltage with overpotential loss.} \tag{7.2g}$$

Here, e^- is an electron, h^+ is a hole, F is Faraday's constant, and n (=2) is the number of electrons exchanged. V_{op} is the operational voltage, η_a, η_c, η_Ω, and η_{sys} are overpotentials associated with anode, cathode, ionic-conductivity, and system losses, respectively.

Alternatively, the half-reaction set can also be written in terms of OH^- ionic exchange:

$$2\gamma \rightarrow 2e^- + 2h^+ \quad \text{Photon-induced electron-hole pair generation} \tag{7.3a}$$

$$2OH^- + 2h^+ \rightarrow H_2O + 1/2O_2(g) \quad \text{OER: half-reaction} \tag{7.3b}$$

$$2H_2O + 2e^- \rightarrow 2OH^- + H_2(g) \quad \text{HER: half-reaction} \tag{7.3c}$$

$$H_2O + 2\gamma \rightarrow H_2(g) + 1/2O_2(g) \quad \text{Net PEC water-splitting}. \tag{7.3d}$$

The corresponding expressions for the Gibbs free energy, the standard reversible potential and the operating voltage are equivalent to those given by (7.2e–g).

The reaction pathways depend on the nature of the electrodes, the electrolyte, and the light-induced energy levels. For example, the half-reaction formulations expressed in terms of H^+ tend to dominate in acidic electrolyte solutions, while the OH^- formulations can dominate in basic solutions, specifically due to the predominance of one ionic species over the other under the different conditions [51]. In either case, it is important to emphasize that each of the half-reactions described in (7.2) and (7.3) is simplifications of more complex multistep electrochemical reaction pathways, and that competing or parasitic reactions are also possible [52, 53]. Without parasitic reactions and losses, the Gibbs free energy of photo-driven water splitting is +237.18 kJ/mol, and the standard reversible potential of PEC water splitting is 1.23 V, as indicated in (7.2) and (7.3). This is independent of the dominant pathways involved in achieving the net reaction. All practical systems will involve losses, such as those represented in the various overpotential terms in the equations; and these losses can be significantly different for different reaction pathways. Including overpotential losses, the energy/voltage needed to split water will inevitably exceed 1.23 V.

7.3.2 PEC Water Splitting Losses

Losses incurred during PEC water splitting include the overpotential losses at the anode and cathode interfaces, the ionic conductivity overpotential loss associated with ion transport in the electrolyte, as well as other solid state and balance of system losses. The anode and cathode overpotential losses include the effects of activation energy, kinetics, and mass-transport of the multistage half reactions. These can be substantial, commonly several tenths of volts, with more severe loss for oxygen evolution, which is the more complex and less facile reaction [54]. In addition to the electrode overpotential losses, the electrochemical losses due to ionic conductivity in the solution can also be severe, depending on electrode geometry and spacing, in addition to solution properties. The splitting of *pure water*, for example, is particularly difficult, since the ionic conductivity, typically less than 0.05 S/m range is prohibitively low. Weak acid or alkaline solutions with conductivities exceeding 10 S/m are typically used to compensate, although this creates a more corrosive environment for the electrodes.

On top of the losses at the interfaces and in solution, there are also significant losses in the solid state of any PEC system. While the ions in solution are migrating through the aqueous media and reacting at the interfaces, the solid-state electrons–holes need to be exchanged between the anode and cathode through a conductive pathway (such as an interconnecting wire in two electrode schemes and the conducting substrate in monolithic single electrode configurations), and therefore some ohmic loss is inevitable. Additionally, there can be severe losses in the bulk semiconductor and solid-state interfaces of the electrodes. These include optical absorption losses, electron–hole pair recombination losses, and interface recombination losses, among others, which also degrade system performance.

The PEC water splitting process is a delicate balancing act, where photon-energized electron–hole pairs under the right conditions can simultaneously drive the electrochemical half-reactions. In steady state, the anodic and cathodic half-reactions need to be sustained at the same reaction rate, or a charge build-up will occur, impeding and eventually stopping the entire process. A similar situation exists with the charge carriers in the solid state. The anodic half-reaction generates two electrons (i.e., "consumes" two holes), while the cathode half-reaction consumes two electrons. These electrons must be shuttled from anode to the cathode via electrical current and steady state cannot be maintained if anode and cathode reaction rates are not the same.

As a result of all the electrochemical, solid-state losses and any balance of system losses occurring at steady state, water splitting cannot occur at the bulk reversible potential of 1.23 V. The operating voltage for water-splitting must exceed this value to compensate for all the losses. In practice, water electrolysis systems typically require operating voltages of 1.6–1.9 V, depending on gas production rates [55, 56]. In other words, to drive the water-splitting process including all solution and electrode losses, the absorbed photons must induce sufficient electrochemical potential to the electron–hole pairs to generate photopotentials in excess of approximately 1.6 eV. The photoelectrolysis *balancing act* can be set into motion *ONLY* if this photopotential requirement is met. As discussed in the following sections, it is the photopotential requirement that may necessitate the use of multijunction approaches for efficient PEC water splitting. If and only if the conditions are right to sustain the water splitting process, the hydrogen evolution will then be proportional to electron consumption rate at the cathode, which represents a *photocurrent*. This is critical to the understanding of STH conversion efficiency.

7.3.3 PEC Solar-to-Hydrogen Conversion Efficiency

During steady-state operations, the solid-state shuttling of charges between anode and cathode represents a photon-induced current, or photocurrent. This photocurrent is integrally tied to the hydrogen-producing performance of the PEC system. Specifically, the PEC half reactions from the previous section illustrated that two electrons are consumed in the evolution of one H_2 molecule. This results in a hydrogen production rate that is half the rate of electron flow, in other words, half the photocurrent:

$$R_{H_2} = \frac{I_{ph}}{2e} = \frac{(J_{ph} \times A)}{2e}, \tag{7.4}$$

where R_{H_2} is the hydrogen production rate (s^{-1}), I_{ph} is the photocurrent (A), e is the charge of an electron (C), A is the area of the illuminated photoelectrode (m^2), and J_{ph} is photocurrent density (A/m^2).

The chemical STH conversion efficiency of a solar-based hydrogen production system is defined as the ratio of the useable chemical energy in the generated hydrogen gas to the total solar energy delivered to the system [57–61]. For steady-state operations, this is equivalent to the ratio of the *power output* to *the power input* of the system. This power ratio can be expressed as:

$$\frac{P_{out}}{P_{in}} = \frac{(\text{hydrogen production rate}) \times (\text{hydrogen energy density})}{\text{solar flux integrated over illuminated area}}. \tag{7.5}$$

Using the hydrogen production rate from (7.4) and the Gibbs energy as the useful energy density of the hydrogen, the STH efficiency for a PEC system can be written as:

$$\text{STH}(\%) = \frac{\frac{\Delta G}{N_A} \times R_{H_2}}{P_{solar} \times A} = \frac{\frac{\Delta G}{N_A}\left(\frac{J_{ph} \times A}{2e}\right)}{P_{solar} \times A} = \frac{\frac{\Delta G}{2eN_A}}{P_{solar}} \times J_{ph}, \tag{7.6}$$

where N_A is Avogadro's number and P_{solar} is the solar flux of energy in W/m^2. The first ratio in (7.6) is generic for any STH production system, while the second term is derived specifically for PEC hydrogen processes using the production rate from (7.4).

The solar flux of radiant energy is comprised of a broad spectrum of energetic photons, which are quantized particles of light, each with discrete energy content. Figure 7.9a shows the standard atmosphere-filtered global solar spectrum (AM1.5G) indicating the range of photon energies and the distribution of energy transmitted by these photons [62]. This AM1.5G spectrum is also frequently presented as a function of photon wavelength, as shown in Fig. 7.9b.

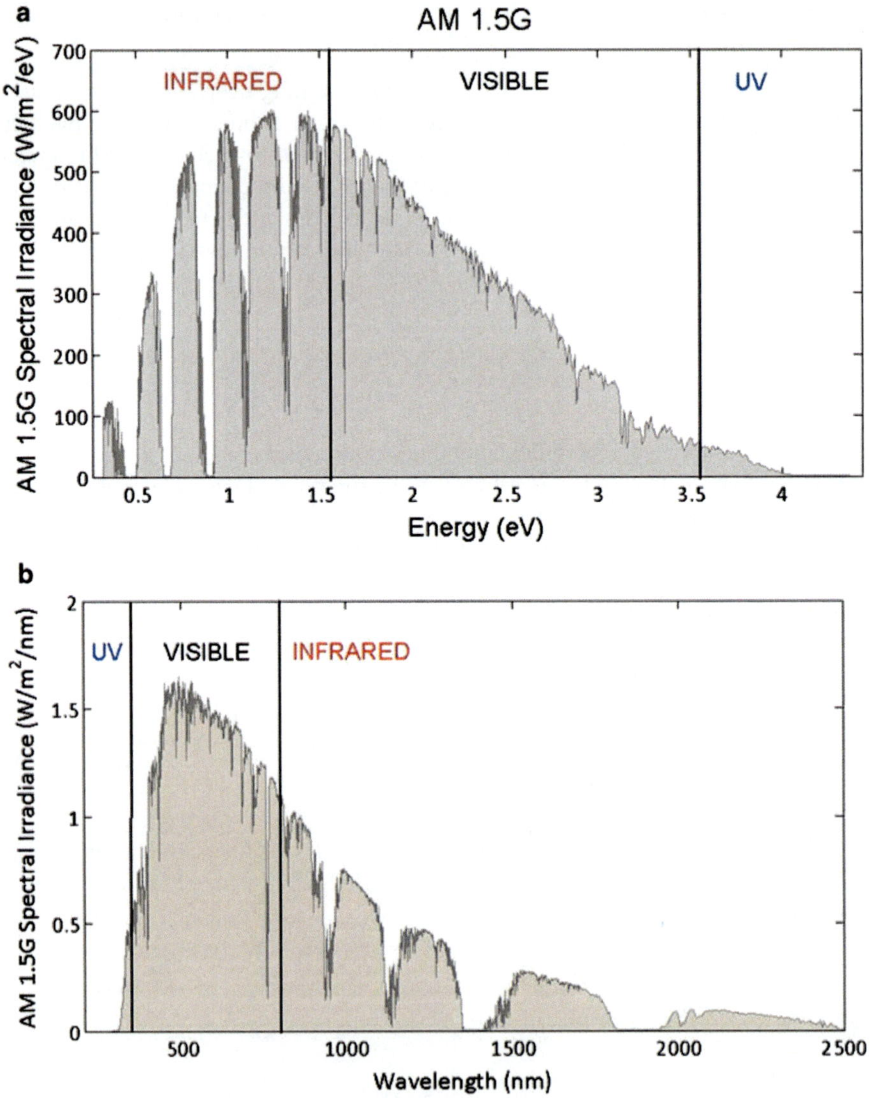

Fig. 7.9 AM1.5 global solar irradiance spectrum: (**a**) as a function of photon energy and (**b**) as a function of photon wavelength, indicating the ultra-violet, visible, and infra-red regions of sunlight

Integrating the AM1.5G solar flux yields a total power density of ~1,000 W/m^2 for sunlight. Using this value for P_{solar} in (7.6), the STH conversion efficiency can be related directly to the photocurrent density of a PEC system:

$$\text{STH}_{\text{AM1.5}}\ (\%) \approx \underbrace{0.123 \times J_{\text{ph}}}_{\text{for } J \text{ in}(\text{A/m}^2)} = \underbrace{1.23 \times J_{\text{ph}}}_{\text{for } J \text{ in}(\text{mA/m}^2)} . \tag{7.7}$$

Equation (7.7) explicitly states the linear relationship between conversion efficiency and photocurrent density calculated under AM1.5G solar illumination. The use of the Gibbs free energy in these equations reflects chemical energy in the hydrogen that can be retrieved using an ideal fuel cell. This in effect, calculates the lower heating value (LHV), which is standard in practical comparisons between different fuels [63].

It is clear from (7.7) that the PEC photocurrent density is the key determining factor for STH efficiency. This is in contrast to solid-state photovoltaic cells, which need to be operated at the *maximum power point* (i.e., maximum product of photocurrent and photovoltage) for the best solar-to-electric conversion efficiency [64]. It is a subtle but important distinction that PEC cells should be operated at maximum photocurrent for best hydrogen-production performance. It is the bandgap-limited *saturated photocurrent density* of a semiconductor that ultimately constrains the hydrogen production rate. For peak STH efficiency, sufficient useable photopotential must be generated in the device to drive the photocurrent as close as possible to the saturation limit. With all the built-in losses in the solid state, in solution, at the interfaces, and in the balance of system, this can be a challenge. In fact, due to limits of useable photopotential generation in a single junction, it may be necessary to resort to multijunction schemes to drive the PEC system into the maximum photocurrent region. There are inherent trade-offs and limitations as well as benefits to this approach. These are explored in detail in the following sections.

7.4 Single Junction PEC Semiconductor Devices

In a semiconductor-based PEC system, photons are absorbed in the semiconductor and the photogenerated charge carriers are separated, and then extracted by the interfacial HER and OER reactions. In photoelectrode-based systems, the charge separation is largely due to the rectifying nature of the semiconductor/electrolyte junction. The charge separation details are somewhat different in the photocatalyst systems [65], but the basic absorption and extraction mechanisms at the solid/liquid interface are the same. In either case, hydrogen production performance is strongly influenced by both semiconductor material properties and interfacial characteristics. An examination of junction formation and operation in the case of photoelectrode configurations reveals much about the key performance parameters and conversion efficiency limitations.

7.4.1 Photoelectrode Junction Formation

In PEC photoelectrode systems, the semiconductor/electrolyte interface can form a *rectifying* junction, similar to the solid-state pn junctions or Schottky diode junctions used in solar cells. Such rectifying junctions exhibit built-in electric fields capable of separating photogenerated charge carriers (i.e., electron–hole pairs) created by absorption of photons in the semiconductor bulk. In PV cells, this charge separation mechanism drives photocurrents to produce electricity, while in the PEC case, the charge separation can drive the HER and OER half-reactions for water splitting. In both cases, illumination creates extra photoexcited electron–hole pairs which need to be separated and extracted before they recombine in order to be effective in the energy conversion process. Extraction of the photogenerated charge carriers with elevated electrochemical potentials in effect converts the solar energy to electricity or hydrogen in PV or PEC systems, respectively. Excellent sources of information are available detailing semiconductor material properties, solid-state junctions and solar cells, and treating the fundamentals of rectifying junction formation and behavior [26, 50, 66–70].

Formation of a rectifying PEC junction at a semiconductor/electrolyte interface follows similar principles, but key concepts from both solid-state physics and electrochemistry need to be combined for a complete description of the process. For reference, there is a wealth of literature on fundamental electrochemical principles [41–46, 50], which are useful, especially in conjunction with the previous citations covering semiconductor physics. Of particular interest to PEC studies are the models developed by Gerischer, which establish important connections between the in-solution electrochemical potentials of electrons and solid-state Fermi levels [47–49]. Using the Gerischer models, descriptions of semiconductor/electrolyte junctions follow closely the solid-state junction analogies. Photoanodes using n-type semiconductors form PEC junctions similar to a Schottky barrier or an np^+ junction, and internal electric fields are set up to drive photogenerated holes toward the electrolyte interface promoting oxygen evolution. In contrast, p-type photocathode junction formation is analogous to a solid-state pn^+ device, and the internal fields promote the injection of photogenerated electrons at the interface to drive the HER [50]. The field directions for both n-type and p-type junctions are shown explicitly in the following illustrations of junction formation.

Specific steps in semiconductor/electrolyte junction formation are illustrated in Fig. 7.10 for both the photoanode (7.10a) and photocathode (7.10b) cases. In Fig. 7.10a (i), the n-type semiconductor and electrolyte are shown in thermal equilibrium prior to contact for the photoanode case. At equilibrium, the Fermi level in the semiconductor (F_n) is close to the conduction band (CB), and the Fermi level in solution (F_s) falls between the redox (reduction/oxidation) levels for hydrogen reduction (H^+/H_2) and water oxidation (H_2O/O_2). After contact, the electrode and electrolyte Fermi levels align to reach thermal equilibrium, as shown in Fig. 7.10a (ii). Since the initial electrode Fermi level is higher than the electrolyte Fermi level, free electrons in the n-type semiconductor will migrate to the solid–liquid interface exposing positively charge fixed donor sites. The electrons form a surface charge layer at the interface which induces a thin Helmholtz double layer in the electrolyte. A *depletion-region*, also known as a *space-charge region* (SCR), forms where free charge carrier diffusion is counter-balanced by the built-in electric field generated by the fixed charges flanking the junction. The charge distributions including the fixed space charges in the solid-state electrode and the Helmholtz layer charges in solution are shown in Fig. 7.10a (iii). Typically, Helmholtz layers are on the

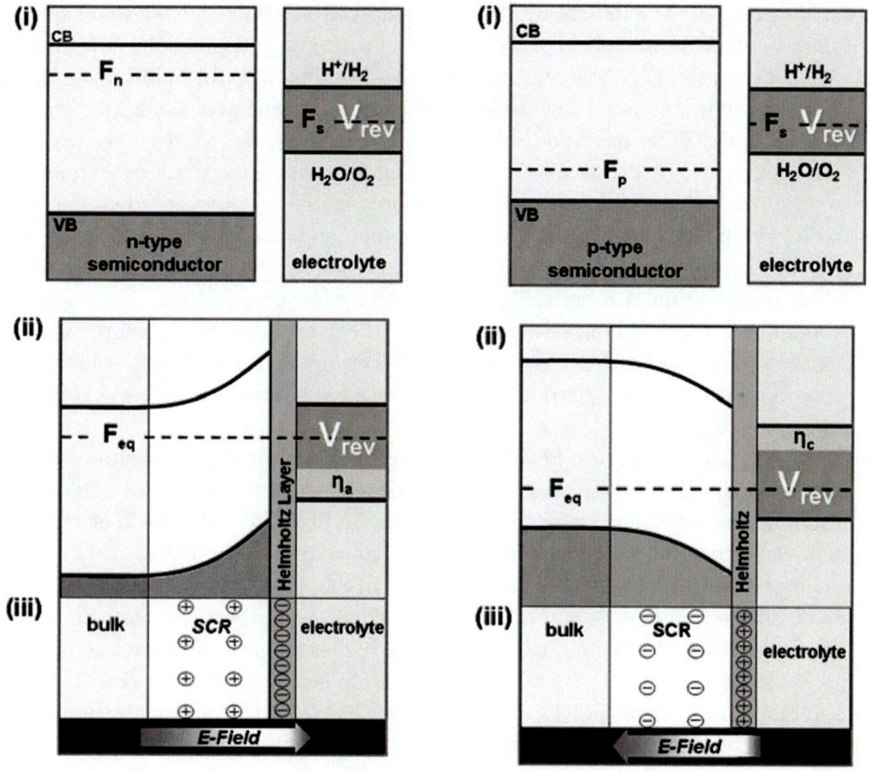

Fig. 7.10 Semiconductor/electrolyte junction formation for (**a**) n-type semiconductor and (**b**) p-type semiconductor junction. Subheadings (**i**), (**ii**), and (**iii**) represent the energy diagrams before contact, the energy diagrams after contact, and the charge distributions, respectively

order of a few nanometers, compared with several micrometers for the semiconductor space-charge region.

For the p-type photocathode case, the equilibrium charge distribution process is the same, but as shown in Fig. 7.10b the charges and band bending are reversed. As illustrated in Fig. 7.10b (i) and (ii), the initial migration of holes toward the interface during contact equilibration creates a space charge region of fixed negative acceptor charges, reversing the direction of the built in electric field compared with the photoanode cases. In both cases, the thermal equilibrium is disturbed upon exposure to light, and illuminated operations need to be described in terms of "quasi-equilibrium" statistics, which generally remain valid for AM1.5G insolation as well as moderate levels of concentrated sunlight [66, 67, 70].

7.4.2 Illuminated Response

Under illumination, the electron–hole pairs generated by photon absorption in the vicinity of the space-charge regions can be separated by the built-in electric fields. Some of the photocharges will be successfully extracted in the water splitting process, while the remainder will recombine, losing energy as radiation or heat. In the photoanode case, the photogenerated holes will be driven by the built-in field toward the electrolyte interface, where, with the appropriate energetics and kinetics, they can drive the OER oxidation reaction. At the same time, photoexcited electrons are driven toward the electrode's back contact, where they can be extracted to solution at counter-electrode surfaces appropriately conditioned to promote the HER. For photocathodes, the charge separation supplies photogenerated electrons to the hydrogen evolution process at the electrolyte interface, and drives photogenerated holes to an OER counter electrode surface. Under these operating conditions, the initial thermal equilibrium has been disturbed by the energy influx of the light, and the band descriptions with Fermi levels are no longer valid. Under conditions of quasi-equilibrium, a useful alternative description can be developed in terms of "quasi-Fermi" levels.

Figure 7.11 illustrates PEC junction response to solar illumination for both photoanodes (Fig. 7.11a) and photocathodes (Fig. 7.11b) using the quasi-Fermi level descriptions. For both configurations in Fig. 7.11, the photoelectrode is immersed in solution, and the back contact is connected by external wiring to a counter electrode also in solution. In addition, the η_a and η_c overpotentials for the OER and the HER in solution have been added to the reversible potential of the redox system to stress the minimum voltage requirement to sustain water splitting.

With the absorption of sunlight in the semiconductor bulk, the original Fermi levels split into separate quasi-Fermi levels for electrons (F_e) and holes (F_h) [47, 66, 67, 70] resulting from the excess concentration of photogenerated electron–hole pairs over the equilibrium populations. For the photoanode case in Fig. 7.11a, the excess hole population significantly alters the minority carrier distribution with respect to the equilibrium populations, while the excess electrons barely affect the majority carrier numbers. As a result, the hole quasi-Fermi level shifts substantially in contrast to insignificant change in the electron quasi-Fermi level. Conversely, in the photocathode case in Fig. 7.11b, there is a significant shift in the electron quasi-Fermi level, but little hole quasi-Fermi level shift. In both cases, near the back contact away from the effects of illumination, the two quasi-Fermi levels converge back to a bulk equilibrium Fermi level.

The back contact potential is tied to the counter electrode potential, which under operating conditions is coupled to the back half reaction. As an important result, the quasi-Fermi separation in the semiconductor determines the *useable* energy for driving the front half reaction. For the illuminated photoanode device shown in Fig. 7.11a, the electron potential at the counter electrode is sufficiently high to drive the hydrogen-reduction half-reaction, including the η_c loss. Simultaneously, the quasi-Fermi hole energy at the solution interface is sufficiently *low* to drive water-oxidation, including the η_a loss. This configuration, in consequence, is capable of sustaining the net PEC water-splitting process, driven by *useable* energy in the quasi-Fermi split.

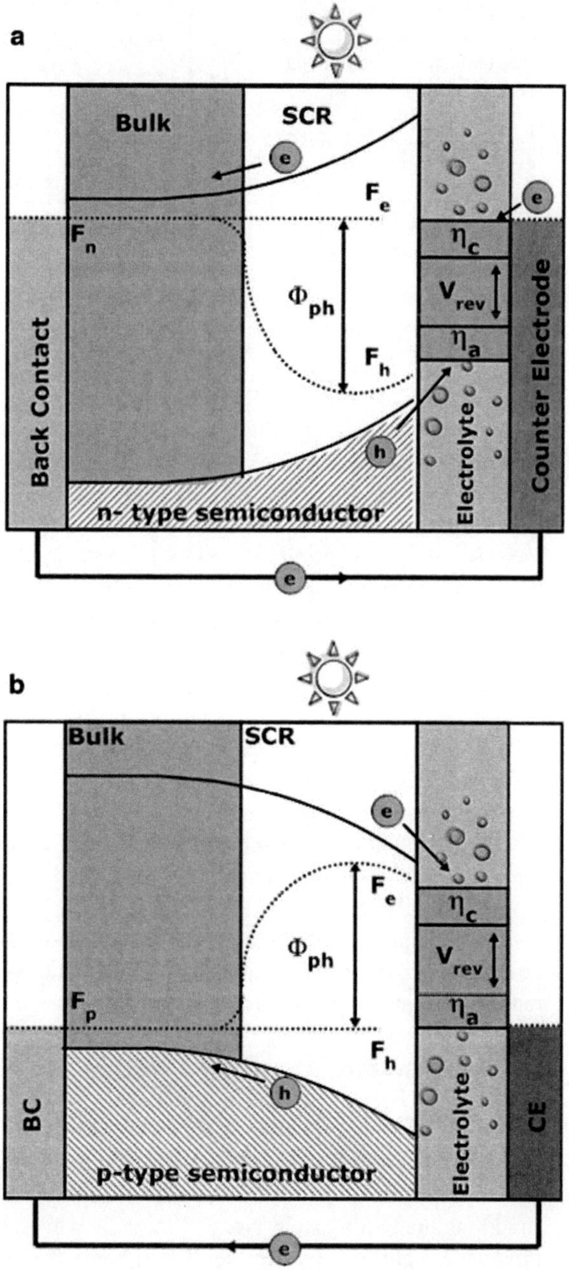

Fig. 7.11 Operational illuminated PEC junctions (**a**) n-type junction with correct hole energy alignment for oxygen evolution and electron energy alignment for hydrogen production, (**b**) p-type junction with correct electron energy alignment for hydrogen production and electron energy alignment for hydrogen evolution

7.4.3 Useable Photopotential

The concept of the quasi-Fermi level split is extremely important to how much *useable* photopotential a semiconductor device can generate, specifically in relation

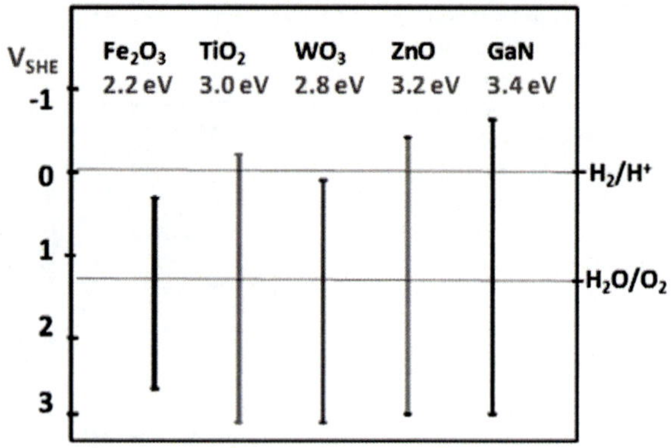

Fig. 7.12 Bandgap position of several common PEC semiconductors relative to the water-splitting redox potentials measured vs. the standard hydrogen electrode (SHE)

to the bandgap energy. In crystalline silicon photovoltaic cells, for example, the bandgap energy is 1.1 eV, while single-junction open-circuit voltages typically range between 0.6 and 0.7 V [26]. Thermodynamically, semiconductor band diagram representations reflect the *internal energy* of electrons and holes, not the *useable energy*. Electricity can be extracted from a PV cell only at potentials below the open-circuit voltage, which is typically 50–70% of the semiconductor bandgap energy. The output voltage limit can be increased using higher bandgap cells, but as a tradeoff, fewer solar-spectrum photons would be absorbed, limiting the saturated photocurrent.

The concept of useable potential is even more important in the case of PEC water-splitting devices. The useable potential of a PEC device must be sufficiently high to drive both half reactions in addition to the overpotential losses. This can typically require over 1.6–1.9 V. To achieve this in a single junction device, semiconductor bandgaps over 3.0 eV are typically necessary, which severely restricts optical absorption, saturated photocurrent, and therefore, STH conversion efficiency.

The bandgap position relative to the redox potentials of the electrolyte must also be considered. It is often reported that semiconductors with bandgaps "straddling" the redox potentials will be able to photosplit water in single junction configurations. While this is a "necessary" condition, it is by no means a "sufficient" condition. The band positions of semiconductor interfaces in aqueous solutions have been reported [5]. Figure 7.12, for example, shows the band edge positions with respect to the redox levels for hydrogen reduction and water oxidation for several commonly studied PEC semiconductor materials. Of these materials, titanium dioxide, zinc oxide, and gallium nitride all have bandgaps greater than 3 eV, and band edges straddling the redox levels. None of these, however, have demonstrated spontaneous PEC water splitting. This further drives home the point that the real energetics and useable photopotential under illumination are dictated by the quasi-Fermi levels, and not by the band edges. In addition, it is important to stress that all the overpotential losses directly subtract form useable potential. As a result, there are many situations where a bandgap straddling the redox potentials will fail to split water.

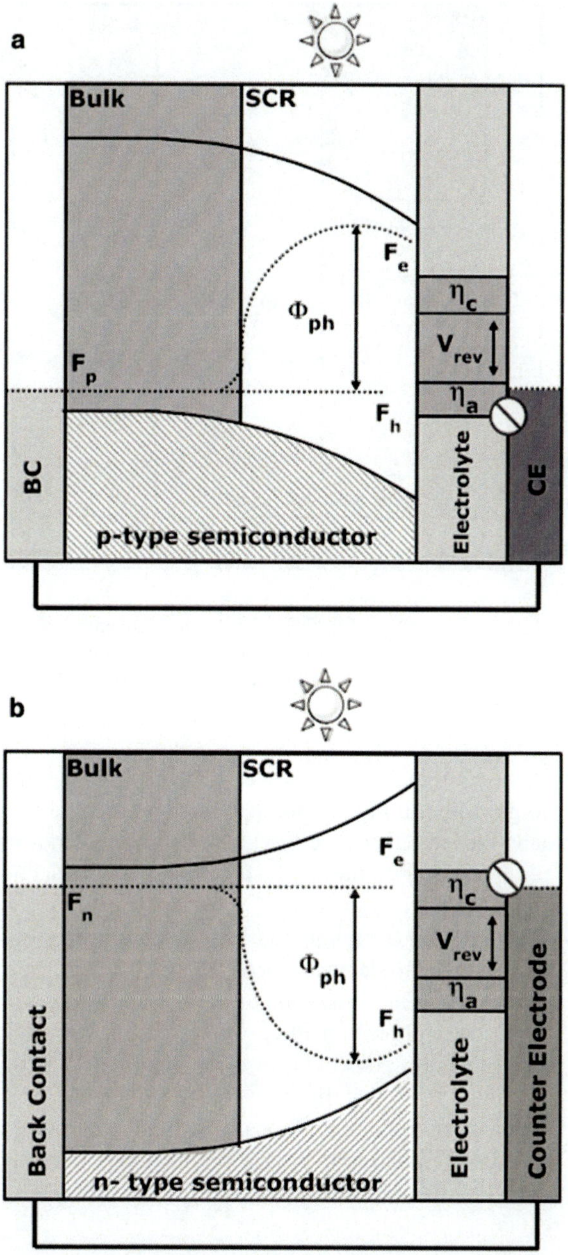

Fig. 7.13 Nonoperational illuminated PEC junctions (**a**) n-type junction with correct hole energy alignment for oxygen evolution but improper electron energy alignment for hydrogen production, (**b**) p-type junction with correct electron energy alignment for hydrogen production but improper electron energy alignment for hydrogen evolution

To illustrate this further, two additional hypothetical devices are illustrated in Fig. 7.13a, b representing photoanode and photocathode junctions, respectively, that are incapable of solar water splitting. In both cases, the conduction and valence band-edges clearly straddle the redox levels, even including the overpotential

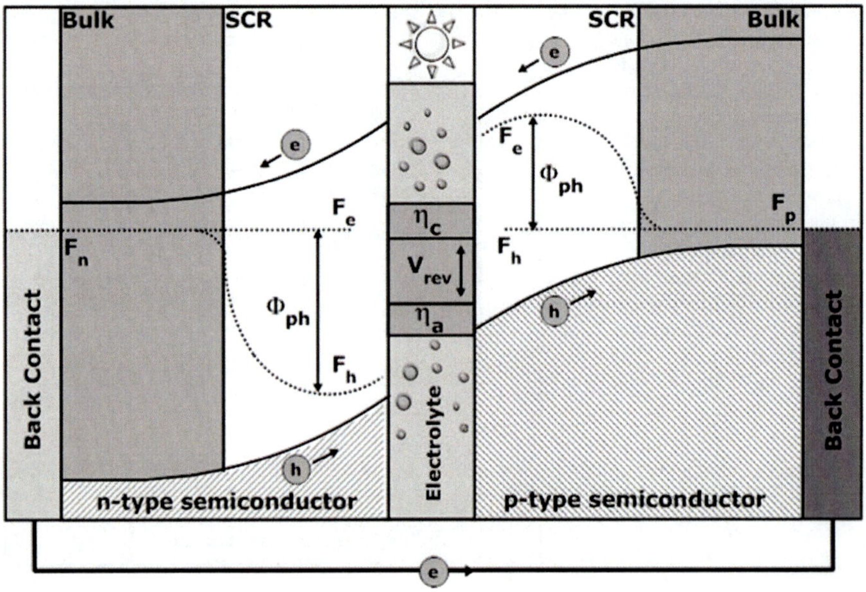

Fig. 7.14 Functional PEC–PEC tandem arrangement combining the nonoperational n-type and p-type electrodes from Fig. 7.13. There is Fermi level alignment at the interconnected back contacts, and the combined photopotentials developed in the two illuminated electrodes is sufficient to split water

losses. In Fig. 7.13a, however, the hole quasi-Fermi level at the PEC interface is too high, while in Fig. 7.13b, the electron potential in the counter electrode is too low and thus in either case PEC water-splitting is not sustainable.

Interestingly, these "dysfunctional" photoelectrodes could, in theory, be combined in tandem to achieve sufficient useable potential for water splitting. This is illustrated in the combined band-diagram representation seen in Fig. 7.14, where the two electrodes have been interconnected to each other, instead of to counter electrodes. This utilization of multiple junctions to enhance useable photopotential is a powerful tool for PEC water-splitting. As discussed in the following sections, multijunction stacking can offer important benefits, but can also introduce new system losses that limit overall STH conversion efficiency. In all single- and multijunction configurations, loss minimization in the device and system designs becomes critical to hydrogen production performance and efficiency.

7.4.4 Loss Minimization

There are numerous losses in PEC devices and systems that can limit STH conversion efficiency, and which need to be minimized. These include losses in solution, at interfaces, and in the semiconductor bulk. The optoelectronic losses in the semiconductor photoelectrode, for example, need to be as low as possible. In single-junction absorbers, photons with energies below the semiconductor bandgap cannot be absorbed or converted. Photons with energies exceeding the bandgap are absorbed, but at rates dependent on the allowed transitions in the semiconductor. Direct bandgap materials, for example, absorb more efficiently than indirect bandgap materials. In all semiconductors, photogenerated electron–hole pairs rapidly thermalize to band-edge energy levels within picoseconds, losing energy to heat. Large bandgap semiconductors generate little photocurrent due to poor absorption, while small bandgap semiconductors can suffer from low conversion efficiencies due to high thermalization losses. At the band-edge energy states, the electron–hole pairs can often survive for several microseconds before recombining, and they must be separated and transported to electrochemical interfaces for extraction during this time for effective energy conversion. This separation is assisted by the electric fields set up by charge distributions in the semiconductor and at the solid/liquid interface. Defects in the bulk and at the interface can adversely affect the separation fields, and also result in poor mobility for charge transport. If wide absorption widths are needed, for example, in indirect semiconductors (i.e., the absorber needs to be relatively "thick"), the charge transport losses can be severe. Thinner devices with back-surface optical reflection may reduce the loss.

To minimize overpotential loss due to interfacial charge extraction and electrochemical product formation, the interface conditions need to be optimized. Ideally, charge is extracted via the water-splitting half-reaction at the solid/liquid interface. The extraction process can be slowed or completely inhibited by poor energetic alignment or poor surface kinetics at the photoelectrode or counter electrode surfaces. Parasitic or corrosion reactions competing with the water-splitting reactions can also result in substantial loss; though surface treatments can be employed to kinetically and/or energetically favor water-splitting over the parasitic processes. Surface incorporation of nanoparticle catalysts is one approach, though light blockage due to such particles must be avoided. Since PEC water-splitting is a low-current density process, typically operating below 20 mA/cm^2, nonprecious metal catalysts can be used. Additionally, nanostructuring of electrode surfaces can increase effective surface area for enhanced charge extraction, although this can also lead to higher surface recombination loss.

The electrolyte is an important factor determining stability, efficiency of the charge-extracting reactions, and the electrochemical byproducts. Splitting seawater, for example, is a challenge since it is difficult to electrochemically suppress the production of chlorine gas from the Cl^- ions [71]. During PEC water-splitting, the evolved hydrogen or oxygen gas must be efficiently removed from the photoelectrode surface to avoid mass-transport losses in the surface reactions, and to minimize adverse optical effects. Surfactants added to the electrolyte have been successful in promoting rapid bubble formation and dissipation. In solution, ionic conductivity losses tend to be a larger problem. High electrolyte concentrations can be used to minimize this loss, but the tradeoff is in higher corrosivity.

Photoelectrode geometry and counter-electrode proximity are critical parameters to the redistribution of ions. In some geometries, gas separating membranes are needed, introducing further ionic transport loss.

Overall, the key condition for PEC hydrogen production is that the quasi-Fermi levels under solar illumination generate sufficient useable photopotential to drive the redox reactions for water splitting after all the solid-state, interfacial and solution overpotential losses have been taken into account. This challenge has severely limited efficiency in all single-junction PEC devices to date.

7.4.5 Single Junction Efficiency Limits

It is difficult to achieve high STH conversion efficiencies in single-junction PEC photoelectrode systems with inherent electrochemical and solid-state losses. Large bandgap semiconductor materials would be needed to generate sufficient useable photopotential. This, however, can severely limit photon absorption, therefore reducing photocurrent and STH conversion efficiency. The bandgap tradeoff between photopotential and photocurrent is particularly detrimental for single junctions. For example, if the minimum water-splitting potential, including the redox separation and overpotentials, amounts to 1.6 V, then the required quasi-Fermi level separation is also 1.6 eV. Even for high-quality semiconductor materials, the quasi-Fermi level separation can only achieve 50–70% of the bandgap level [72]. In this case, the minimum bandgap for the onset of photoelectrolysis would be 2.7–3.2 eV. For increasing levels of hydrogen production, the system losses will also increase, including the current-dependent surface overpotentials as well as electrical and ionic conductivity losses. Even higher bandgaps would then be required.

An upper bound for STH conversion efficiency can be established as a function of semiconductor bandgap based on fundamental optical absorption limits [50, 73]. Figure 7.15 plots the maximum attainable AM1.5G photocurrent densities in a semi conductor as a function of bandgap based on optimal light absorption and carrier extraction. The derivation assumes that every photon in the solar spectrum with an energy exceeding the bandgap will create an electron–hole pair, and that all of these electron–hole pairs are converted to photocurrent. For bandgaps greater than 3.2 eV, the photocurrent density is limited to approximately 1 mA/cm^2. This places an upper limit of 1.23% on the STH efficiency, as calculated using (7.4). The STH values corresponding to the maximum achievable photocurrents are listed in parentheses on the right vertical axis of Fig. 7.15, but these only apply to junctions capable of spontaneous water-splitting. Strictly speaking, any photoelectrode system incapable of sustaining photoelectrolysis will not produce photocurrent and thus have 0% STH efficiency.

Fundamental thermodynamic limits for the conversion efficiency of single- and multijunction semiconductor PEC devices have also been established in numerous studies [74–81]. The thermodynamic single-junction limit under ideal conditions, i.e., including thermalization losses but with no overpotential losses, has been

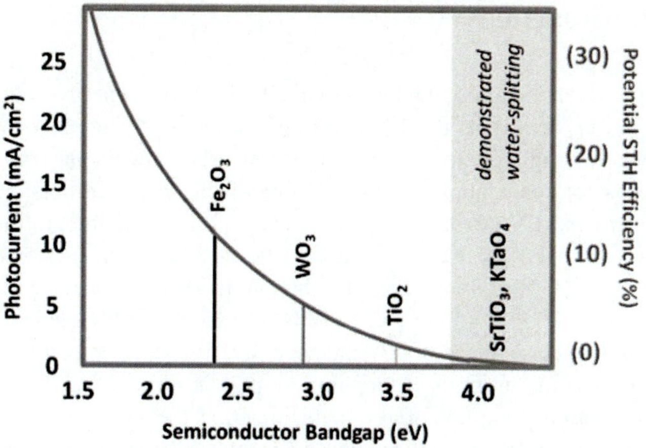

Fig. 7.15 Maximum achievable photocurrent density levels and potential solar-to-hydrogen conversion efficiencies for single-junctions as a function of bandgap based on optical absorption limits. Highlighted are the high-bandgap materials that have demonstrated spontaneous PEC water splitting but at correspondingly low efficiencies

calculated to be approximately 30% STH. With recovery of some thermalization losses through advanced multi-exciton device schemes, this limit is modestly increased to 32% STH [74]. More thorough analyses have been reported which include overpotential losses in addition to the optical and thermalization losses. These have indicated that the best single junction PEC cells for AM1.5G water-splitting will likely to be limited to STH efficiencies below 12%, even using the highest quality semiconductor and catalyst materials currently available [82].

7.4.6 Single Junction Examples

To date, the only demonstrations of single-junction water-splitting have utilized very high bandgap materials such as $SrTiO_3$ and $KTaO_4$ [83, 84]. Based on poor photon absorption, the demonstrated STH values have been well below 1%, consistent with the values in Fig. 7.15. Commonly studied semiconductor PEC materials have included iron oxide (Fe_2O_3), tungsten trioxide (WO_3), and titanium dioxide (TiO_2). The theoretically achievable photocurrent densities highlighted in Fig. 7.15 look encouraging, especially those for iron oxide. However, none of these materials can develop enough useable photopotential under sunlight to split water, not even the TiO_2 with a bandgap over 3.0 eV. To enhance the photopotential, these and other promising semiconductors can be incorporated in multijunction PEC schemes. This is consistent with the multijunction device enhancements that are well known in the PV community. For PEC devices, the same approach can be taken to enhance the photopotential and also to increase the absorption efficiency, as described in the following sections.

7.5 Multijunctions

Multiple-junction semiconductor devices for solar energy conversion are well-known, especially to the PV research community. The use of multijunction approaches for high photovoltaic conversion efficiency is the basis of current technologies for space applications, and one of the major cornerstones for future *Third Generation* PV devices [85]. In theory and practice, integration of multiple junctions in stacked solar cells enhances the absorption of photons in the solar spectrum [64, 85, 86]. In fact, multijunction III–V solar cells with efficiencies greater than 40% are commercially available [87, 88]. Efficiencies in PEC solar conversion devices can also be enhanced by improved spectral absorption in multijunction configurations. More importantly, multijunction PEC devices can develop increased photopotentials, better meeting the voltage demands of water splitting. The tradeoffs are reduced photocurrent, which adversely affects hydrogen production rates, and added device complexity. Overall, a PEC multijunction device design must strike the right balance to maximize STH conversion efficiency without adding too much cost. The multijunction approach, with its benefits, limitations, and tradeoffs are further discussed in the following sections.

7.5.1 Multijunction Stacks

The concept of stacking multiple photorectifying cells is illustrated in Fig. 7.16. Though the figure specifically depicts a triple cell stack, the illustration can be generally extended to represent any number of stacked cells. Each cell in the stack represents an independent photoconverting junction, such as a semiconductor *pn junction*, or a semiconductor/electrolyte PEC junction. Electrically and optically the cells are stacked together in a series-connected fashion. Incident sunlight on the "top" cell is partially absorbed, generating a photovoltage and photocurrent in the process. The remaining filtered light reaches the "middle" cell, where further absorption and photoconversion takes place. Any remaining sunlight can be absorbed and converted in the "bottom" cell. Since the cells are series-connected, the individual photovoltages will add, but the net current will be the minimum of the individual cell photocurrents. This is because the net current through the device must be continuous, and will be bottlenecked by the minimum component cell current. Any excess carrier generation in the other cells will be lost internally to recombination. Since the individual cells are all powered by some fraction of the incident solar photons, particularly the buried cells, they operate at reduced photocurrents compared with full-sun levels. As a result, overall device current is reduced by the bottleneck effect. Typically, then, a multijunction device based on a given semiconductor material system will generate lower photocurrent compared with a single-cell device of the same material. In general, multijunction devices offer higher voltage with lower current. As illustrated in later sections, over-all conversion efficiencies can be enhanced due to the extended range of photons effectively absorbed using multiple bandgaps in an integrated device.

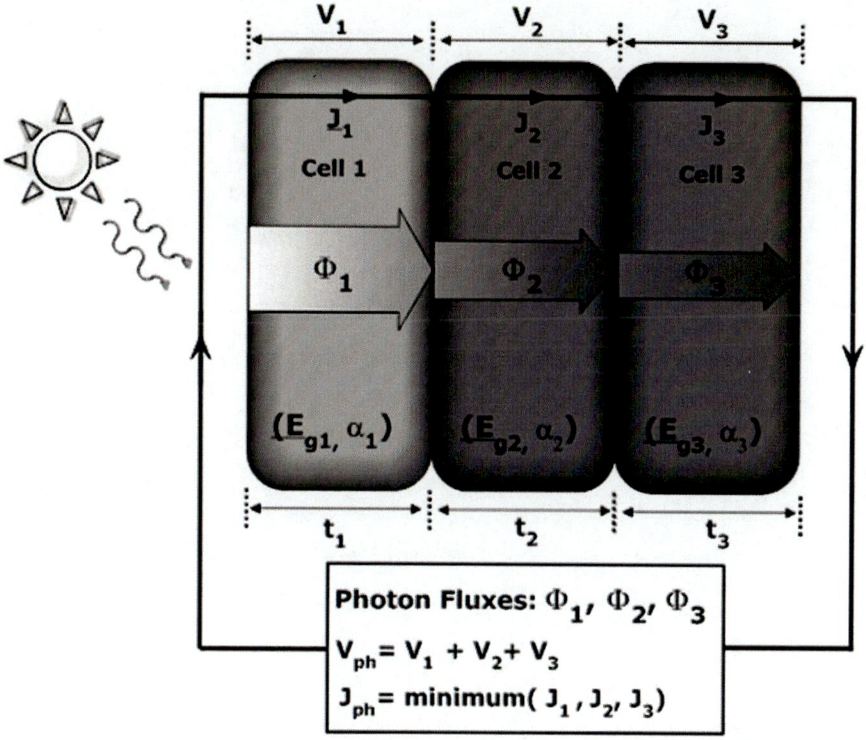

Fig. 7.16 General schematic of a three-cell stack of three absorber materials with bandgap E_g and absorption coefficient α. Indicated are the photon fluxes in each cell (Φ), along with the generated photovoltages (V) and photocurrent densities (J). The total voltage in the series-connected stack is the sum of the individual cell voltages, while the stack current is the minimum of the individual cell currents

7.5.2 Optical Considerations

Semiconductor photorectifying cells absorb photons in the solar spectrum, and can convert some of the absorbed photon energy into electricity or chemical products, such as hydrogen. Different semiconductors are specifically sensitized to different parts of the solar spectrum dependent on optoelectronic properties related to bandgap and absorptivity. In general, the macroscopic light absorption process in a material follows the Beer–Lambert Law [89], here expressed as a function of photon wavelength, λ:

$$I_\lambda(x) = I_{0,\lambda}\, e^{-\alpha_\lambda x}. \tag{7.8}$$

For each photon wavelength λ in (7.8), $I_{0,\lambda}$ is the intensity of the incident flux (e.g., in W/m^2), α is the absorption coefficient (in m^{-1}), and x is the length of the optical pathway through the material (in m). I_λ is the unabsorbed flux intensity emerging out the back of the material, as illustrated in Fig. 7.17.

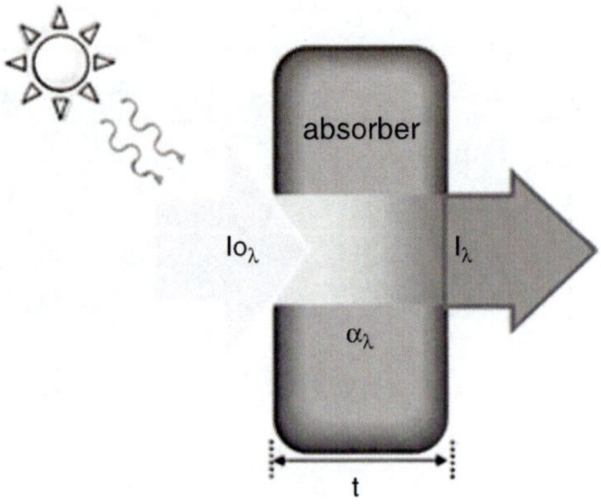

Fig. 7.17 Illustration of the Beer–Lambert Law for wavelength-dependent photon absorption in a material with absorption coefficient α

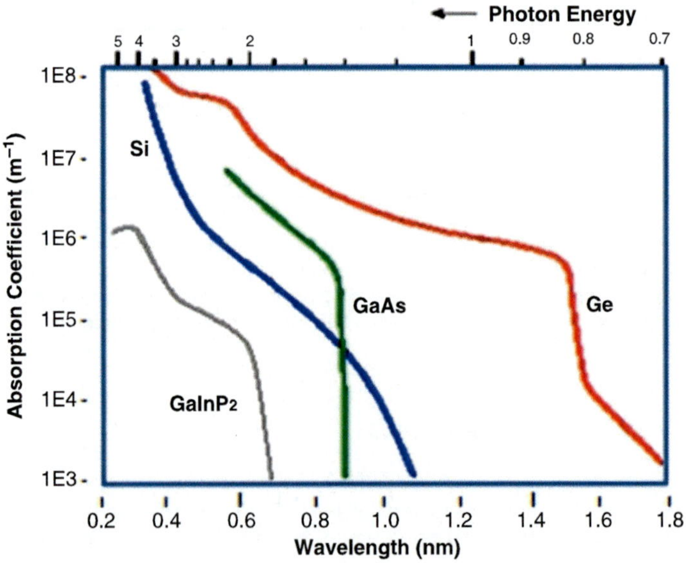

Fig. 7.18 Absorption coefficient spectra of $GaInP_2$GaAs, Ge, and Si

Important information about the spectral-sensitivity of a given semiconductor material is incorporated in the absorption coefficient α, which is strongly dependent on material bandgap and allowable energy transitions. Figure 7.18, for example, depicts typical absorption coefficient curves for four different crystalline semiconductors: (1) gallium indium phosphide ($GaInP_2$), a direct band transition semiconductor with bandgap 1.86 eV (corresponding to a photon wavelength of 667 nm); (2) gallium arsenide (GaAs), a direct band transition semiconductor with bandgap 1.43 eV (868 nm); (3) germanium (Ge), an indirect transition material with bandgap 0.66 eV (1,880 nm); and (4) silicon (Si), an indirect transition material with bandgap 1.12 eV (1,110 nm). For all four cases, the absorption coefficient vanishes at the gap energy, consistent with nonabsorption of photons with energies below the bandgap.

For photons with energies above the bandgap, the direct transition materials absorb more efficiently. This is reflected in the steeper slopes and correspondingly higher values for α in $GaInP_2$ and GaAs near their band energies, specifically compared with Ge and Si. Solar energy conversion devices based on direct-transition materials such as GaAs can be made as thin as a few micrometers and still absorb most of the available photons. In contrast, silicon solar cells need to be made much thicker, typically 100 μm or more to absorb most of the sunlight. A key point relevant to multijunction device design is that photon absorption and photocurrent generation in the individual component cells can be tailored by the selection of semiconductor bandgap and absorption properties. They can also be adjusted by varying cell thickness, as indicated in the Beer–Lambert law.

Another important parameter describing the optoelectronic response in a photoactive semiconductor device is the *quantum efficiency* (QE) (more precisely, the external quantum efficiency or EQE), which is critical to both single- and multijunction device designs. For each wavelength of light, the "external" QE expresses the percentage of photons incident on the device surface that results in an extracted electron–hole pair. This parameter is therefore an accurate measure of the device's ability to generate photocurrent, and convert solar energy. The QE ratio and the corresponding relationship to photocurrent generation in a given semiconductor device can be expressed as:

$$\mathrm{QE} = \frac{\#\mathrm{charges/s}}{\#\mathrm{photons/s}} = \frac{I/e}{P_{\mathrm{total}}(\lambda)/E_{\mathrm{photon}}(\lambda)}, \tag{7.9}$$

where e is the elementary charge, I is the current, $P_{\mathrm{total}}(\lambda)$ is the total power of incident light at a wavelength λ, and $E_{\mathrm{photon}}(\lambda)$ is the energy of a single photon at wavelength λ.

The quantum efficiency is closely related to the absorption coefficient, but incorporates additional information regarding the efficiency of extracting photo-charges created in the absorption process. Figure 7.19 shows representative QE spectra for $GaInP_2$, GaAs, and Ge devices, with sensitized regions of the AM1.5G included above. It is evident that the different semiconductors are sensitized to specific portions of the solar spectrum. Ge exhibits a broad bandwidth in its spectral sensitivity, but there is a relatively low level of solar photon flux across this bandwidth. In contrast, both $GaInP_2$ and GaAs have more narrow bandwidths, but each is efficient at converting photons in high-flux portions of sunlight. It is important to emphasize that QE is dependent not only on absorption and charge separation properties of device, but also on device operating conditions, such as voltage biasing, which alter

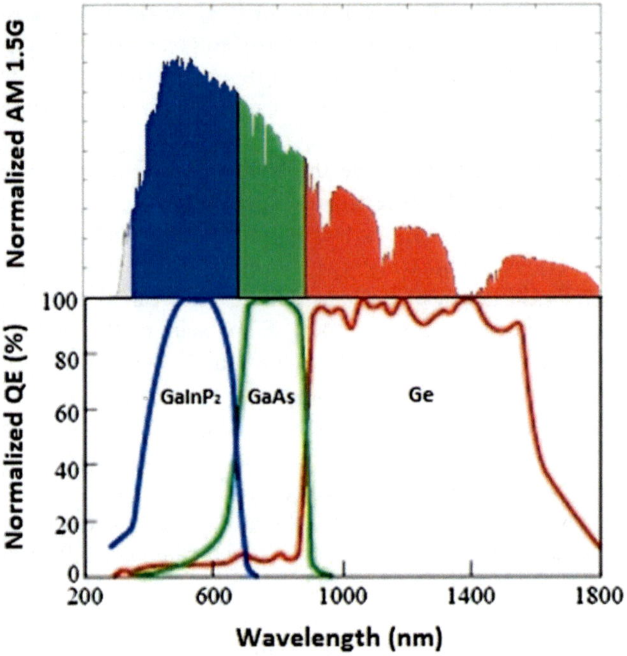

Fig. 7.19 Normalized QE curves of $GaInP_2$, GaAs, and Ge shown with the regions of spectral absorption for each material highlighted in the AM1.5G spectrum above

charge distributions and electric fields. These operating conditions need to be specified in reported QE curves to be useful in device design.

The solar energy conversion in a single bandgap semiconductor device is ultimately limited by the spectral width of the quantum efficiency curve. Also, since photogenerated electron–hole pairs rapidly thermalize to near band-edge energy states, energy conversion is most efficient for near-bandgap wavelengths. Absorbed photons with energies in excess of the bandgap will lose this excess to thermalization, reducing the conversion efficiency of light at the corresponding photon wavelength. Referring to the QE curves in Fig. 7.19, individual $GaInP_2$, GaAs, and Ge devices, for example, can convert photons in the 300–700, 700–900, 900–1,800 nm ranges, respectively, with optimal conversion near the 667, 868, and 1,800 nm bandgaps, respectively. By extension, a triple-junction stacked device using all three materials could absorb over a broader range of wavelengths from 300 to 1,800 nm, with optimal conversion efficiency at all the component cell bandgaps of 667, 868, and 1,800 nm. This illustrates an important avenue for enhanced efficiencies in multicell device schemes, relevant to both PV and PEC systems. The strategy has been studied extensively in the PV research and development community, and "full spectrum" absorber devices based on stacked multiple junctions are envisioned as a future pathway to ultrahigh solar-to-electric conversion efficiencies [90]. As discussed in following sections, photovoltaic efficiencies in triple-junction III-V devices have already exceeded 40%.

7.5.3 Current and Voltage Relationships

The external quantum efficiency spectrum indicates the efficiency of photogenerated charge extraction and current generation as a function of photon wavelength. The QE specified at the device operating condition coupled with the incident photon flux provide sufficient information for determining the photocurrent of the device. Specifically, the theoretical carrier extraction rate is determined by integrating the product of QE and flux over the light spectrum. The photocurrent density is then equal to the extraction rate scaled by the ratio of electron charge/photon energy. The photocurrent equation can be explicitly written as an integral over photon wavelengths in the incident irradiation:

$$J_{\mathrm{ph}} = \int_{\mathrm{irradiance}} \Phi_\lambda \left(\frac{e\lambda}{hc} \mathrm{QE}_\lambda \right) \mathrm{d}\lambda. \tag{7.10}$$

In (7.10), J_{ph} is the photocurrent density (e.g., in $\mathrm{A/m^2}$ or $\mathrm{mA/cm^2}$) at device operating conditions, h is Planck's constant, and c is the speed of light. Wavelength-dependent parameters include the incident irradiance flux, Φ_λ, and the quantum efficiency, QE_λ, which is specified at the operating conditions.

As an example of J_{ph} determination from QE and solar flux, Fig. 7.20 illustrates the QE response at short-circuit conditions for a typical monocrystalline silicon PV single-junction cell along with the corresponding responsive regions of the AM1.5G spectrum. Integrating QE according to (7.10) using AM1.5G as the input irradiance yields 35 $\mathrm{mA/cm^2}$, consistent with the short-circuit photocurrent levels of commercial single-junction silicon PV devices [64]. As seen in Fig. 7.20, silicon, with a bandgap of 1.12 eV, has a broad spectral response for absorbing sunlight. Efficient energy conversion with low thermalization loss in this single-junction device, however, is limited to absorbed photons with energies near the bandgap energy (wavelengths near 1,100 nm). As a result, compared with the PV efficiencies exceeding 40% in multijunction III–V devices, efficiencies are limited to about 20% using single-junction silicon.

The useable photovoltage (equivalently, the photopotential) generated in a semiconductor device under illumination can be represented by the split in the quasi-Fermi levels as described in Sect. 7.4. This voltage is a function of semiconductor bandgap, but it is always less than the internal energy separation between conduction and valence bands represented by the bandgap. In PV cells, the maximum useable potential, expressed by the open-circuit voltage, is typically 50–70% of the bandgap energy [64, 91]. For example, silicon has a bandgap of 1.1 eV, and

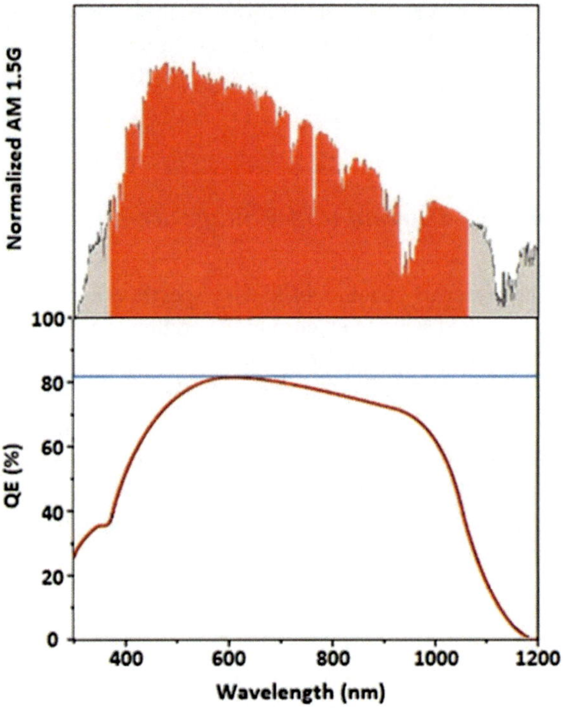

Fig. 7.20 QE curve of a typical monocrystalline silicon solar cell the region of spectral absorption for the device highlighted in the AM1.5G spectrum above

typical open circuit voltages of single junction crystalline silicon cells are around 0.6 V. Similarly, high-quality GaAs has a bandgap of 1.4 eV, and device open circuit voltages of approximately 1.0 V. As the photoconversion device is asked to deliver current to an external load, the operating voltage will start to decrease below the maximum potential. The photovoltage and photocurrent levels are integrally related. Specifically, the maximum photovoltage is logarithmically related to the current-generating capability of the device, as derived from junction analysis [64, 92]. For photovoltaic cells, this is expressed as a relationship between the open-circuit voltage, V_{oc}, and the short-circuit current-density, J_{sc}:

$$V_{oc} = V_T \ln\left(\frac{J_{sc}}{J_0} + 1\right), \tag{7.11}$$

where V_T is the temperature-dependent thermal voltage of the device, and J_0 is the device's saturation current density.

The current and voltage relationships coupled with the optical parameters described in the previous section for single material systems provide powerful insights into the voltage, current, and conversion efficiency properties of single junctions, and also of stacked semiconductor device configurations. Another important performance parameter representing the quality of photorectifying junctions is the "fill factor," or FF. The FF expresses how severely voltage will drop below the maximum useable potential under current loading. This is illustrated in Fig. 7.21, showing photoconversion device performance curves with "good" and "poor" fill

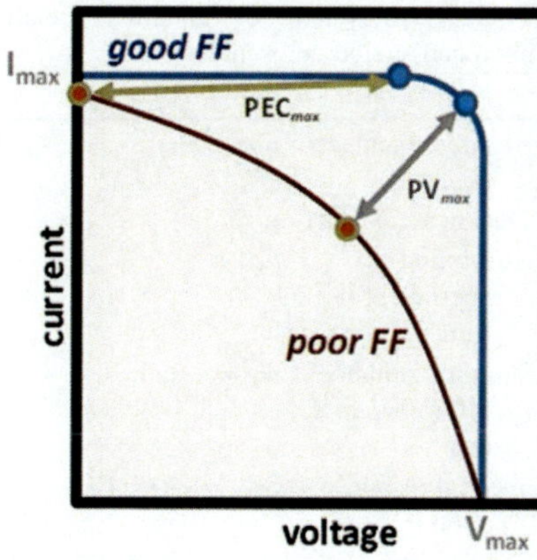

Fig. 7.21 Current/voltage response curves of photoconversion devices with good and poor fill factors (FF), showing maximum voltage, current, and power points for each curve

factors. In PV cells, the maximum power point, found at the "knee" of the performance curve, is greatest for high fill factors. For PEC hydrogen production cells, it is necessary to maintain a high photopotential to drive the water splitting reactions while maintaining high photocurrent to maximize production rates. In other words, it is desirable to operate a PEC cell at maximum possible current with minimum voltage drop below the device's maximum useable potential. A good fill factor is also critical in this situation. A high solar conversion efficiency in PV and PEC devices requires material systems and devices with maximum-photovoltage, maximum-photocurrent and fill factor all optimized.

The current and voltage relationships for individual single-junction cells can be combined to model of multijunction devices. The generic triple-junction scheme shown in Fig. 7.16, for example, can be modeled by the set of equations summarized in Table 7.1. These equations specifically highlight the dependencies of device photovoltage and photocurrent on the individual cell absorption coefficients and quantum efficiencies. This multijunction equation set is explicit to a three-cell device, but readily generalized to devices with any number of stacked cells. Additionally, the relationships expressed are applicable whether the individual cell junctions are photovoltaic, PEC, or a combination. If the device stack is designed to generate electricity, the PV conversion efficiency is determined at the "maximum power point," as illustrated in the response curves of Fig. 7.21. Conversely, if the stack is designed to utilize solar energy for purposes of splitting water, the PEC conversion efficiency depends on the maximum achievable electrochemical photocurrent. In either case, optimizing voltage, and even more importantly, maximizing current is desirable for achieving high efficiency. Some key take-away points from the relationships in Table 7.1 relating to multijunction device design and optimization include the following:

- *Photocurrents are decreased in multijunctions*
  - Photocurrent in a component cell is dependent on the physical and optoelectronic properties of that cell, such as thickness, bandgap, and QE, but also on the available photon flux.

Table 7.1 System of equations relating voltage, current, and spectral responses in photoconversion cells and stacks

	Cell 1	Cell 2	Cell 3
Absorber bandgap (E_g) (eV)	E_{g1}	E_{g2}	E_{g3}
Thickness (t) (μm)	t_1	t_2	t_3
Absorption coefficient (α) (μm^{-1})	$\alpha_1(\lambda)$	$\alpha_2(\lambda)$	$\alpha_3(\lambda)$
Photon irradiance (Φ) (mW/cm^2/nm)	$\Phi_1 = 1$ sun	$\Phi_2 = \Phi_1\,\mathrm{e}^{-\alpha_1 t_1}$	$\Phi_3 = \Phi_2\,\mathrm{e}^{-\alpha_2 t_2}$
Spectral response (CR) (C/J)	$\mathrm{SR}_1 = \frac{q}{E_g(\lambda)}\mathrm{QE}_1$	$\mathrm{SR}_2 = \frac{q}{E_g(\lambda)}\mathrm{QE}_2$	$\mathrm{SR}_3 = \frac{q}{E_g(\lambda)}\mathrm{QE}_3$
Short-circuit current (J_{ph}) (mA/cm^2)	$J_{ph1} = \int_{\lambda g1}^{\infty} \Phi_1 \mathrm{SR}_1\,\mathrm{d}\lambda$	$J_{ph2} = \int_{\lambda g2}^{\lambda g1} \Phi_2 \mathrm{SR}_2\,\mathrm{d}\lambda$	$J_{ph3} = \int_{\lambda g3}^{\lambda g2} \Phi_3 \mathrm{SR}_3\,\mathrm{d}\lambda$
Open-circuit voltage (V_{oc}) (mV)	$V_{oc1} = V_T \ln\left(\frac{J_{sc1}}{J_{01}} + 1\right)$	$V_{oc2} = V_T \ln\left(\frac{J_{sc2}}{J_{02}} + 1\right)$	$V_{oc3} = V_T \ln\left(\frac{J_{sc3}}{J_{03}} + 1\right)$
Total device V_{oc}	$V_{oc} = V_{oc1} + V_{oc2} + V_{oc3}$		
Total device J_{ph}	Minimum (J_{ph1}, J_{ph2}, J_{ph3})		

  - Photon flux available to a component cell is filtered, therefore reduced, by absorption in preceding cells in the stack. This optically couples photocurrent generation in all the cells.
  - The output current of the stack is limited to the smallest of the photocurrents produced by any of the constituent cells, and optical coupling effects between cells become important in device design.

- *Photovoltages are enhanced in multijunctions*
  - Photovoltages of the components cells are additive in the stacked device, providing higher useable photopotential.
  - Photovoltage in a component cell is logarithmically dependent on the photocurrent, so current reduction due to filtering will decrease voltage, but not significantly.

- *Optics are important*
  - Available flux in a component cell is determined by light absorption in previous cells, but light reflection is also important. Minimizing reflection at the front surface and at each interface while maximizing back-surface reflections to redirect light back through the device can significantly enhance efficiency.

- *Fill factor is important*
  - Although fill factor is not represented explicitly in the equation set, optimal fill factor is important in each cell to ensure maximum output photocurrents with minimal loss in photovoltage.

Using these general points as guidelines, it is possible to develop practical designs for highly efficient multijunction devices. There are still a number of general design challenges for practical implementations that need to be considered.

7.5.4 General Design Challenges

In effective designs, maximum current in a multijunction photoconversion device is achieved by "current-matching" the individual cells in the stack [93]. For current-matching, the component cell materials and thicknesses can be tailored to ensure all component cell photocurrents are approximately the same. The goal is to minimize internal current losses. Since current must be continuous across the entire device, the lowest component cell photocurrent will bottleneck the output. If current excess is generated in any of the other component cells, it will be lost through internal recombination. If all junction currents are the same, this recombination loss is minimized, and net current is maximized. As an alternative viewpoint of current matching, the cells in the stack are essentially "splitting" all of the incident solar energy photons that can be used for photocurrent generation. Again, since the cell with the lowest current dictates the device current, an equal split will be optimal. Since photocurrent generation is sensitive to the incident flux of photons as well as physical and optoelectronic properties of all the cells, the process of designing an optimal current-matched device is complex. A general approach involves stacking semiconductors with decreasing bandgaps. Top layers absorb higher energy photons while transmitting lower energy photons that are subsequently absorbed by following layers of the device. After a component cell's bandgap and quantum efficiency have been established, cell thickness can be adjusted to alter the number of absorbed photons, and tune the photocurrent.

An important challenge in the design of multijunction devices is the limited selection of availability of semiconductor materials with appropriate bandgaps that offer high efficiency through low defect densities. Examples of PV material systems that have been successful or have shown promise for multijunction configurations include the III–V crystalline materials, amorphous hydrogenated silicon films, and copper chalcopyrite thin-films [86, 94, 95]. In each case, bandgap tuning is available through modification of alloy composition. A broader class of low-cost thin-film PEC material systems for multijunction water-splitting applications is under investigation, including novel metal oxides and oxi-nitride compounds covering a broad range of bandgaps [96–99]. In the PEC junction case, there are additional material challenges associated with the potential and band-edge requirements for water splitting, as well as stability in solution.

In addition to the material selection challenge, interface design for low optical and defect recombination losses in stacked cells is complicated, and highly dependent on the materials being interfaced. In crystalline semiconductor multijunction devices, lattice-matching between adjacent layers is critical for avoiding defects and significant interfacial recombination losses [100, 101]. In polycrystalline and amorphous thin film material systems, stress and strain at the various film interfaces must be minimized to avoid defects and even possible delamination of the device. In semiconductor material systems where heavy doping can be readily achieved, thin "tunnel junctions" are the preferred method of forming low-loss connection between adjacent cells [101]. Precise process control is essential in successful tunnel junction implementations. In other material systems where tunnel junction formation is not possible, the use of conductive transparent oxides can be used to interface adjacent cells, but optical and contact resistance losses can be introduced. To retain the efficiency advantages of multijunctions, all interface losses need to be minimized.

Broader challenges exist in the processing and manufacture of monolithic multijunction designs. In crystalline, polycrystalline, and amorphous semiconductor processing, precisely controlled process conditions are essential to the formation of device quality materials; and these conditions are highly material specific. Developing a process sequence to successfully stack numerous semiconductor layers can be extremely difficult. During the process, fabrication conditions at later stages, particularly high-temperature processes, can adversely affect the quality of all preceding layers. Low temperatures are advantageous for process compatibility and for lower manufacture cost, but it is often difficult to obtain the necessary material quality. The trade-offs boil down to enhanced performance vs. added complexity and manufacture cost. In PEC water-splitting applications, the increased photopotentials developed in multijunction devices is an added benefit to the mix. In the PV industry, the multijunction approach has been a clear winner in applications when the highest conversion efficiencies are absolutely required, despite higher cost.

7.5.5 Photovoltaic Multijunctions

The multijunction approach for achieving enhanced conversion efficiencies in PV devices have been studied for several decades. Since the early 1980s, solar-to-electric conversion efficiencies in production scale multijunction PV cells have been increased from just over 15% to over 40% today [86, 102]. The most efficient present-day multijunction photovoltaic cells are based on high-quality, and high cost, crystalline III–V material systems. For example, commercial cells have been on the market for several years, primarily for space applications, utilizing the $GaInP_2$, GaAs, and Ge materials discussed in Sect. 7.5.2. These specific materials were selected based on their collective ability to absorb photons over a broad range of the solar spectrum, as was seen in Fig. 7.19. $GaInP_2$, with a bandgap energy of 1.85 eV, absorbs short wavelength ultraviolet and visible photons in the solar spectrum; GaAs, with a bandgap of 1.42 eV, absorbs near-infrared light; and Ge, with a bandgap of 0.67 eV absorbs the lower energy photon energies in the infrared. In the device fabrication process, layers of crystalline GaInP, GaAs, and Ge layers are monolithically grown on a Ge substrate using epitaxial methods [103]. The completed triple-junction device scheme, containing over 20 layers, is illustrated in Fig. 7.22.

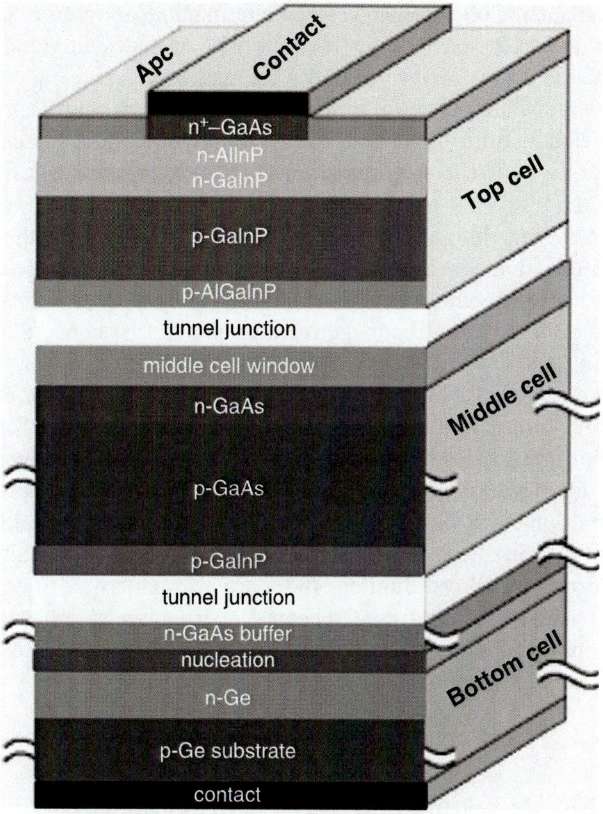

Fig. 7.22 Device structure of a high efficiency $GaInP_2$/GaAs/Ge triple-junction solar cell developed at NREL and commercialized by Spectrolab

Several critical design features were necessary to enable high performance in these devices. For example, current matching among all three component cells was essential for maximizing photocurrent output. To avoid losses due to interface defects, the semiconductor layers in this configuration had to be lattice matched. Additionally, low-loss electrical contact between adjacent cells was achieved using optimized tunnel junctions, comprised of a ministack of highly doped layers producing an effective potential barrier for both minority carriers. Furthermore, a broadband dual-layer dielectric stack antireflection coating was employed. The resulting multijunction device achieves PV efficiencies over 34% under concentrated sunlight [104]. More recent variations on the device scheme utilizing a $Ga_{0.44}In_{0.56}P/Ga_{0.92}In_{0.08}As$/Ge stack have achieved over 40% PV efficiency under concentrated sunlight with open-circuit voltages over 3.0 V and operating voltages near 2.75 V [87]. These efficiencies represent world records in the field of photovoltaic solar energy conversion, but this comes at a price. The crystalline III–V material systems are prohibitively expensive, especially for any large-scale deployment.

As a lower cost alternative, multijunction devices incorporating thin-film semiconductors have also been developed. As an example, triple junction thin film solar cells based on amorphous silicon/germanium alloys have been developed with photovoltaic efficiencies over 10% and with open-circuit voltages up to 2.3 V [54]. The fundamental design strategies for optimizing efficiency in these lower cost devices are similar to those in the III–V designs, including current matching, tunnel junction optimization, and minimization of optical losses. In a typical device, the thin-film silicon–germanium alloy bandgaps are adjusted to approximately 1.8, 1.55, and 1.4 eV in the top, middle, and bottom cells, respectively, to achieve current matching and broad spectral response. Also in the design, front-surface antireflection coatings are incorporated, while back surface reflector layers are used to reflect photons back into the device. Compared with single-junction devices using amorphous silicon–germanium material systems, the multijunction approach fully utilizes more of the solar spectrum. Higher conversion efficiencies are therefore achieved, but again at the expense of added device complexity and cost. Rigorous efforts in loss minimization are critical to ensure that the enhanced performance in the PV multijunctions offer worthwhile advantages over simpler single-junction devices. Efforts to achieve even higher efficiencies in economical thin-film multijunction technologies are on-going; not only in amorphous silicon devices, but also in alternative, enhanced absorbers based on bandgap-adjustable copper chalcopyrite alloys. Similar efforts are also underway to develop efficient multijunction water-splitting PEC schemes incorporating a broader class of low-cost semiconductor materials, such as metallic oxide alloys. Additional design considerations apply in the PEC case.

7.5.6 PEC Multijunction Design Considerations

Multijunction device designs can combine PV cells to improve solar-to-electricity conversion efficiency, or they can incorporate a combination of PV and PEC cells to enhance solar hydrogen production via water splitting. In both cases, the stacked junctions can be beneficial, providing photovoltage enhancement, as well as the broadened solar absorption width. The primary tradeoff is added device complexity and cost. In the PEC case, an additional tradeoff is reduced photocurrent, which limits hydrogen production rates. To make multijunctions worthwhile, the PEC device design must strike the right balance to maximize conversion efficiency and minimize expense. The general design considerations discussed in preceding sections still apply, including current matching and interface loss minimization. There are, moreover, additional considerations specifically applicable to a PEC component cell, which could be incorporated into a multijunction stack as photoanode (OER electrode) or a photocathode (HER electrode). These considerations include the following:

1. Low loss semiconductor PEC electrode materials with appropriate bandgap, absorption properties, and device quantum efficiency are needed for efficient photon absorption and charge extraction.
2. Sufficient useable potential must be developed (i.e., overall split in quasi-Fermi levels for holes and electrons) across the stack under illuminated operating

conditions to overcome overpotential losses and drive the OER and HER reactions.

3. Suitable band-edge position at the semiconductor/electrolyte interface need to be established to facilitate the proper alignment of the quasi-Fermi levels under illumination.
4. High surface activities for the OER/HER reactions need to be established to minimize the interface overpotential losses.
5. High surface passivation, based on energetics and/or kinetics, needs to be established to mitigate corrosion and all other parasitic reactions.
6. There needs to be enough flexibility in selection of semiconductor bandgap, absorption properties and thickness to allow for current-matching in the stack design.
7. Losses associated with electrolyte conductivity and oxygen/hydrogen product gas management (related to relative electrode placement, ionic pathways between electrodes, possible membrane separators, and other related design features) need to be minimized.

Design optimizations, for all the component cells and for the integrated stack, are essential for realizing enhanced efficiency in multijunction devices. Decades of research and development in PV multijunctions has resulted in solar-to-electric conversion efficiencies exceeding 40% in triple junctions, approaching theoretical limits. In contrast, PEC multijunction research is at a far less mature level. Though STH conversion efficiencies exceeding 10% have been demonstrated in best PEC tandem devices based on high-quality III–V crystalline semiconductor materials, these "First Generation" devices lack long-term stability and affordability. More stable and affordable "Second Generation" thin-film PEC multijunction devices have been demonstrated, but only in the 3–5% STH range of performance levels. The following section examines the theoretical limits for any "First-," "Second-," or even "Third-" Generation multijunction systems, with specific emphasis on attainable STH efficiency.

7.5.7 *Multijunction Efficiency Limits*

As with the single-junction PV or PEC devices, efficiency bounds can be placed on multijunction configurations based on optical absorption limits. Figure 7.23 is a two-dimensional extension of Fig. 7.15 for tandem devices, where maximum photocurrent, and the corresponding STH levels for PEC devices, are calculated as a function of both top- and bottom-junction bandgaps. The assumptions included in the derivation of this graph are as follows:

1. The top cell absorbs all photons with energies exceeding the top-cell bandgap.
2. The bottom cell is illuminated with the top-cell-filtered light, and absorbs all of the remaining photons with energies exceeding its bandgap.

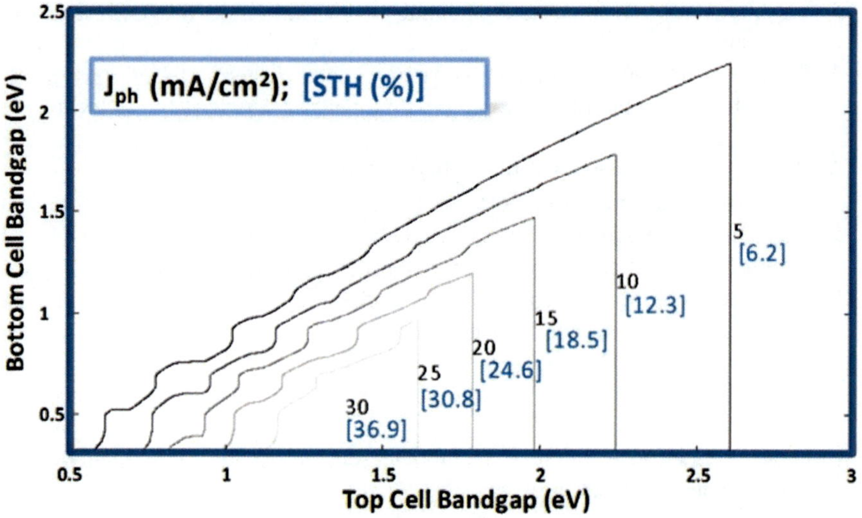

Fig. 7.23 Maximum achievable photocurrent densities in tandem semiconductor devices as a function of top and bottom cell bandgaps based on theoretical optical absorption limits. Included in brackets are the corresponding PEC STH efficiencies achievable in devices capable of water splitting

3. For every photon absorbed, an electron–hole pair is generated and extracted as photocurrent.
4. The total photocurrent of the stack is then simply the smaller of the two photocurrents produced by each individual cell.
5. For the case of PEC cells, the STH efficiencies, indicated in brackets in the plot, are calculated using (7.7), These ONLY apply to systems with sufficient photopotentials to split water.

The results from Fig. 7.23 are definitely encouraging for the viability of tandem photovoltaic electricity production as well as PEC hydrogen production. Despite the reduced photocurrents inherent in the tandem arrangement, a wide range of bandgap combinations yield high enough photocurrents for STH conversion efficiencies over 10%. An important key is to identify the possible bandgap combinations capable of the necessary photopotential levels for water splitting. The useable potential of a single cell ranges from 50 to 70% of the semiconductor bandgap, with 70% achievable only in the highest quality materials. In addition, the tandem device photopotential needs to exceed the water-splitting potential in addition to overpotentials, typically requiring around 1.6 V. The combined bandgap of the two component semiconductor materials would then need to be somewhere between 2.3 and 3.2 eV, with the high end of the range for most materials, and the low end only for the highest quality materials. There are many pitfalls and extra losses in real-world multijunction implementations, so the general rule is no guarantee of successful water splitting.

It is important to emphasize that even if the tandem device develops sufficient potential to split water, the STH efficiency will always be restricted based on the lowest photocurrent in the stack, usually in the highest bandgap cell. As an illustration, TiO_2 with bandgaps of 3.1 eV could be combined with a low bandgap semiconductor such as silicon (1.1 eV) in a hybrid tandem to meet the photopotential requirement for PEC water splitting. As seen in Fig. 7.17, however, independent of the bottom junction, the device performance will never exceed approximately 2.5% STH. In this example, and in any series-connected tandem configuration, the performance is bound by the photocurrent limits of the highest bandgap PEC semiconductor. As an alternative hypothetical example for achieving STH efficiencies above 10%, a high-quality top cell semiconductor with a bandgap of 2.0 eV can be coupled with a 1.0 eV semiconductor for the bottom cell. This combination satisfies the combined bandgap requirement, and is within the 12–19% STH range, in fact, approaching 19% as seen in Fig. 7.23.

As discussed in Sect. 7.3.1, fundamental thermodynamics limits conversion efficiency for both single- and multijunction semiconductor PEC devices have also been established in numerous studies. The thermodynamic tandem limit under ideal conditions, i.e., including thermalization loss, but with no overpotential loss, has been calculated at approximately 40% STH. This is consistent with the optical limits shown in Fig. 7.23. It is also substantially higher than the 30% STH limit established for the single-junction case. Moreover, employing advanced multi-exciton device schemes which eliminate part of the thermalization loss, the theoretical tandem limit is further increased to 46% STH [58].

As also discussed in Sect. 7.3.1, more detailed analyses of multijunction PEC devices including optical, thermalization, and overpotential losses have been reported. Figure 7.24 shows the maximum achievable photocurrent density and possible associated STH conversion efficiency results from the Rocheleau and Miller analysis [82] for tandem devices of high-quality III–V semiconductors. This analysis, which focuses on devices operating in the high-efficiency region seen in the optical limits plot of Fig. 7.23, includes estimates for solid-state and electrochemical losses. As expected, the STH efficiency limits are somewhat lower than the optical limits, though they still do exceed 25% over a limited range of top and bottom cell bandgaps. The high-quality factor of the modeled semiconductors clearly plays a large role in maintaining high levels of performance. For lower quality thin-film materials, larger efficiency hits would be expected. Nevertheless, there should be adequate margins for achieving over 10% in thin film tandem devices under some circumstances [82].

As an illustration of this, the analysis of efficiency limits for triple junctions based on lower quality amorphous silicon materials were also reported by Rocheleau and Miller. In that analysis, a bottom cell, with a bandgap of 1.45 eV was assumed, and the middle and top cells were modeled based on the available range of bandgaps in thin-film silicon materials. The resulting current density and STH efficiency limits are shown in Fig. 7.25. Interestingly, even with the lower quality thin film materials, and despite the further photocurrent limits in the triple junction case, STH efficiencies in the 10–13% STH range appear readily attainable.

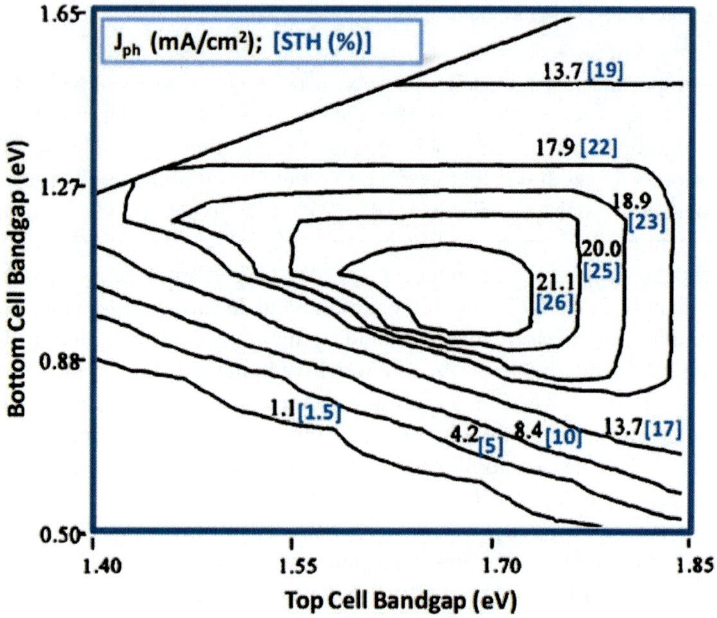

Fig. 7.24 Maximum achievable photocurrent densities and PEC STH efficiencies in tandem III-V semiconductor stacks as a function of top and bottom cell bandgaps based on analyses of Rocheleau and Miller [82] which included solid-state and electrochemical losses

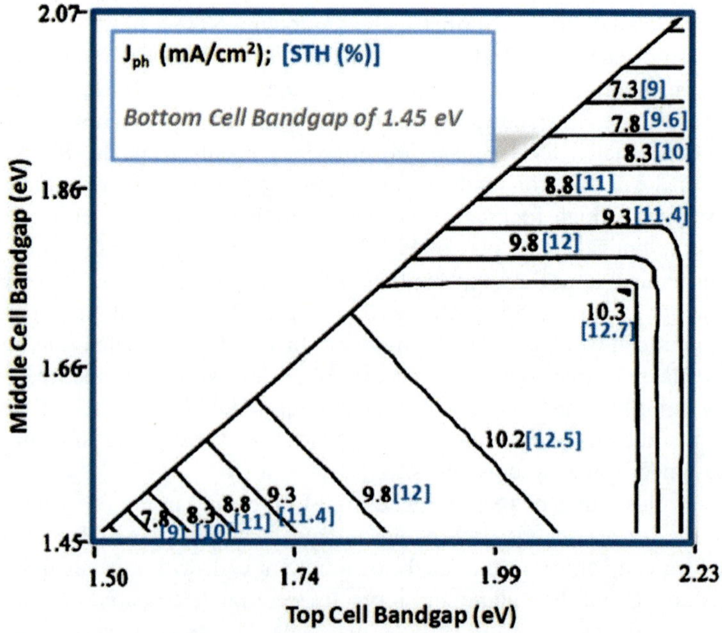

Fig. 7.25 Maximum achievable photocurrent densities and PEC STH efficiencies in triple junction amorphous silicon-based semiconductor devices as a function of top and middle cell bandgaps with a bottom cell gap of 1.45 eV. These results are based on the analyses of Rocheleau and Miller [82] which included solid-state and electrochemical losses

Unfortunately, achieving high STH efficiency levels in laboratory prototypes of thin-film multijunction devices has been elusive to date. Challenges in identifying the most appropriate semiconductor materials to incorporate into a stack, and in developing sufficient solid–solid and solid–liquid interfaces have limited progress. Still, some encouraging research pathways using different device scheme approaches are ongoing.

7.6 PEC Multijunction Device Schemes

A number of different multijunction device schemes incorporating a combination of PEC cells, or a combination of PEC and PV cells have been explored in attempts to enhance solar hydrogen production via water splitting [9, 39, 105]. Some of the device approaches which have demonstrated varying levels of success to date include the following:

1. PV-electrolysis devices with multijunction solid-state PV cells coupled to eletrolyzer systems.
2. Photoelectrode-based PEC–PEC tandem devices incorporating a coupled photoanode and photocathode.
3. Hybrid photoelectrode PV–PEC tandem devices comprising a PEC photoanode or photocathode integrated with a "buried" single-junction PV cell.
4. Hybrid photoelectrode PV–PV–PEC triple-junction devices comprising a PEC photoanode or photocathode integrated with a "buried" double junction PV cell.
5. Photocatalyst Z-scheme systems incorporating hydrogen-evolution and oxygen-evolution photocatalysts in single or dual bed configurations.

In the discussions of these approaches below, schematics of representative device configurations are included, accompanied by the basic device band structures to indicate the energetic driving forces behind the photolytic water splitting. Some examples of implementation in laboratory-scale prototypes are also presented.

7.6.1 Multijunction PV-Electrolysis

Establishing an interesting reference case for multijunction solar water splitting, multijunction PV cells with adequate useable photopotential to directly drive electrolysis have been coupled with electrolyzer systems, which can be separate or fully integrated. This approach is compatible with the "PV-electrolysis" reactor types described in Sect. 7.2.4. However, since the PV output of the multijunction cell is designed to have direct compatibility with electrolysis, no power conditioning units are required. A device level schematic is shown in Fig. 7.26 for the case of a triple-junction PV cell driving the electrolyzer reactions. The useable potential

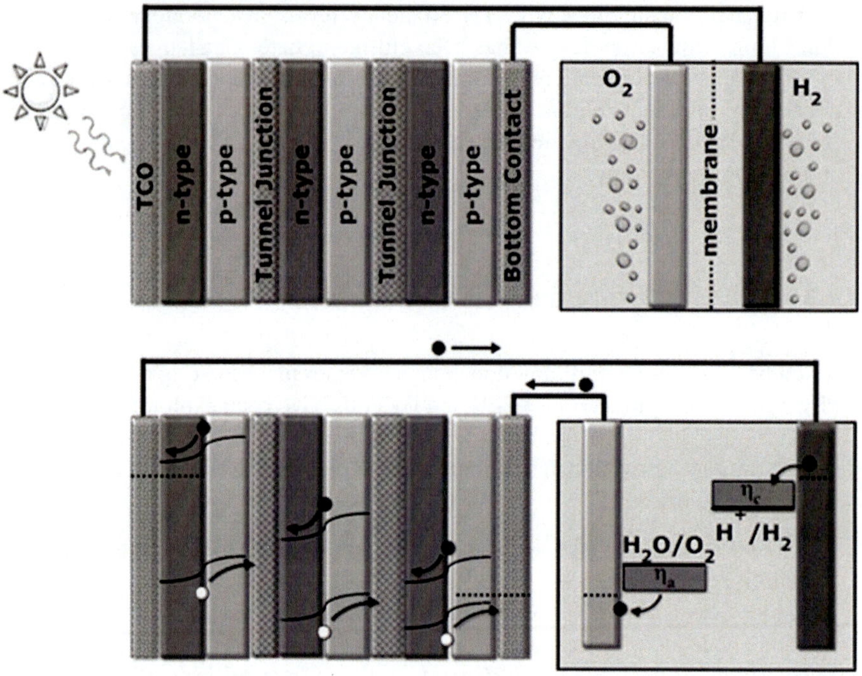

Fig. 7.26 Multijunction PV cell coupled to an electrolyzer system with band diagram showing photogenerated electrons (*black circles*) and holes (*white circles*)

developed by the triple-junction is capable of simultaneously driving photoholes at the electrolyzer anode surface and photoelectrons at the electrolyzer cathode surface. A unique set of design considerations apply in this case. The standard design strategies of current and lattice matching are necessary in the PV semiconductor stack. Additionally, the PV output voltage and current need to be sized to match the optimal operating conditions of the electrolyzer system. The electrolysis process can be independently optimized with appropriate electrodes and catalysts. Since there are no PEC cells in the device stack, any issues of semiconductor durability in solution are avoided. Auxiliary system losses including the lateral collection of current from the PV system and transmission of electrons through external wiring are inevitable, but can be minimized. Also inevitable, cost remains an issue in this approach. As discussed in Sect. 7.2.5, using commercialized PV and electrolyzer technologies, the hydrogen production cost would be greater than \$10/kg, far exceeding the US DOE targets of \$2–4/kg.

As a potentially lower cost implementation of the PV-electrolysis approach, triple junction amorphous silicon solar cells have been fully integrated with on-board electrolysis units. The amorphous silicon cells generate photovoltages over 2 V, and can convert sunlight to electricity with a stabilize PV efficiency of approximately 8–10%. At the same time, the electrolysis system has been optimized to convert electricity to hydrogen at efficiencies near 70%. Net STH efficiencies from 5 to 7% are then achievable, as reported by Kelly and Gibson. [106]. The approach is viable, but cost estimates have not been fully developed to evaluate how pragmatic it could be.

Higher efficiency, higher cost multijunction solar cells, such as the III–V materials systems capable of 20–40% PV conversion efficiency are also commercially available. Using these materials, 14–28% STH would be achievable using PV-electrolysis. In today's market, however, such systems would be prohibitively expensive for any large-scale deployment. The 5–7% mark for lower cost amorphous silicon technology is a more appropriate near-term benchmark for practical solar hydrogen production. Alternative PEC-based schemes need to meet or exceed this benchmark to be viable.

7.6.2 PEC–PEC Tandem Photoelectrodes

PEC–PEC tandem photoelectrode systems for solar water splitting have been investigated as one alternative. These comprise separate photoanodes and photocathodes that are optically stacked in series and electrically interconnected, for example, using external wires. The two-photoelectrode approach is compatible with the photoelectrode-based reactor schemes, and subject to the associated reactor-level considerations described in Sect. 7.2.4. A device level schematic of the photoelectrodes is shown in Fig. 7.27, depicting a photoanode stacked in front of a photocathode. The electrode order can be reversed, however, if required by design consideration. For example if a higher bandgap photocathode material is available, this should be the top junction for optical reasons. As seen in the band diagram of Fig. 7.27, the back contacts of two electrodes are directly interconnected, and therefore their electronic levels align. Under illumination, the level of photoelectrons is driven "up" in the photocathode, and the level of photoholes is driven "down" in the photoanode, resulting in sufficient separation to split water.

There are a number of significant design challenges in this approach. The PEC–PEC arrangement requires two different stable PEC semiconductors, one n- and p-type, with added demands of optical compatibility for current matching, and band alignment compatibility for efficient photovoltage generation. As an extra consideration, the front junction must be fabricated on a transparent substrate coated with a transparent conductive layer to enable both light transmission and electron transfer through to the back junction. This adds complication, optical losses, electrical losses, as well as material cost in the photoelectrode fabrication. Additional conductivity losses can be associated with electrode spacing and gas separation. Due to these challenges, efficient PEC/PEC tandem systems have not been demonstrated to date. Proof of concept experiments have been reported, though. For example, a zinc-doped p-type Fe_2O_3 thin film photocathode has been used with a n-type Fe_2O_3 photoanode with reported efficiencies of 0.11%

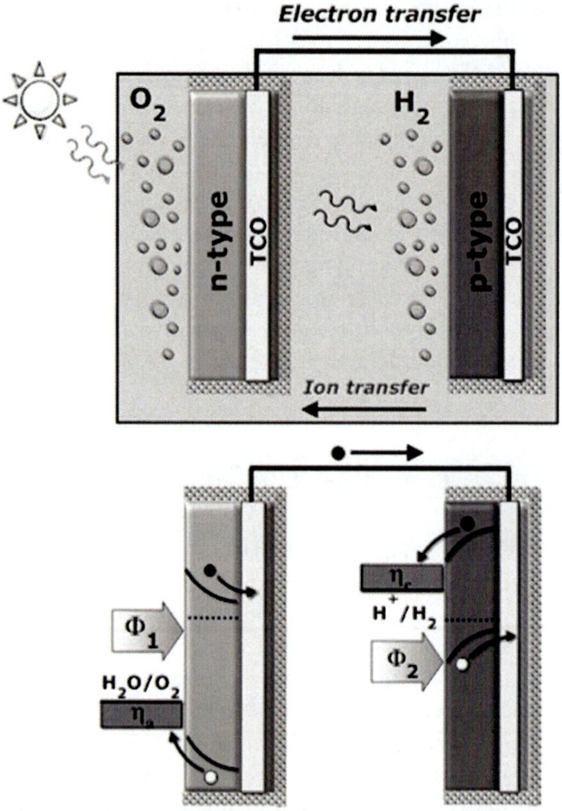

Fig. 7.27 PEC–PEC tandem water splitting device with band diagram showing photogenerated electrons and holes

STH in 0.1 M H_2SO_4 electrolyte [107]. An interesting recent innovation in this area has been the exploration of novel, futuristic tandem nanorod electrode structures using silicon and metal oxides [105].

7.6.3 PV–PEC Tandem Hybrid Photoelectrodes

An alternative tandem device being developed for solar water splitting is the PV/PEC hybrid photoelectrode. In this approach, a PEC top cell is monolithically stacked with a solid-state single-junction PV back cell. Such a photoelectrode device is compatible with both the one- and two-electrode reactor configurations described in Sect. 7.2.4, and the reactor-level considerations described in that section apply. Figure 7.28 shows a device level schematic for a specific two-electrode implementation incorporating a hybrid photocathode and a separate counter electrode. Possible alternative variations include the use of a hybrid photoanode instead of the photocathode, and the integration of the counter and

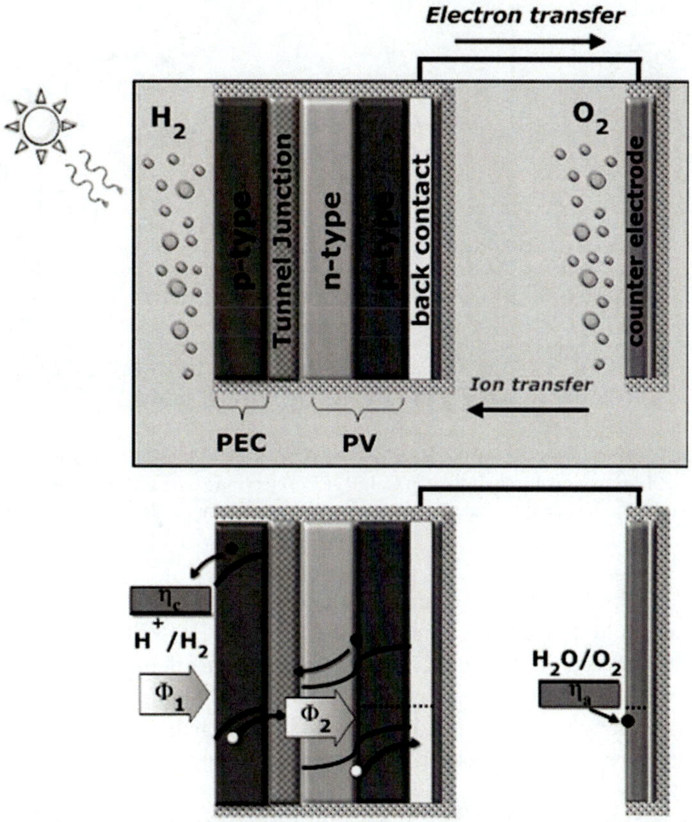

Fig. 7.28 PEC–PV monolithically stacked hybrid water-splitting device

photoelectrodes into a single monolithic device. As seen in the band diagram of Fig. 7.28, the back contact of the PEC cell and the front contact of the PV cell are electronically interconnected via a tunnel junction. In addition, the back contact of the PV cell is wired to the counter electrode, and therefore electronically aligned. As illustrated in the band diagram, under illumination the photoelectron levels at the electrolyte interface and the photohole levels at the counter electrode are separated sufficiently to split water.

In contrast to the PEC–PEC tandem, the two junctions are stacked onto the same substrate, so no transparent substrate is needed. Also, in single-electrode implementations of this approach, electron transmission through wires is eliminated, reducing ohmic losses. There are some clear benefits, but also some specific design challenges associated with the PV–PEC tandem. Appropriately matched PV and PEC semiconductor materials systems are needed, which is challenging enough. In addition, process compatibility in fabricating the PEC cell directly onto the PV cell can be problematic; for example, the necessary restrictions on processing temperature can adversely affect device quality and efficiency. As an alternative strategy for avoiding the process compatibility issues, the PV and PEC cells can be

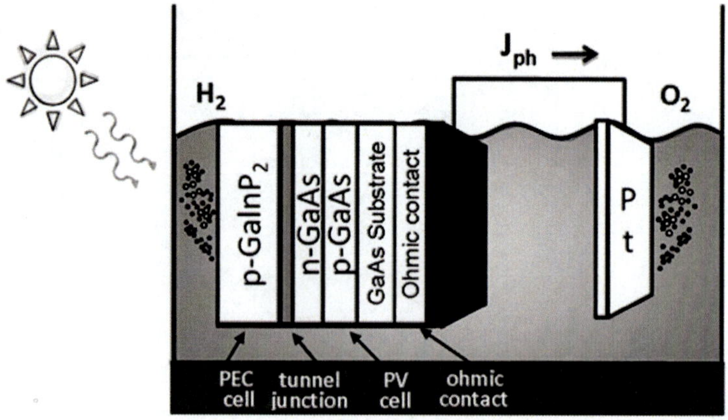

Fig. 7.29 Device schematic of the NREL GaAs/GaInP$_2$ hybrid tandem device submersed in the electrolyte. Light enters through the top GaInP$_2$ PEC electrode then filters down to the GaAs solar cell. This device has demonstrated spontaneous water splitting under illumination with over 12 STH%

separately fabricated and mechanically stacked. This, however, adds manufacturing complexity and new losses. Similar to the tandem PEC–PEC case, stability of the PEC interface is a primary concern in the PV–PEC tandem. There are also similar conductivity losses associated with electrode spacing and gas separation. Overall, it is felt by many that the benefits can outweigh the challenges.

In evidence, the hybrid tandem approach has been under investigation for more than a decade [1, 73, 108, 109]. The best laboratory-scale demonstration to date is unquestionably the NREL's III–V tandem hybrid photoelectrode structure, which still holds the world-record STH efficiency for PEC water-splitting [1, 2]. This hybrid device, shown in Fig. 7.29, consists of a p-type GaInP$_2$ PEC electrode interconnected by a tunnel junction to a buried GaAs p/n PV cell grown on a GaAs wafer substrate. The basic band structure is equivalent to the one shown in Fig. 7.28. In the NREL device, the top junction GaInP$_2$ bandgap is 1.83 eV and the back junction GaAs bandgap is 1.44 eV. Under illumination, hydrogen is evolved at the front photocathode surface and oxygen is evolved at the counter electrode. Under concentrated 11 sun illumination, and in 3 M sulfuric acid electrolyte, prototypes of this device have demonstrated over 12% STH efficiencies for up to 20 h. Subsequent drops in photocurrent and efficiency were reported due to corrosion of the GaInP$_2$. Circumventing the stability limitation, the basic PV structure of this device has also been integrated in tandem-PV-electrolysis prototypes, demonstrating STH efficiencies exceeding 16%. In attempts to enhance the tandem PV–PEC implementation, fundamental research to stabilize the GaInP$_2$ PEC interface is ongoing [97]. Though lacking long-term stability, and reliant on high-cost III–V materials, the results of the NREL demonstrations affirm the potential for high efficiency in multijunction PEC systems. As a "sanity check" for the high-performance levels

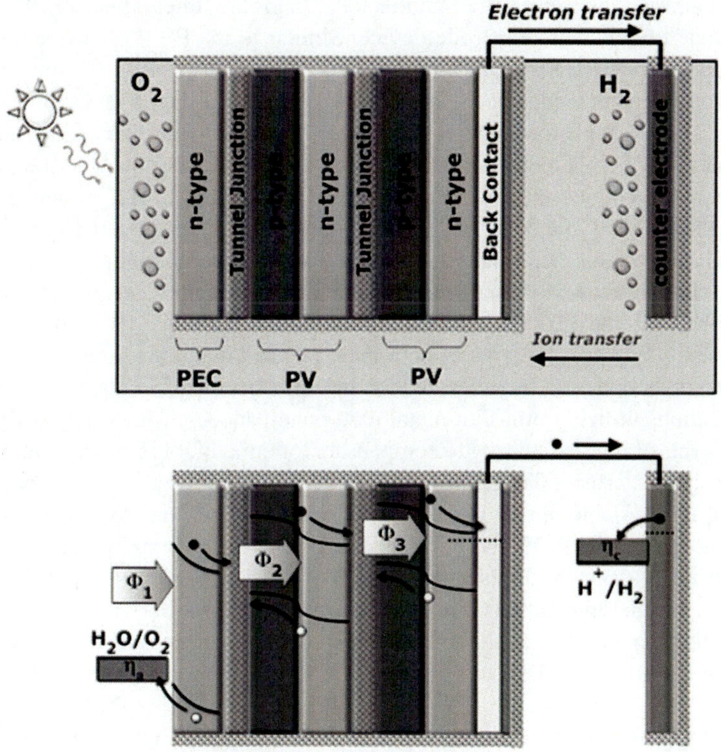

Fig. 7.30 PEC–PV–PV device schematic and band diagram

achieved, it is interesting to note that, according to Fig. 7.23, the STH efficiency limit for a III–V 1.83 eV/1.44 eV tandem stack, is about 18% STH, consistent with the experimental results.

7.6.4 PV–PV–PEC Triple Junction Hybrid Photoelectrodes

A further extension of the PV–PEC hybrid photoelectrode device approach is the PV–PV–PEC triple-junction hybrid photoelectrode. The triple-junction variation is capable of developing higher photopotential, but at the cost of reduced photocurrent and added device complexity. In this configuration, a PEC top cell is monolithically stacked on a solid-state double-junction PV cell. Again, the resulting device is compatible with both the one- and two-photoelectrode configurations in reactor types described in Sect. 7.2.4. A specific two-electrode implementation with a PV–PV–PEC hybrid photoanode and a separate counter electrode is schematically illustrated in Fig. 7.30. Of course, alternative variations include the use of a hybrid photocathode instead of the photoanode, and the integration of the counter electrode into a single electrode device. Similar to the PV–PEC tandem case, the back contact of the PEC cell and the front contact of the PV–PV double cell are electronically interconnected. Also, the back contact of the PV–PV cell is electronically aligned with the counter electrode. As seen in the band diagram, under illumination the photoelectron levels at the electrolyte interface and the photohole levels at the counter electrode are separated sufficiently to split water. Compared with the PV–PEC tandem, additional photopotential is generated in the PV–PV buried cell of this triple-junction device, offering compatibility with a broader range of lower performing PEC materials. The same design considerations described for the PV–PEC tandem hold, though the extra PV cell adds further complexity and cost in terms of both materials and manufacturing.

The triple-junction hybrid approach has also been well studied, particularly in applications utilizing thin-film metal oxide photoanodes, which frequently require high levels of internal biasing to compensate for unfavorable band positioning at the electrolyte interface. Tungsten trioxide (WO_3), for example, has a proven track record in the stable photo-oxidation of water. However, due to its band structure, biasing in excess of 1.0 V has been needed to achieve water splitting [110]. This has necessitated using a double-junction PV buried cell in triple-junction hybrid photoelectrode applications. At the University of Hawaii (UH) at Manoa, n-type WO_3 thin-film electrodes have been mechanically stacked on top of amorphous silicon (a-Si) tandem PV cells into a PV–PV–PEC hybrid device, as illustrated in Fig. 7.31a [7]. The "load-line" analysis of this device [54] is shown in Fig. 7.31b, superimposing the independently measured two-electrode response curves of both the WO_3 PEC (in 0.33 M H_3PO_4) cell and the a-Si tandem PV cell under AM1.5G illumination [57]. Consistent with the operating point indicated by the intersection in the load-line plot, the device successfully split water, with a stabilized photocurrent of 2.5 mA/cm^2, corresponding to a STH conversion efficiency of approximately 3.1%. An interesting and important insight from the load-line plot is that the limiting factors to high efficiency in this specific device are the WO_3 onset potential, saturation photocurrent, and fill factor. If these could be adequately addressed, such a triple-junction approach using low cost thin film semiconductor materials has clear potential for exceeding 5% STH. Based on the optical limits, as seen in Fig. 7.25, successful bandgap reduction in the WO_3 would be needed to exceed the 5% mark. Alternatively, new lower bandgap thin-film materials may emerge through research for this application.

In addition to the amorphous silicon implementation, several interesting variations of the PV–PV–PEC hybrid device based on WO_3 photo-oxidation cells have been demonstrated focusing on low-cost material components. Most notably is the impressive work using tungsten oxide PEC cells in conjunction with dye-sensitized solar cells. The Grätzel–Augustynski "Tandem Cell" is the best example of this [5], though some more recent work has been reported by the Arakawa group in Japan [111]. Efficiencies in these all-PEC configurations are reported in the 2.5–4.5% STH range. Moreover, there has been commercial interest in this approach.

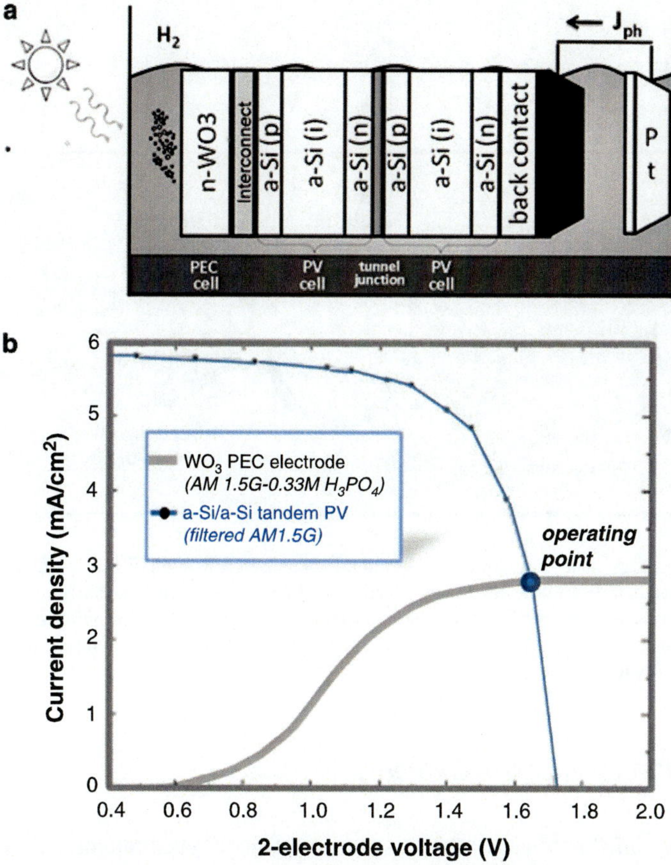

Fig. 7.31 (**a**) Device schematic of the UH WO_3/a-Si/a-Si hybrid triple junction device submerged in the electrolyte. Light enters through the top WO_3 PEC electrode then filters down to a-Si/a-Si tandem solar cell. This device has demonstrated spontaneous water splitting under illumination with over 3.0 STH%. (**b**) Load-line analysis of the device, showing operating point of 2.5 mA/cm^2 based on superposition of the PV and PEC responses

Another interesting example has been the development of an all thin-film silicon triple-junction hybrid device by MVSystems, Incorporated [109, 112–114]. In this device, hydrogenated amorphous silicon-carbide alloy thin-films are used for all three junctions, including the PEC junction and both buried PV cells. Bandgap tuning through alloy composition variations is used for current matching. Compatible with the triple-junction efficiency analysis shown in Fig. 7.25, such a device has theoretical potential for exceeding 10% STH. Laboratory prototypes, though, have been limited by high-loss potential barriers formed at the PEC interface, and demonstrations to date of spontaneous solar water splitting have performed closer to 1% STH. Ongoing work to reduce the interfacial barrier issues offers much promise for achieving commercialized PEC hydrogen production based on thin-film silicon alloys.

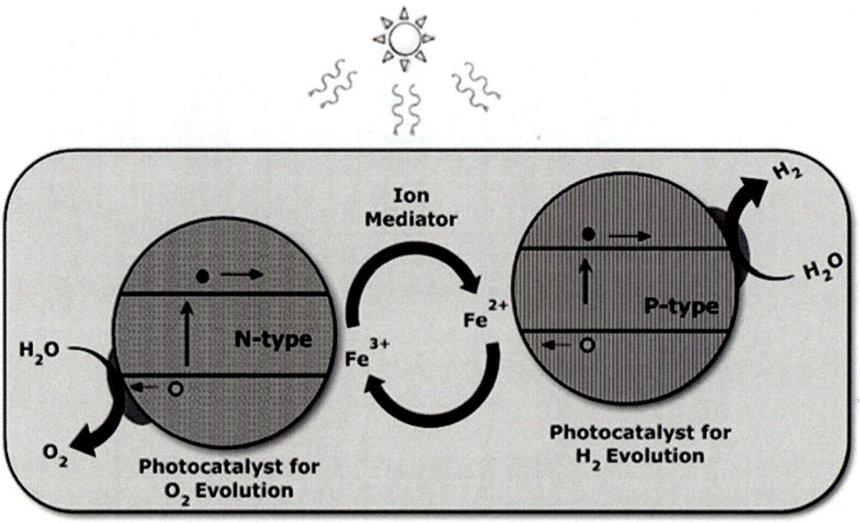

Fig. 7.32 Energy band diagram of the Z-scheme showing OER and HER photocatalyst particles with functionalized surface sites for the oxygen- and hydrogen-evolution reactions, respectively. Concentrations of both particle types are suspended together in solution with an ion-shuttling mediator, such as the Fe^{3+}/Fe^{2+} couple, effectively coupling the gas evolution reactions in tandem to photosplit water

7.6.5 Photocatalyst Z-Schemes

The "Z-scheme" approach, which is receiving much recent attention, particularly in the Japanese research community, is essentially a photocatalyst version of the tandem PEC–PEC photoelectrode system. As shown in the schematic representation in Fig. 7.32, n-type photocatalyst particles drive the hydrogen evolution reaction (HER) and p-type photocatalyst particles drive the oxygen evolution reaction (OER), achieving water splitting in a tandem fashion. In the Z-scheme, electrons are exchanged between the photoanode and photocathode particles via an ion mediator in solution (such as the Fe^{3+}/Fe^{2+} couple), rather than through wires, as in the PEC–PEC photoelectrode case. Slurries of the two photocatalysts can be mixed together in single-bed reactors, or alternatively, slurries of the HER and OER photocatalysts can be separated into two beds, connected with an ionic bridge, consistent with the one- and two-bed reactors discussed in Sect. 7.2.4.

As indicated in the DTI techno-economic analysis report, the particle-based systems offer some of the best hope for low-cost hydrogen production on a large scale, but there are a number of unique design challenges posed by this approach. For example, charge separation within each catalyst particle needs to be enhanced. The photocathode particle has to simultaneously reduce hydrogen and oxidize the mediator at separate surface sites. Similarly, the photoanode particle must both oxidize water and reduce the mediator. Functionalization of surface sites, for example, with reaction-specific nanoparticle catalysts, is necessary to draw apart photogenerated electrons and holes, and to enable the simultaneous reactions.

As a particular challenge in the single bed reactor system using Z-scheme photocatalysts, both hydrogen and oxygen are evolved in the same reactor bed, with associated safety hazards. Safe gas separations techniques could be employed, but at substantial additional cost. The dual bed system avoids any possible explosive hydrogen/oxygen gas mixtures, but the greater land area requirements and conductivity losses in the ion bridge reduce the STH performance. Both the single- and dual-bed systems have their advantages and disadvantages, but either way, new HER and OER photocatalysts need to be identified with sufficient efficiency and stability for practical hydrogen production.

The search for effective HER and OER photocatalyst materials for Z-scheme systems is ongoing. Extensive libraries of multicomponent metal oxides and oxinitrides are being developed and screened [9, 96–99]. Specific material systems have demonstrated spontaneous water-splitting under visible light, for example, $SrTiO_3$:Rh combined with $BiVO_4$ in the presence of an Fe^{3+}/Fe^{2+} redox couple [115], but STH efficiencies remain extremely low. Visible light water splitting has also been accomplished using a RuO_2–TaON and Pt–TaON mixture with I_3^{-}/I^{-} redox couple, but again with low STH efficiency [116]. A recurring theme in PEC research is that new materials continue to offer new hope.

7.6.6 Many Options

Compared with the photovoltaics industry, the field of multijunction PEC hydrogen production is relatively new, but there has been substantial exploratory research in recent years. In addition to the above examples in *PV-electrolysis, multijunction photoelectrolysis*, and *Z-scheme photocatalysis*, numerous variations which mix and match concepts and components of all three are conceivable. In the final analysis, commercialization will depend on balances between performance, complexity, and cost. The simpler the system, the better. Single-junction photocatalyst or photoelectrode systems offer the greatest simplicity, but the significant difficulties in achieving high efficiencies in these systems necessitate a serious look at the multijunction alternatives. Efficient tandem systems are the next best hope. Analysis has validated the potential for high STH conversion efficiencies in tandem double- and triple-junction schemes and there has been experimental verification, though still limited, on the laboratory scale. Real challenges remain in the identification of low-cost material systems capable of reproducing the laboratory performance levels on a commercial scale. As the research community advances in its ability to engineer high-efficiency absorber materials and functionalized interfaces, and as it learns to integrate these in optimal multijunction designs, the opportunities for success will continue to grow.

7.7 In Summary

There have been significant R&D efforts over the past several decades advancing multijunction device technologies, particularly in the PV field. These have resulted in the highest solar-to-electric conversion efficiencies to date, exceeding 40% in high quality III–V cells being commercialized for space applications. These ultrahighly efficient cells are, however, expensive, and not practical for large-scale deployment in solar power production plants. The multijunction approach continues to make strides in improving performance in lower cost thin-film solar cells, including, for example, amorphous silicon and copper chalcopyrite devices, though the commercial benefits have not yet been fully realized. It is nevertheless clear that low cost and high performance are the keys to the success of transformative scale solar electricity. Multijunction thin film devices offer hope in this area, and this hope extends to PEC hydrogen production.

There are more recent efforts to develop multijunction devices and systems for solar hydrogen production, though this field is still in its infancy in comparison with PV technologies. STH conversion efficiencies of 12–16% have been demonstrated on the laboratory scale in tandem devices using high-quality III–V materials, but with limited durability. Multijunction PEC and PV-electrolysis systems using lower cost thin-film materials have demonstrated better stability, but at conversion efficiencies limited in the 3–8% range. Ultimately, durable, low cost systems with STH conversion efficiencies over 10% are needed for commercial deployment of solar hydrogen technologies. Fortunately, continued development of multijunction schemes offers great promise, specifically in terms of enhanced photopotentials and broadened spectral bandwidth. Analyses of tandem and triple-junction PEC schemes indicate that efficiencies exceeding 10% STH and in special cases up to 25% STH, are in fact possible. Technical barriers have stood in the way of achieving such performance levels, but these are solvable if the fundamental problems are broken down and appropriately addressed.

7.7.1 What Is Needed?

Based on PEC research efforts to date, it has become increasingly clear that new, innovative material systems, interfaces, and devices are needed to achieve high-performance, low-cost solar water splitting. Breaking this down into specifics relative to the multijunction approach:

- Low-cost semiconductor materials with appropriate bandgap and sufficient opto-electronic quality need to be identified and optimized to enable efficient solar energy absorption and charge extraction.
- Effective PEC interfaces activated for the gas evolution reactions and passivated for corrosion and other parasitic reactions need to be engineered, using, for example, heterojunction and/or catalytic coating and/or dispersions.

- Low loss solid-state interfaces need to be engineered for bridging component cells in a multijunction device structure, for example, using tunnel junctions or intermediate transparent conductive oxides layers.
- Innovative multijunction designs combining the best materials and interfacing schemes into efficient photoelectrode or photocatalyst device structures will be the key.
- Innovative synthesis and fabrication techniques will be needed to enable cost-effective large-scale manufacture of multijunction photoelectrodes or photocatalysts.

Leveraging the knowledge gained from the PV community in the successful development of high-efficiency multijunction device designs will be integral to success. Key advances, however, are needed specifically in the development of the PEC material components and interfaces, since these remain the greatest technical barriers.

7.7.2 *What Is Happening*

In efforts to address the technical barriers, the PEC research community over the past several years has been developing new theoretical, synthesis and characterization tools using improved state-of-the-art methodologies [57]. These tools are being employed in several parallel strategies for developing the innovative materials and interfaces necessary to the success of PEC hydrogen production systems.

One specific approach is the further development of the traditional PEC semiconductor materials in thin-film and nanostructured forms for higher efficiencies. There have been significant scientific efforts in this approach, including continued development of the following materials classes:

- Tungsten-oxide and related bandgap-reduced compounds [6, 117–121].
- Iron-oxide and related enhanced mobility compounds [122–126].
- Vanadates, including bismuth vanadate and related compounds [127–129].
- Titanium dioxide and related bandgap-reduced compounds [61, 130].

This list is by no means comprehensive, but does represent a large part of the research endeavors to advance well-established PEC semiconductors.

Another important strategy is the adaptation of efficient PV semiconductor thin-films and nanostructures for effective use in PEC applications. Recent research in this area has focused on material classes with inherent bandgap tuning capabilities such as the amorphous silicon compounds (including silicon carbides and nitrides) [109, 112–114, 131, 132], and polycrystalline copper chalcopyrite compounds [133–137].

Some of the most recent innovative approaches have included the development of entirely new materials classes and structures for use in photoelectrode or photocatalyst devices. Examples include quantum-confined WS_2 and MoS_2

nanoparticle photocatalysts embedded in a conductive scaffolding [138] and silicon nanorods embedded in membrane structures [105, 139]. Also cutting-edge are efforts in the development of breakthrough synthesis technologies to reduce the cost and enhance the stability of the high-performance III–V multijunction structures that have demonstrated the best efficiencies to date, both in PV and PEC [4, 87]. Future progress in all these approaches, which will be integral to the ultimate success of PEC hydrogen production, will depend on the best use of the most advanced research tools.

7.7.3 The Path Forward

For all PEC systems, and for multijunction water-splitting systems in particular, the fundamental solutions reside in *MATERIALS* and *INTERFACES*. Good solid-state light absorber materials with efficient charge transport properties that can be synthesized inexpensively are the cornerstone. Low loss interfaces are the next major step. For example, low-defect and low-recombination solid-state interfaces are essential between cells in a multijunction stack, and at back contacts. These are achievable, as demonstrated through years of R&D in the PV research community. More challenging are the solid–liquid electrochemical interfaces in PEC multijunction schemes. Functionalizing surfaces to produce the right energetic and kinetics for promoting the water-splitting half reactions while inhibiting parasitic reactions and recombination loss will be the key to unlock the full potential of PEC hydrogen production.

The lessons learned to date in the PEC research and development community, coupled with of PV knowledge and experience amassed over decades could be extremely fruitful, both for solar hydrogen and solar electricity production. Some questions to consider for the pathways forward include the following:

- Can functions of light absorption, charge extraction, reaction catalysis, and production management be effectively separated, and independently optimized in PEC devices? If so, it opens up the design space considerably. For example, a good absorber material would not need to be catalytic, but could be coupled with a good catalyst material with the design of an efficient interface.
- Can cost-effective thin film PV materials (e.g., amorphous silicon or more efficient copper chalcopyrites) and multijunction devices be adapted through surface modifications to split water? If so, a wealth of PV manufacturing experience would be available for enabling commercial scale deployment.
- Can new low-cost synthesis routes be devised for ultrahigh efficiency III–V materials and devices? If so, the barriers to large-scale deployment of solar electricity are largely gone. Moreover, if appropriate surface modifying materials and interfaces are also developed, then the commercial viability of PEC hydrogen production would be assured.

- Can new forms of earth-abundant materials, including oxides, nitrides, and sulfides, be discovered and developed with sufficiently good light absorbing and charge transport properties for solar energy conversion? This, with the development of efficient solid-state and electrochemical interfaces, and with the design of effective multijunction photoelectrode or photocatalyst systems, could revolutionize both PV and PEC technologies!

These questions encompass a broad vision and provide a coarse roadmap for paths forward. The pathways to success of PEC hydrogen production are along the road of technical collaboration; and the trail will be blazed by the new generation of scientists and engineers.

Acknowledgments The authors would like to acknowledge and express their great admiration for all members of the international PEC research and development community; with special nods of appreciation to the US Department of Energy's PEC Working Group supported by the Fuel Cells Technologies Office, and to Annex-26 of the International Energy Agency's Hydrogen Implementing Agreement. They also thank the members of the Thin Films Laboratory at the University of Hawaii at Manoa's Hawaii Natural Energy Institute, including Drs. Nicolas Gaillard, Bor Yann Liaw, Yuancheng Chang, and Richard Rocheleau, as well as Jess Kaneshiro, Jeremy Kowalczyk, Xi Song, and Brett Ikei for their encouragement and support to this effort.

References

1. Khaselev, O., Bansal, A., Turner, J.A.: High-efficiency integrated multijunction photovoltaic/electrolysis systems for hydrogen production. Int. J. Hydrogen Energy **26**, 127–132 (2001)
2. Khaselev, O., Turner, J.A.: A monolithic photovoltaic photoelectrochemical device for hydrogen production via water splitting. Science **280**, 425–427 (1998)
3. Andreev, V.M.: GaAs and high-efficiency space cells. In: Markvart, T., Castañer, L. (eds.) Practical Handbook of Photovoltaics: Fundamentals and Applications. Elsevier, New York (2003)
4. Deutsch, T.G., Koval, C.A., Turner, J.A.: III − V nitride epilayers for photoelectrochemical water splitting: GaPN and GaAsPN. J. Phys. Chem. B **110**, 25297–25307 (2006)
5. Grätzel, M.: Photoelectrochemical cells. Nature **414**, 338 (2001)
6. Marsen, B., Miller, E.L., Paluselli, D., Rocheleau, R.E.: Progress in sputtered tungsten trioxide for photoelectrode applications. Int. J. Hydrogen Energy **32**, 3110–3115 (2007)
7. Gaillard, N., Chang, Y., Kaneshiro, J., Deangelis, A., Miller, E.L.: Status of research on tungsten oxide-based photoelectrochemical devices at the University of Hawai'i. Proc. SPIE **7770**, 77700V–77701V (2010)
8. Rocheleau, R.E., Miller, E.L., Misra, A.: High-efficiency photoelectrochemical hydrogen production using multijunction amorphous silicon photoelectrodes. Energy Fuels **12**, 3–10 (1998)
9. Miller, E.L., Gaillard, N., Kaneshiro, J., DeAngelis, A., Garland, R.: Progress in new semiconductor materials classes for solar photoelectrolysis. Int. J. Energy Res **34**, 1215–1222 (2010)
10. Kuang, C.: Fast Company. http://www.fastcompany.com/blog/cliff-kuang/design-innovation/start-aims-clean-energys-holy-grail-freeing-hyrodgen-water, 16 Apr 2009. Accessed 29 Mar 2011
11. Li, Y., Zhang, J.Z.: Hydrogen generation from photoelectrochemical water splitting based on nanomaterials. Laser Photonics Rev. **4**, 517–528 (2010)

12. Bush, G.W.: State of the Union. Presented in Washington, DC, USA, 28 January 2003
13. Rifkin, J.: The Hydrogen Economy. Tarcher (2003)
14. Romm, J.J.: The Hype About Hydrogen. Island, New York (2004)
15. Bromaghim, G., Gibeault, K., Serfass, J., Serfass, P., Wagner, E.: Hydrogen and Fuel Cells: The U.S. Market Report. National Hydrogen Association, 22 March 2010
16. Collodi, G., Bressan, L., Ruggeri, F., Uncuoglu, D.: Hydrogen Production for Upgrading Projects in Refineries. Foster Wheeler Italiana S.p.A, Via Caboto 1, 20094 Corsico – Milan – Italy. http://www.fwc.com/publications/tech_papers/files/Hydrogen%20Production%20for%20Upgrading%20in%20Refineries.pdf (2009). Accessed 29 Mar 2011
17. U.S. Department of Energy: Natural Gas Reforming. http://www1.eere.energy.gov/hydrogenandfuelcells/production/natural_gas.html. Accessed 29 Mar 2011
18. Ball, M., Wietschel, M.: The Hydrogen Economy: Opportunities and Challenges. Cambridge Press, New York (2009)
19. Yürüm, Y.: Hydrogen energy system: production and utilization of hydrogen and future aspects. Kluwer Academic Publishers, Dordrecht (1995)
20. Turner, J.A.: A realizable renewable energy future. Science **285**, 687–689 (1999)
21. Energy Information Administration: World Proved Reserves of Oil and Natural Gas, Most Recent Estimates. 3 March 2009. http://www.eia.doe.gov/emeu/international/reserves.html. Accessed 29 Mar 2011
22. Energy Information Administration: World Petroleum Consumption, 1960–2008. http://www.eia.doe.gov/aer/txt/ptb1110.html. Accessed 29 Mar 2011
23. Energy Information Administration: International Energy Outlook 2007: Petroleum and Other Liquid Fuels. http://www.eia.doe.gov/oiaf/archive/ieo07/pdf/oil.pdf. Accessed 29 Mar 2011
24. Safina, C.: Testimony to the House Subcommittee on Energy and Environment. 21 May 2010
25. U. S. Department of Energy, Office of Science: Basic Research Needs for Solar Energy Utilization. Washington (2005)
26. Green, M.A.: Solar cells: Operating Principles, Technology, and System Applications. Prentice-Hall, Inc, Kensington, NSW (1982)
27. U. S. Department of Energy, Energy Information Administration: International Energy Outlook 2008 (DOE/EIA-0484). Washington (2008)
28. Greentech Media and the Prometheus Institute: PV Technology, Production and Cost, 2009 Forecast: The Anatomy of a Shakeout. Cambridge (2008)
29. Solarbuzz: Marketbuzz 2009: Annual World Solar PV Market Report. San Francisco (2009)
30. Bauman, R.P.: Modern Thermodynamics with Statistical Mechanics. Macmillan Publishing Company, New York (2003)
31. Akkerman, I., Janssen, M., Rocha, J., Wijffels, R.H.: Photobiological hydrogen production: photochemical efficiency and bioreactor design. Int. J. Hydrogen Energy **27**, 1195–1208 (2002)
32. Zaborsky, O.R.: Biohydrogen. Plenum, New York (1998)
33. Funk, J.E., Reinstrom, R.M.: Energy requirements in production of hydrogen from water. Ind. Eng. Chem. Process Des. Dev. **5**, 336–342 (1966)
34. Minggu, L.J., Daud, W.R.W., Kassim, M.B.: An overview of photocells and photoreactors for photoelectrochemical water splitting. Int. J. Hydrogen Energy **35**, 5233–5244 (2010)
35. James, B.D., Baum, G.N., Perez, J., Baum, K.N.: Technoeconomic Analysis of Photoelectrochemical (PEC) Hydrogen Production. Directed Technologies, Inc. https://www1.eere.energy.gov/hydrogenandfuelcells/pdfs/pec_technoeconomic_analysis.pdf (2009). Accessed 11 Mar 2011
36. Ruth, M., Laffen, M., and Timbario, T.A.: NREL technical report (NREL/BK-6A1-46676). Hydrogen Pathways: Cost, Well-to-Wheels Energy Use, and Emissions for the Current Technology Status of Seven Hydrogen Production, Delivery, and Distribution Scenarios. September 2009

37. NREL Technical Report (NREL/TP-6A1-46612): Current (2009) State-of-the-Art Hydrogen Production Cost Estimate Using Water Electrolysis. September 2009
38. DOE EERE Fuel Cells Technologies Program: Multi-Year Research, Development and Demonstration Plan: Planned Program Activities for 2005–2015. http://www1.eereenergy.gov/hydrogenandfuelcells/mypp/. April 2009
39. Kelly, N.A., Gibson, T.L.: Solar energy concentrating reactors for hydrogen production by photoelectrochemical water splitting. Int. J. Hydrogen Energy **33**, 6420–6643 (2008)
40. Mavroides, J.G., Kafalas, J.A., Kolesar, D.F.: Photoelectrolysis of water in cells with $SrTiO_3$ anodes. Appl. Phys. Lett. **28**, 241–243 (1976)
41. Bard, A.J., Faulknerk, L.R.: Electrochemical Methods: Fundamentals and Applications. Wiley, New York (2000)
42. Bockris, J.O.M., Reddy, A.K.N., Gamboa-Aldeco, M.E.: Modern Electrochemistry: Fundamentals of Electrodics, vol. 2a. Springer, New York (2001)
43. Memming, R.: Semiconductor Electrochemistry. Wiley-VCH, Weinheim (2001)
44. Lipkowski, J., Ross, P.N.: Electrochemistry of Novel Materials. VCH Publishers, New York (1994)
45. Gellings, P.J., Bouwmeester, H.J.M.: The CRC Handbook of Solid State Electrochemistry. CRC, Boca Raton (1997)
46. Nozik, A.J., Memming, R.: Physical chemistry of the semiconductor–liquid interface. J. Phys. Chem. **100**, 13061–13078 (1996)
47. Gerischer, H.: Solar photoelectrolysis with semiconductor electrodes. In: Seraphin, B.O. (ed.) Solar Energy Conversion, Solid-State Physics Aspects, pp. 115–172. Springer-Verlag, New York (1979)
48. Gerischer, H.: Physical Chemistry: An Advanced Treatise, vol. 9A. Academic, New York (1970)
49. Gerischer, H.: The impact of semiconductors on the concept of electrochemistry. Electrochim. Acta **35**, 1677–1690 (1990)
50. Miller, E.L.: Solar hydrogen production by photoelectrochemical water splitting: the promise and challenge. In: Vayssieres, L. (ed.) On Solar Hydrogen and Nanotechnology, pp. 3–35. Wiley, Asia (2009)
51. Lee, K., Nam, W.S., Han, G.Y.: Photocatalytic water-splitting in alkaline solution using redox mediator. 1: Parameter study. Int. J. Hydrogen Energy **29**, 1343–1347 (2004)
52. Marcus, R.J.: Chemical conversion of solar energy. Science **123**, 399–405 (1965)
53. Bockris, J.O.M.: Kinetics of activation controlled consecutive electrochemical reactions: anodic evolution of oxygen. J. Chem. Phys. **24**, 817–827 (1956)
54. Kanan, M.W., Nocera, D.G.: In Situ Formation of an oxygen-evolving catalyst in neutral water containing phosphate and Co^{2+}. Science **321**, 1072–1075 (2008)
55. Dutta, S.: Technology assessment of advanced electrolytic hydrogen production. Int. J. Hydrogen Energy **15**, 379–386 (1990)
56. LeRoy, R.L.: Industrial water electrolysis: present and future. Int. J. Hydrogen Energy **8**, 401–417 (1983)
57. Chen, Z., Jaramillo, T.F., Deutsch, T.G., Kleiman-Shwarsctein, A., Forman, A.J., Gaillard, N., Garland, R., Takanabe, K., Heske, C., Sunkara, M., McFarland, E.W., Domen, K., Miller, E.L., Turner, J.A., Dinh, H.N.: Accelerating materials development for photoelectrochemical (PEC) hydrogen production: Standards for methods, definitions, and reporting protocols. J. Mater. Res. **25**, 3–16 (2010)
58. Parkinson, B.: On the efficiency and stability of photoelectrochemical devices. Acc. Chem. Res. **17**, 431–437 (1984)
59. Dohrmann, J.K., Schaaf, N.S.: Energy conversion by photoelectrolysis of water: determination of efficiency by in situ photocalorimetry. J. Phys. Chem. **96**, 4558–4563 (1992)
60. Heller, A.: Electrochemical solar cells. Solar Energy **29**, 153–162 (1982)
61. Khan, S.U.M., Al-shahry, M., Ingler Jr., W.B.: Efficient photochemical water splitting by a chemically modified n-TiO_2. Science **297**, 2243–2245 (2002)

62. Emery, K.: Measurements and characterization of solar cell modules. In: Luque, A., Hegedus, S. (eds.) Handbook of Photovoltaic Science and Engineering, pp. 701–752. Wiley, New York (2003)
63. NIST Chemistry WebBook: NIST Standard Reference Database Number 69 http://webbook.nist.gov/chemistry/. Accessed 29 Mar 2011
64. Luther, J.: Motivation for photovoltaic application and development. In: Luque, A., Hegedus, S. (eds.) Handbook of Photovoltaic Science and Engineering, pp. 45–60. Wiley, New York (2003)
65. Asahi, R., Morikawa, T., Ohwaki, T., Aoki, K., Tage, Y.: Visible-light photocatalysis in nitrogen-doped titanium oxides. Science **293**, 269 (2001)
66. Sze, S.M.: Physics of Semiconductor Devices. Wiley, New York (2006)
67. Neamen, D.A.: Semiconductor Physics and Devices: Basic Principles. McGraw-Hill, New York (2002)
68. Balandin, A.A., Wang, K.L.: Handbook of Semiconductor Nanostructures and Nanodevices (5-Volume Set). American Scientific Publishers, Stevenson Ranch (2006)
69. Muller, R.S., Kamins, T.I.: Device Electronics for Integrated Circuits. Wiley, New York (2002)
70. Yu, P.Y., Cardona, M.: Fundamentals of Semiconductors: Physics and Materials Properties. Springer, New York (2004)
71. Mussini, T., Longhi, P.: Chlorine. In: Bard, A.J., Parsons, R., Jordan, J. (eds.) Standard Potentials in Aqueous Solution, pp. 70–77. IUPAC, New York (1985)
72. Tan, M.X., Kenyon, C.N., Krulger, O., Lewis, N.S.: Behavior of Si photoelectrodes under high level injection conditions. 1. Steady-state current–voltage properties and quasi-fermi level positions under illumination. J. Phys. Chem. B **101**, 2830–2839 (1997)
73. Miller, E.L., Paluselli, D., Marsen, B., Rocheleau, R.: Optimization of hybrid photoelectrodes for solar water splitting. Electrochem. Solid-State Lett. **8**, A247–A249 (2005)
74. Hanna, M.C., Nozik, A.J.: Solar conversion efficiency of photovoltaic and photoelectrolysis cells with carrier multiplication absorbers. J. App. Phys. **100**, 074510 (2006)
75. Ross, R.T., Hsiao, T.L.: Limits on the yield of photochemical solar energy conversion. J. Appl. Phys. **48**, 4783–4785 (1977)
76. Bolton, J.R., Haught, A.F., Ross, R.T.: Photochemical energy storage: an analysis of limits. In: Connolly, J.S. (ed.) Photochemical Conversion and Storage of Solar Energy, pp. 297–330. Academic, New York (1981)
77. Bolton, J.R., Strickler, S.J., Connolly, J.S.: Limiting and realizable efficiencies of solar photolysis of water. Nature **316**, 495–500 (1985)
78. Weber, M.F., Dignam, M.J.: Splitting water with semiconducting photoelectrodes – efficiency considerations. Int. J. Hydrogen Energy **11**, 225 (1986)
79. Archer, M.D., Bolton, J.R.: Requirements for ideal performance of photochemical and photovoltaic solar energy converters. J. Phys. Chem. **94**, 8028–8036 (1990)
80. Bolton, J.R.: Solar photoproduction of hydrogen: a review. Solar Energy **57**, 37 (1996)
81. Licht, S.: Multiple band gap semiconductor/electrolyte solar energy conversion. Phys. Chem. B **105**, 6281–6294 (2001)
82. Rocheleau, R.E., Miller, E.L.: Photoelectrochemical production of hydrogen: engineering loss analysis. Int. J. Hydrogen Energy **22**, 771–782 (1997)
83. Ellis, A.B., Kaiser, S.W., Wrighton, M.S.: Semiconducting potassium tantalate electrodes. J. Phys. Chem. **80**, 1325–1328 (1976)
84. Green, M.A.: Third Generation Photovoltaics: Advanced Solar Energy Conversion. Springer-Verlag, Heidelberg (2003)
85. Yamaguchi, M.: Super-high-efficiency multi-junction solar cells. Prog. Photovolt. Res. Appl. **13**, 125 (2005)
86. King, R.R. et al: Advances in High-Efficiency III-V Multijunction Solar Cells. Adv. Opto-Electr. Article ID 29523, 8 pages (2007)

87. Press Release: Spectrolab solar cell breaks 40% efficiency barrier. 7 December 2006. http://www.insidegreentech.com/node/454. Accessed 29 Mar 2011
88. Guter, W., et al.: Current-matched triple-junction solar cell reaching 41.1% conversion efficiency under concentrated sunlight. Appl. Phys. Lett. **94**, 223504 (2009)
89. Swinehart, D.F.: The Beer–Lambert law. J. Chem. Educ. **39**, 333 (1962)
90. López, N., Reichertz, L.A., Yu, K.M., Campman, K., Walukiewicz, W.: Engineering the electronic band structure for multiband solar cells. Phys. Rev. Lett. **106**, 028701 (2011)
91. Baruch, P., De Vos, A., Landsberg, P.T., Parrott, J.E.: On some thermodynamic aspects of photovoltaic solar energy conversion. Solar Energy Mater. Solar Cells **36**, 201–222 (1995)
92. Fonash, S.: Solar Cell Device Physics. Academic, New York (1982)
93. Smestad, G.P.: Optoelectronics of Solar Cells. SPIE, Bellingham (2002)
94. Yang, J., Yan, B., Guha, S.: Amorphous and nanocrystalline silicon-based multi-junction solar cells. Thin Solid Films **487**, 162–169 (2005)
95. Nishiwaki, S., Siebentritt, S., Walk, P., Lux-Steiner, M.C.: A stacked chalcopyrite thin-film tandem solar cell with 1.2 V open-circuit voltage. Prog. Photovolt. Res. Appl. **11**, 243–248 (2003)
96. Arai, T., Konishi, Y., Iwasaki, Y., Sugihara, H., Sayama, K.: High-throughput screening using porous photoelectrode for the development of visible-light-responsive semiconductors. J. Comb. Chem. **9**, 574–581 (2007)
97. Kusama, H., Wang, N., Miseki, Y., Sayama, K.: Combinatorial search for iron/titanium-based ternary oxides with a visible-light response. J. Comb. Chem. **12**, 356–362 (2010)
98. Jianghua, H., Parkinson, B.A.: A combinatorial investigation of the effects of the incorporation of Ti, Si, and Al on the performance of α-Fe_2O_3 photoanodes. J. Comb. Chem. **13**(4), 399–404 (2011)
99. Woodhouse, M., Parkinson, B.A.: Combinatorial approaches for the identification and optimization of oxide semiconductors for efficient solar photoelectrolysis. Chem. Soc. Rev. **38**, 197–210 (2009)
100. Burnett, B.: The Basic Physics and Design of III-V Multijunction Solar Cells. NREL, Golden (2002)
101. Yamaguchi, M.: III–V compound multi-junction solar cells: present and future. Solar Energy Mater. Solar Cells **75**, 261–269 (2003)
102. Wolf, M.: Limitations and possibilities for improvement of photovoltaic solar energy converters. Proc. Inst. Radio Eng. **48**, 1246–1263 (1960)
103. Poortmans, J., Arkhipov, V.: Thin film solar cells: fabrication, characterization and applications. Wiley, Hoboken, NJ (2006)
104. The Basic Physics and Design of III-V Multijunction Solar Cells http://photochemistry.epfl.ch/EDEY/NREL.pdf. Accessed 6 Oct 2011
105. Walter, M.G., Warren, E.L., McKone, J.R., Boettcher, S.W., Mi, Q.X., Santori, E.A., Lewis, N.S.: Solar water splitting cells. Chem. Rev. **110**, 6446–6473 (2010)
106. Gibson, T.L., Kelly, N.A.: Predicting efficiency of solar powered hydrogen generation using photovoltaic-electrolysis devices. Int. J. Hydrogen Energy **35**, 900–911 (2010)
107. Ingler, W.B., Khan, S.U.M.: A self-driven p/n-Fe_2O_3 tandem photoelectrochemical cell for water splitting. Electrochem. Solid State Lett. **9**, G144–G146 (2006)
108. Miller, E.L., Rocheleau, R.E., Deng, X.M.: Design considerations for a hybrid amorphous silicon/photoelectrochemical multijunction cell for hydrogen production. Int. J. Hydrogen Energy **28**, 615–623 (2003)
109. Zhu, F., Hu, J., Kunrath, A., Matulionis, I., Marsen, B., Cole, B., Miller, E.L., Madan, A.: a-SiC:H films used as photoelectrodes in a hybrid, thin-film silicon photoelectrochemical (PEC) Cell for progress toward 10% solar-to hydrogen efficiency. Sol. Hydrogen Nanotechnol. Proc. SPIE **6650**, 66500S (2007)
110. Santato, C., Ulmann, M., Augustynski, J.: Photoelectrochemical properties of nanostructured tungsten trioxide films. J. Phys. Chem. B **105**, 936–940 (2001)

111. Arakawa, H., Shiraishi, C., Tatemoto, M., Kishida, H., Usui, D., Suma, A., Takamisawa, A., Yamaguchi, T.: Solar hydrogen production by tandem cell system composed of metal oxide semiconductor film photoelectrode and dye-sensitized solar cell. Proc. SPIE **6650**, 665003 (2007). doi:10.1117/12.773366
112. Hu, J., Zhu, F., Matulionis, I., Kunrath, A., Deutsch, T., Kuritzky, L., Miller, E.L., Madan, A.: Solar-to-hydrogen photovoltaic/photoelectrochemical devices using amorphous silicon carbide as the photoelectrode. 23rd European Photovoltaic Solar Energy Conference, Valencia, Spain, 1–5 September 2008
113. Matulionis, I., Zhu, F., Hu, J., Gallon, J., Kunrath, A., Miller, E.L., Marsen, B., Madan, A.: Development of a corrosion-resistant amorphous silicon carbide photoelectrode for solar-to-hydrogen photovoltaic/photoelectrochemical devices. SPIE Solar Energy and Hydrogen Conference, San Diego, USA, 10–14 August 2008
114. Stavrides, A., Kunrath, A., Hu, J., Treglio, R., Feldman, A., Marsen, B., Cole, B., Miller, E.L., Madan, A.: Use of amorphous silicon tandem junction solar cells for hydrogen production in a photoelectrochemical cell. SPIE Optics & Photonics Conference, San Diego, USA, 13–17 August 2006
115. Higashi, M., Abe, R., Ishikawa, A., Takata, T., Ohtani, B., Domen, K.: Z-scheme overall water splitting on modified-TaON photocatalysts under visible light ($\lambda < 500$ nm). Chem. Lett. **37**, 138–139 (2008)
116. Arakawa, H., Zou, Z., Sayama, K., Abe, R.: Direct water splitting by new oxide semiconductor photocatalysts under visible light irradiation. Pure Appl. Chem. **79**, 1917–1927 (2007)
117. Miller, E.L., Marsen, B., Cole, B., Lum, M.: Low-temperature reactively sputtered tungsten oxide films for solar-powered water splitting applications. Electrochem. Solid State Lett. **9**, G248–G250 (2006)
118. Yan, Y., Wei, S.-H.: Doping asymmetry in wide-bandgap semiconductors: origins and solutions. Phys. Stat. Sol. B **245**, 641 (2008)
119. Alexander, B.D., Kulesza, P.J., Rutkowska, I., Solarska, R., Augustynski, J.: Metal oxide photoanodes for solar hydrogen production. J. Mater. Chem. **18**, 2298–2303 (2008)
120. Cole, B., Marsen, B., Miller, E.L., Yan, Y., To, B., Jones, K., Al-Jassim, M.M.: Evaluation of nitrogen doping of tungsten oxide for photoelectrochemical water splitting. J. Phys. Chem. C **112**, 5213–5220 (2008)
121. Honga, S.J., Juna, H., Borsea, P.H., Lee, J.S.: Size effects of WO_3 nanocrystals for photooxidation of water in particulate suspension and photoelectrochemical film systems. Int. J. Hydrogen Energy **34**, 3234–3242 (2009)
122. Miller, E.L., Paluselli, D., Marsen, B., Rocheleau, R.E.: Low-temperature reactively sputtered iron oxide for thin film devices. Thin Solid Films **466**, 307–313 (2004)
123. Duret, A., Grätzel, M.: Visible light-induced water oxidation on mesoscopic α-Fe_2O_3 films made by ultrasonic spray pyrolysis. J. Phys. Chem. B **109**, 17184–17191 (2005)
124. Hu, Y.-S., Kleiman-Shwarsctein, A., Forman Hazen, A.J., Park, J.N., McFarland, E.W.: Pt-doped α-Fe_2O_3 thin films active for photoelectrochemical water splitting. Chem. Mater. **20**, 3803–3805 (2008)
125. Kleiman-Shwarsctein, A., Hu, Y.-S., Forman, A.J., Stucky, G.D., McFarland, E.W.: Electrodeposition of α-Fe_2O_3 Doped with Mo or Cr as Photoanodes for Photocatalytic Water Splitting. J. Phys. Chem. C **112**, 15900–15907 (2008)
126. Kay, A., Cesar, I., Grätzel, M.: New benchmark for water photooxidation by nanostructured α-Fe_2O_3 films. J. Am. Chem. Soc. **128**, 15714–15721 (2006)
127. Berglund, S.P., Flaherty, D.W., Hahn, N.T., Bard, A.J., Mullins, C.B.: Photoelectrochemical oxidation of water using nanostructured $BiVO_4$ films. J. Phys. Chem. C **115**, 3794–3802 (2011)
128. Liang, Y., Kleijn, S.J., Mooij, L.P.A., Van de Krol, R.: Defect properties and photoelectrochemical performance of $BiVO_4$ photoanodes. 216th ECS Meeting, Abstract #1172 (2009)

129. Enache, C.S., Lloyd, D., Damen, M.R., Schoonman, J., Van de Krol, R.: Photo-electrochemical properties of thin-film $InVO_4$ photoanodes: the role of deep donor state. J. Phys. Chem. C **113**, 19351–19360 (2009)
130. Chen, X., Liu, L., Yu, P.Y., Mao, S.S.: Increasing solar absorption for photocatalysis with black hydrogenated titanium dioxide nanocrystals. Science **331**, 746 (2011)
131. Yae, S., Kobayashi, T., Abe, M., Nasu, N., Fukumuro, N., Ogawa, S., Yoshida, N., Nonomura, S., Nakato, Y., Matsuda, H.: Solar to chemical conversion using metal nanoparticle modified microcrystalline silicon thin film photoelectrode. Solar Energy Mater. Solar Cells **91**, 224–229 (2007)
132. Sebastian, P.J., Mathews, N.R., Mathew, X., Pattabi, M., Turner, J.: Photoelectrochemical characterization of SiC. Int J. Hydrogen Energy **26**, 123–125 (2001)
133. Repins, I., Contreras, M.A., Egaas, B., DeHart, C., Scharf, J., Perkins, C.L., To, B., Noufi, R.: 19.9%-efficient ZnO/CdS/$CuInGaSe_2$ solar cell with 81.2% fill factor. Prog. Photovolt. Res. Appl. **16**, 235 (2008)
134. Bär, M., Weinhardt, L., Pookpanratana, S., Heske, C., Nishiwaki, S., Shafarman, W., Fuchs, O., Blum, M., Yang, W., Denlinger, J.D.: Depth-dependent band gap energies in Cu(In, Ga) (S, Se)2 thin films. Appl. Phys. Lett. **93**, 244103 (2008)
135. Bär, M., Bohne, W., Röhrich, J., Strub, E., Lindner, S., Lux-Steiner, M.C., Fischer, Ch-H: Determination of the band gap depth profile of the penternary $Cu(In_{(1-X)}Ga_X)(S_YSe_{(1-Y)})_2$ chalcopyrite from its composition gradient. J. Appl. Phys. **96**, 3857 (2004)
136. Bär, M., Weinhardt, L., Heske, C., Nishiwaki, S., Shafarman, W.: Chemical structures of the Cu(In, Ga)Se2/Mo and Cu(In, Ga)(S, Se)2/Mo interfaces. Phys. Rev. B **78**, 075404 (2008)
137. Marsen, B., Cole, B., Miller, E.L.: Photoelectrolysis of water using thin copper gallium diselenide electrodes. Solar Energy Mater. Solar Cells **92**, 1054–1058 (2008)
138. Jaramillo, T.F., Jørgensen, K.P., Bonde, J., Nielsen, J.H., Horch, S., Chorkendorff, I.: Identifying the active site: atomic-scale imaging and ambient reactivity of MoS_2 nanocatalysts. Science **317**, 100–102 (2007)
139. Maiolo, J.R.I.I.I., Atwater, H.A., Lewis, N.S.: Macroporous silicon as a model for silicon wire array solar cells. J. Phys. Chem. C **112**, 6194–6201 (2008)

Part II
Materials Properties and Synthesis

Chapter 8
Economic and Business Perspectives

Julian Keable and Brian Holcroft

8.1 Introduction

Mankind faces the biggest challenge in the history of energy use – to find how to store natural energy in a way that would replace the vast store created by great nature herself, a store of which we have used a half in the last 50 years [1].

This is also one of the biggest opportunities ever to arise. The market is vast for a fuel that can be conveniently stored and transported, provided from a sustainable, low environmental impact source, at an affordable price, to meet the inevitable shortfall after "peak oil."

There are many possible ways forward. One generic way is to split water, and store the resulting hydrogen. Photoelectrochemical (PEC) hydrogen production is not yet a commercial solution for this, and for it to be considered there are basic material improvements required, together with robust system implementation. Materials with parameters in the range of photocurrents of 8 mA/cm^2 in "1 Sun" with 0.8 V anode to cathode bias would open the way to exploitation of PEC-derived hydrogen, and the possibility of a cost/m^2 under \$80, excluding PV bias costs with lifetimes $>$10 years. An early demonstration model of such a system is shown in Fig. 8.1.

8.2 The Need

Our entire world economy has grown out of the availability of abundant and cheap energy supplies. The cost of energy is now under threat, and the future is by no means secure.

J. Keable (✉) • B. Holcroft
Stored Solar Ltd, Reading, UK
e-mail: julian@jk-p.demon.co.uk

R. van de Krol and M. Grätzel (eds.), *Photoelectrochemical Hydrogen Production*,
Electronic Materials: Science & Technology 102, DOI 10.1007/978-1-4614-1380-6_8,

Fig. 8.1 Demonstration model of a 1-m^2 PEC array

8.2.1 Increased Demand/Reduced Supply = Rising Cost

In the 4 years up to 2007 [2],

- Oil consumption increased by 3 million barrels/day
- Natural gas increased by 8 billion cubic feet/year
- Coal by 1.4 thousand short tons/year
- Electricity by 2.1 billion kWh/year

Projections show that demand continues to grow – and that the transport sector is elastic to price, there being no available alternative.

Oil production per capita fell from 5.26 barrels in 1980 to 4.73 barrels in 2006. The DOE/EIA give a price projection from [2]:

- \$60/barrel in 2009 to
- \$110 in 2015
- \$115 in 2020
- \$130 in 2030

8.2.2 *Growth in Energy Demand Arises from Several Different Causes*

These causes are

- Population growth – mainly in the developing world
- Increased standards of living within upper income sectors
- Increased size of middle income groups

The figures quoted above hint at the size of this growth in energy demand. The EIA forecast an 80% increase in consumption in non-OECD Asia, largely due to transport needs [3]. Of special interest is the ever increasing proportion of energy use in the transport sector generally, in particular road vehicles and air travel. Of total primary energy use, some 40% is used in the transport sector. This is met by the use of oil in different grades (petrol, diesel, jet fuel), and there is little prospect of a change to other energy sources here. There is an absolute need for transport fuel to be convenient and transportable – on board – which means that the use of electricity is itself dependent on storage, in batteries of some sort.

8.2.3 *"Peak Oil"*

Peak oil was first predicted by M. King Hubbert in 1956 [4]. He showed that every well increased its production up to a peak, coinciding with the depletion of half the total capacity of the well, then slowly declined once past this peak. He showed that this is just as true for the overall capacity of all wells. Dr. C.J. Campbell, a trustee of the Oil Depletion Analysis Centre in London, spelt out the consequences in some detail in his 1997 book "The coming Oil Crisis" [5]. In 1980, oil production first surpassed new oil discoveries. Therefore, there is a need for an alternative to augment oil and preferably from a low impact source due to climate change issues.

Renewable sources are attractive for sustainability – e.g., wind, hydro, tidal, solar etc., and geographic factors will determine where each is appropriate. However, the sheer size of overall demand points to solar – the solar resource far outstrips all others. With a global usable energy flux of 120 PW, it is some 8,000 times the current total energy use, and able to supply growing demand as far ahead as we can see [6].

There has been a huge emphasis on the production of electricity from alternative sources, and for many, "solar cells" mean photovoltaic cells. Wind, hydro, tidal, and atomic sources all tend to equate to the supply of electricity. As noted above, electricity needs to be stored if it is to meet the demands of transport, and, in the case of wind and solar, periods when the source is unavailable.

8.3 The Commercial Challenges

Energy must be affordable, but the cost of energy in different forms (with and without tax) varies widely.

8.3.1 Price Distortions

Governments distort true prices by giving subsidies on the one hand, and applying taxes on the other. The 2005 Energy Bill in the USA gave $6 billion for oil and gas, $9 billion for coal, and $12 billion for nuclear [7]. President Obama is seeking to remove those subsidies. Tax on petrol varies by country. In UK 67% is duty, in Russia 0%. In the USA, tax varies by state, with Alaska at 18 cents/gallon, and New York at 61 cents [8].

8.3.2 The Climate Change Factor

Despite the growing awareness and acceptance that climate change is real, and that it poses huge threats to our whole existence, there has been little appetite by Governments to address the issue in a robust way. The conclusion drawn from this may be that carbon-free alternatives must compete without relying on government support or coercion.

8.3.3 The Need for Clear Focus

Research effort, and the subsequent process of bringing research results to a successful product and market, is costly. It is important, therefore, to focus on well-formulated objectives. Set out below are a number of such objectives, all having in view the aim of finding a viable way to store solar energy, as an alternative to mining the natural resources we have inherited.

8.3.4 Definition of Aims

Development of an energy supply which is

- As cheap as existing "conventional" supplies
- As convenient in use
- Less environmentally damaging
- Reasonably safe in use (all energy can be dangerous!)

8.4 Storage of Solar Energy: The New Challenge, and The Hydrogen Solution

Throughout history, mankind has existed by harvesting and mining natural resources (wood, coal, oil, and gas) for his energy needs. Only with the coming of the atomic age can it be said we have ourselves invented or created energy, and even now we have still to find anything comparable with the stored energy created by nature. Her "invention" of the tree, which covers itself with disposable solar collectors during the sunny season, converts sunlight to a form of stored energy, then discards those leaves when winter comes – how elegant!

Wood, coal, oil, and gas can and do store energy for any amount of time, ready for instant use.

This is our real challenge, for whether "peak oil" has already come, or whether it is still 20 years away ("tomorrow morning"), once we are on the far side of the mining curve nothing can replace that golden resource.

So we come to our subject here – the possibility of splitting water using sunlight, to provide a source of stored energy at an affordable price. PEC is such a possibility, but many hurdles remain.

8.4.1 Can We Learn from Nature?

The sunlight/photosynthesis/fossil fuel process has been working at efficiencies well below 1%, thus needing vast land areas – land areas now committed to other uses where they are fertile, or lacking either water or sun elsewhere. There are many ways to approach this problem, PEC being only one, and this is not the place to examine the alternatives. Each has advantages, disadvantages, risks, and state of readiness.

8.4.2 There Are A Number of Ways to Split Water and Yield Hydrogen

Direct thermal splitting is one such way – and there is work going on looking at efficient catalysts for this [9, 10]. Electrolysis is another and well-understood way – the use of electric power to split water. It may be driven by electricity from any source – conventional (fossil fuelled), wind power, hydro or wave power, nuclear or PV [11]. Steam methane reformation (SMR) is at present the cheapest way to produce hydrogen, and relies on fossil fuel (natural gas). This provides a cost per kg and kWh of H_2 burnt split between capital and electrical power cost [12, 13].

The DOE target for hydrogen production from natural gas (NG) was \$3/kg H_2 by 2005, and \$1.50/kg by 2010 [2]. The volatile price of NG is the main factor here.

If NG is traded at \$1/kg equivalent, and if conversion from NG to H_2 is at a ratio of 1–2.4 (when carbon capture and storage is included) then even at very large scale the cost of H_2 is likely to be some \$2.4/kg. Thus the \$3/kg target, not yet reached, is likely to remain the lowest.

Although meeting a target cost of twice this figure using PEC is already very challenging, this aim must be kept in sight, and achievements measured in the light of such goals.

8.4.3 End Use of Hydrogen

The end use of hydrogen is, ideally, by direct combustion. Process heat, cooling by the absorption cycle, combustion in IC vehicle engines, cooking, and water heating all make the best use of hydrogen. To generate electricity, while perfectly possible, is to lose efficiency overall. Fuel cells have yet to be proven, and have yet to show cost benefits. Hydrogen has an important role to play without the need for deployment in that way.

8.4.4 EU/US Targets for H_2 Evolution

The US DOE target has been shifting gradually upward, from \$1.50/kg delivered H_2 (2001) to \$2/kg at production site, to \$3/kg. EU targets twice the untaxed petrol cost, on the basis that petrol taxation will make H_2 competitive if untaxed.

8.5 PEC Issues

Each element of the PEC process has a bearing on the final cost of hydrogen production, and hence of commercial viability.

8.5.1 PEC Costs for Hydrogen

It is constructive to consider the optimistic cost of hydrogen produced today using PEC compared with the target prices. In this calculation, typical values of photocurrent for modified iron oxide of 2 mA/cm^2 at 1 "Sun" illumination with a bias of 1.2 Volts have been used and costs for 17% efficient PV at \$2/W peak incorporated. Clearly, the cost of hydrogen per kg produced is dependent on the location and a site in Southern Spain has been assumed with an averaged monthly insolation of 4.8 $kWh/m^2/day$. To produce 1 kg of hydrogen per day would require some

50 m^2 of PV and 300 m^2 of PEC with minimal overall losses and inefficiencies. Assuming that the PEC costs are around \$200/m^2 (PV costs are \$300/m^2) then this yields a cost per kg of hydrogen of \$20 (assuming a 10-year lifetime) for the capital costs alone. Even if the PEC costs were halved to \$100/m^2 then the hydrogen costs are still \$12/kg. At this lower price the PV costs are around half the PEC costs due to the large area of PEC required.

This cost of \$20/kg is clearly too high and around a factor of 10 greater than the target price and does not include additional costs of land, maintenance, installation, transport, the balance of plant costs, and profit. Clearly, an improvement in PEC efficiency would be advantageous to limit the land usage. This suggests that significant improvements in performance of the PEC device are required to meet commercial demands.

We should not ignore competitive means of producing hydrogen from a solar input via electrolysis. Given the likely timescale for improvements in PEC, we should use a comparison with other methods under development. Direct comparison with low cost PEM electrolyzers powered by PV is instructive. The US targets for low cost PEM are \$400/kW with 67% efficiency by 2012 and \$125/kW with 74% efficiency by 2017. The reported status in 2007 was \$987/kW with a cell efficiency of 67% [14]. Using this data for the site in Southern Spain with PV costs of \$2/W peak we can estimate hydrogen costs for the capital alone with a 10-year payback if the systems were implemented of \$8/kg using the 2007 figures, reducing to around \$5/kg of hydrogen with the 2017 figures. The reduction in cost is not as marked as expected from the reduction in PEM electrolyzer costs due to the influence of the PV costs which have been assumed not to reduce below \$2 per watt peak over the period. The PV costs would be 94% of the capital costs in such a system, and using the 2017 figures it would suggest this system would require around 60 m^2 of 17% efficient PV.

Using the 2012 figures the PV accounts for 80% of the cost of a PEM system, with the PEM electrolyzer capable of producing 1 kg of hydrogen per day in Spain. The estimated capital cost of some \$4,000 yields a cost per kg of hydrogen of \$6 assuming a 10-year payback.

8.5.2 *Efficiency and Improved Photocurrent*

It is widely held that the successful application of PEC to the water-splitting process must reach or exceed an efficiency of 10% (i.e., for a given incident solar energy, the energy content of the liberated hydrogen must be 10% or more). Put simply, at "1 Sun" (1,000 W/m^2), the aim must be for 1 m^2 of device to liberate hydrogen having an energy content of 100 W. In the following sections, we set out to expand on materials parameters that support this.

The efficiency of PEC is directly related to the magnitude of the photocurrent produced by exposure to light. Clearly, doubling the efficiency or photocurrent halves the area of PEC required to produce a given mass of gas.

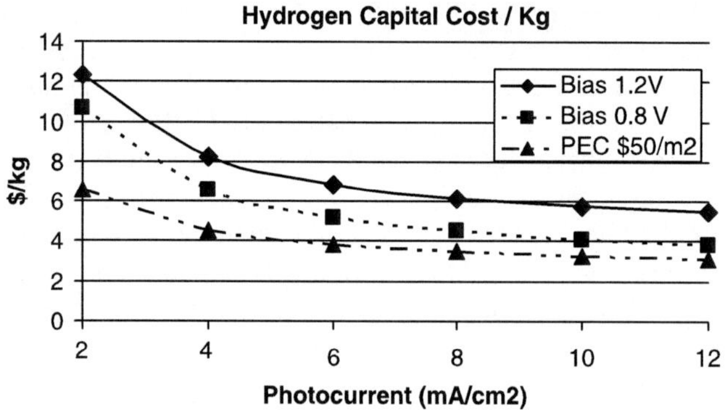

Fig. 8.2 Capital cost of hydrogen per kg as a function of photocurrent

The solid line in Fig. 8.2 represents the capital cost per kg of hydrogen produced with a 10-year payback assuming a PEC device requiring a bias of 1.2 V that can be manufactured for \$100/m^2. To match the low-cost PEM equivalent price of \$6/kg would require an improvement in efficiency to yield a "1 Sun" photocurrent in excess of 8 mA/cm^2. At this value the area of PEC required in Spain would be around 75 m^2 compared to 50 m^2 of PV and 60 m^2 of PV for a low-cost PEM. As can be seen increasing the photocurrent further provides decreasing gains in lower cost hydrogen as the PV costs, which do not decrease with increasing PEC photocurrent, begin to dominate the total costs. At a "1 Sun" photocurrent of 12 mA/cm^2 the PV costs, assuming \$2/W peak, would be \$15,000 compared to \$5,000 for the PEC. The most advantageous method to reduce cost further is to reduce the bias requirements and so reduce the amount of PV required.

8.5.3 Voltage Issues and Reduced Bias

Increasing the efficiency reduces the cost of PEC to produce a given mass of gas but not the cost of PV. Since PV dominates the cost overall, then lower cost of PEC does not itself produce a significant saving. How can we reduce the amount of PV power needed? This can only be achieved by reducing the bias voltage as the mass of hydrogen produced is proportional to the current.

Water splitting by electrolysis has a threshold minimum of 1.23 V and in a PEC device this is effectively reduced by the photocatalyst, the reduction being in part related to the band gap of the material and the positioning of the energy bands relative to the hydrogen and oxygen evolution potentials. However, an overvoltage is observed for realistic devices. The size of this overvoltage is critical and must be minimized.

By comparison of the solid line for a PEC device manufactured for \$100/m^2 with a bias of 1.2 V and the single dashed line for a PEC device with a bias of 0.8 V (Fig. 8.2) it can be seen that the hydrogen cost per kg can be reduced below that of a low-cost PEM at reasonable photocurrent values above 5 mA/cm^2. Efforts to reduce the bias are essential to achieve a meaningful advantage over low-cost PEM and ideally PEC systems requiring no external bias would be highly attractive.

It is important to note that with devices exhibiting "1 Sun" photocurrents of 8 mA/cm^2, currents of the order of 80 A for a single m^2 can be expected. A 1 mΩ resistance in a pathway carrying 80 A would result in a resistive voltage drop of 0.08 V which is significant when efforts to reduce the bias are hard won. Hence, very low electrode resistances and interconnections are crucial to a PEC system. Also sources of electrode overpotential, losses due to gas separation membranes, and reactor designs all play a part in minimizing the external bias requirements for a PEC device.

From a system view point anything that loses voltage, e.g., electrode resistance, interconnections, over potentials, reactor design, electrolyte, H_2 evolving electrode, etc., is important.

8.5.4 Scale-Up Problems

Research is typically carried out using small samples on which to try alternative coatings or application processes. By this means, many great experiments can be made with a relatively small use of materials and other resources. However, in the context of PEC development, the question of electrical resistances must not be overlooked.

At small scale, only small travel distances will be involved – the question of resistance may not well arise. But once a promising process comes to be scaled up to commercially realistic units, current flows then rise from milliamps to amps, and the need for a low resistance pathway may rule out the concept of a semitransparent conducting coating on glass (for example) – only highly conductive metallic electrodes which are nontransparent may suffice.

8.5.5 "Real Estate" Issues

There are many ways in which the area of land required per unit of energy harvested can affect cost. These include location (level of insolation to be expected), urban or off-grid siting, the need for support infrastructure, and other competing land uses.

For example, if we compare the 2012 low-cost PEM figures with the PEC estimates based on performance today we see that the PEC system would require around 20% less area of PV and hence the cost of PV. However, the land area used by the PEC system today would be dominated by the 300 m^2 of PEC which is some five times the area of a compact PEM system in which the PEM electrolyzer does not need exposure to light and can be located beneath the PV. The capital cost of hydrogen produced per kg is around double from the PEC system assuming it can be manufactured at costs of \$100/m^2. This would suggest that to be competitive with proposed low-cost PEM it is essential to reduce the area of PEC required through increased efficiency.

8.5.6 Solar Spectrum Considerations

In general, it is clear that having PV and PEC side by side is less efficient than using the same area to harvest different parts of the spectrum for the two requirements, so light harvesting is itself an issue relating to cost.

The tandem/hybrid approach seeks to address this question. A collaboration between EPFL and the University of Geneva leads to a proposal for a "Tandem Cell" which sought to use different parts of the spectrum of white light for different purposes. A patent was granted for this idea, based on harvesting energy from the blue end of the spectrum in a PEC application, and the red end of the spectrum for generating the required bias to achieve water splitting, using PV.

This elegant solution appeared to offer a way of "taking two bites from the same cherry," and did function at very small scale. Care with such designs must be taken not to limit the transmitted light through the PEC device and into the underlying PV, otherwise PV outputs will be reduced which will result in increased costs of the PV. The need for highly conductive substrates at any reasonable photocurrent density and area of device within PEC devices will severely limit the light transmission and limit the commercial viability of this approach.

8.5.7 Durability

Products must be durable, and pay-back calculations will be influenced by the expected product life-time. As can be seen in Fig. 8.2, even with an external bias of 0.8 V and a PEC manufacturing cost of \$50/m^2, which is one sixth that of current PV and a "1 Sun" photocurrent of 12 mA/cm^2 the hydrogen cost of the capital costs alone are still in excess of the EU and US targets assuming a 10-year pay back. These targets can only be met by increasing the durability of the PEC to give lifetimes approaching 20 years. To put this in perspective a 20-year lifetime operating 9 h per day is in excess of 65,000 h, which is challenging when you consider that a 10,000-h auto engine test would be considered impressive.

Although there may well be markets for shorter life-time products, in the longer term there will be an absolute requirement for products capable of operating for 15 or 20 years, comparable with other core technologies.

For these reasons, durability testing should be built into the overall PEC development program from the earliest stages. In some cases, it will be found that durability cannot be achieved for reasons of fundamental material science, thus in such cases there will be little reason to pursue an otherwise interesting line.

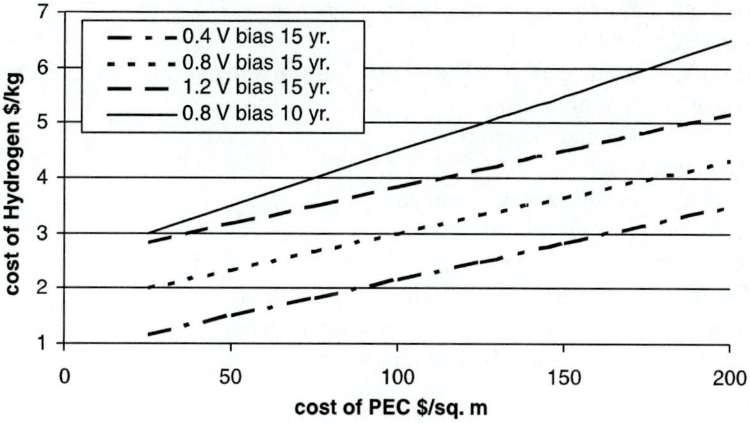

Fig. 8.3 Hydrogen costs per kg (capital costs only) for PEC with 1 Sun photocurrents of 8 mA/cm^2 in Southern Spain assuming \$2/W$_{peak}$, 17% efficient PV

8.5.8 Commercially Attractive Targets

Figure 8.3 highlights that for PEC with a 1 Sun photocurrent of 8 mA/cm^2 and a bias of 0.8 V, capital costs of hydrogen per kg less than \$3/kg, with a 10-year payback, cannot be achieved (solid line). Equally, PEC at similar photocurrents with 1.2 V bias cannot achieve sub-\$3/kg hydrogen costs even with a 15-year payback (large dashed line). To achieve sub-\$3/kg hydrogen capital costs for PEC with photocurrents of 8 mA/cm^2, the bias must be at least as low as 0.8 V with a 15-year payback and PEC costs below \$100/m^2 (small dashed line). If the bias can be reduced further to 0.4 V then with a 15-year payback PEC costs as high as \$150/m^2 can be sustained, which is around half the current cost of PV.

Clearly, the performance of PEC today is not ready for commercial exploitation. However, with defined improvements it offers a route to the production of hydrogen on a large scale which would meet EU and US targets and be comparable with current methods of hydrogen production. To achieve this, "1 Sun" photocurrents in excess of 8 mA/cm^2 are required, using materials that require less than 0.8 V bias, exhibiting much greater than 10-year lifetimes, that can be deposited and incorporated into reactors which cost less than \$100/m^2.

8.5.9 Material Parameters

Several determinants for the choice of materials will contribute to a commercially viable device. These include

- High photocurrent leading to high efficiency
- High photon capture

- High photon-to-electron conversion efficiency
- Low electron–hole loss
- 1.6 eV < band gap < 2.2 eV matching hydrogen and oxygen evolution potentials to minimize bias
- Low cost of source material
- Abundant supply
- Durability in aqueous environment
- Resistance to high solar irradiance
- Resistance to high temperature
- Electrode structures supporting high current flows
- Avoidance of poisonous or contaminating materials
- Electrolyte choice, ideally, commonly available, nontoxic, noncorrosive

8.6 Other Issues

Only issues of relatively direct impact on PEC research are touched on here. These include the following.

8.6.1 Hydrogen Storage

Sometimes seen as an "Achilles heel," the problem of how to store and transport hydrogen remains to be solved. Historically, heavy steel cylinders have been and are still used to contain the gas at a range of pressures which ensure that the volume is not unacceptably large.

Only a small number of companies specialize in this field, and the development of much lighter weight containers based on carbon fibers, special resins, and so on is now well advanced, ready to supply the needs of the automotive industry, where the overall weight of gas plus container is a constraint.

Carbon nanotubes are being investigated, and metal hydrides are already available. Other solutions based on storing hydrogen in another form are also under review. Ammonia has its advocates, being a material which is handled in bulk by the fertilizer industry, and is composed of just two elements – N and H. However, transforming the hydrogen into and then back out of this chemical is troublesome, and other solutions are sought.

One interesting alternative is formic acid (H_2CO_2), as the CO_2 component may be usefully abstracted from power plants.

8.6.2 System Costs

To achieve the stated aim of competing with existing energy sources, without reliance on subsidy or tax hedge, the relevant bench mark is that of hydrogen production by SMR. This is a moving target, as the source for SMR is natural gas, itself a hydrocarbon under pressure from rising demand, and dwindling supply. However, we can compute the following using some assumptions:

- A 10% efficient device with bias requirements less than 0.8 V
- A 15-year minimum life
- A capital cost for the system (PEC + PV) of \$160/m^2

With these assumptions PEC hydrogen could cost less than \$3/kg. A clear statement of aims such as these is essential for the development of a commercially convincing product.

8.6.3 Solar Concentration and Tracking

Modest solar concentration, in the 3× to 10× range, may be helpful in boosting output with a low capital cost increase, provided a nonsolar tracking configuration is adopted.

High concentration schemes are needed for high-temperature applications, but in the context of PEC cannot be recommended. The extra cost of the tracking component is likely to put the overall energy cost beyond an acceptable limit. Remember too that solar tracking implies clear beam radiation, thus reducing the number of locations where such applications can be sited.

8.6.4 The Issue of Intellectual Property ("IP")

The world of "risk capital" funds, hedge funds, and the like places great reliance on the protection of IP by patents, and typically a family of patents is sought, covering all aspects of the given technology. Without seeking to detract from this viewpoint, it may be more important to rely on "know-how" and practical experience. For example, the original "dye-cell" patent has already expired, but the dye-cell team is busier than ever with applications for their know-how. However, there is a conflict between the wish for researchers to publish their work, and the need to maintain secrecy until patent cover has been applied for.

8.6.5 Further Issues

Maintenance/support in operation – the simplicity or otherwise of the PEC solution will affect the ongoing costs – what elements will need cleaning, and at what intervals? What elements will need replacement and at what frequency?

Hydrogen transportation from point of production to point of use will be an issue, especially since every pipe joint represents a potential leak of this finest of all gases.

8.7 Route to Market

Section 8.2 showed the huge sector represented by transport energy. For transport fuel, large scale and large company development is needed to make an impact. This will only happen when the requirements of cost, convenience, and safety have been met. At this point all the other uses of conventional fuels will be opened too.

That said, fleet vehicles tied to a refueling base (taxis, buses, delivery vehicles) avoid some of the "chicken-and-egg" problems associated with the eventual need for fully distributed infrastructure, while small fleet vehicles (fork-lift trucks, airport tugs) are even more readily serviced.

8.7.1 Other Markets

There are as many potential markets as there are activities in this world. The broad divisions noted above can also be subdivided into "grid connected" and "remote" or "off-grid" locations. The latter is of interest in that the price of energy at remote sites is already much higher than where grid connection exists. Niche markets alone do not justify great expenditure, so there is a need to think big – once a competitive process has been developed.

8.7.2 A Comment on "growth"

The idea that we need our world economy to grow for it to be healthy must be treated with caution. If population grows, then growth of supplies is needed. Apart from that, it would seem that the economic and banking crisis of 2008–2009 shows us that growth is chiefly needed to satisfy greed, and leads to collapse. *Sustainability is a much more relevant criterion.*

8.8 Conclusions

The opportunity to find a way to store solar energy poses huge challenges, but offers a huge reward – not just financial, but deeply satisfying in terms of achievement in an entirely new field for humanity. Clear targets must be set, ideally leading to system costs that are easy to accept in being no more than the fossil fuel alternative. Here, proposed is a figure of under $3/kg of hydrogen produced. To achieve this, we see that

- More efficient materials are needed, and a target of 8 mA/cm^2 is set (to reach >10% efficiency)
- Bias requirements must be reduced, with minimal resistive voltage drops
- Durability must be increased to 15, or better 20 years
- System costs must be reduced, and a price not exceeding $160/m^2 for both PEC and PV is set

By splitting water directly – in one step –the need for increased independent electricity generation is avoided, as is the need for grid connection. More importantly, the way would be opened to a real reduction in CO_2 emissions at the most basic level. The tasks involved are large, and higher levels of funding are likely to be needed, especially if results are to be achieved in the near term.

References

1. Leggett, J.: Half Gone: Oil, Gas, Hot Air and the Global Energy Crisis. Portobello Books Ltd, London (2006)
2. Energy Information Administration, Office of Integrated Analysis and Forecasting, U.S. Department of Energy: International Energy Outlook 2009. http://www.eia.doe.gov/oiaf/ieo/index.html. Accessed 1 Nov 2010
3. Energy Information Administration, Office of Integrated Analysis and Forecasting, Chapter 7: Transportation Sector Energy Consumption. U.S. Department of Energy: International Energy Outlook 2009. http://www.eia.doe.gov/oiaf/ieo/index.html. Accessed 1 Nov 2010
4. Wikipedia: Peak Oil. http://en.wikipedia.org/wiki/Peak_oil. Accessed 1 Nov 2010
5. Campbell, C.J.: The Coming Oil Crisis. Multi-Science Publishing Co. Ltd. & Petroconsultants S.A, Essex (1997)
6. National Renewable Energy Laboratory (NREL): Renewable Resource Data Center – the solar resource. http://www.nrel.gov/rredc/solar_resource.html. Accessed 1 Nov 2010
7. Wikipedia: The Energy Policy Act of 2005. http://en.wikipedia.org/wiki/Energy_Policy_Act_of_2005. Accessed 1 Nov 2010
8. CNN Money.com: Gas prices around the world. http://money.cnn.com/pf/features/lists/global_gasprices/. Accessed 1 Nov 2010
9. Kogen, A.: Direct solar thermal splitting of water and on-site separation of the products. II. Experimental feasibility study. Int. J. Hydrogen Energy **23**, 89–98 (1998)
10. Perkins, C., Welmer, A.: Likely near-term solar-thermal water splitting. Int. J. Hydrogen Energy **29**, 1587–1599 (2004)

11. Wikipedia: Electrolysis of Water. http://en.wikipedia.org/wiki/Electrolysis_of_water. Accessed 1 Nov 2010
12. National Research Council: The Hydrogen Economy – Opportunities, Costs, Barriers, and R&D Needs. The National Academies Press, Washington DC (2004)
13. Lindsay, I., Lowe, C., Reddy, S., Bhakta, M., Balkenende, S.: Designing a climate friendly hydrogen plant. Energy Proc. **1**, 4095–4102 (2009)
14. Hamdan, M.: Low cost, high pressure hydrogen generator. 2008 Hydrogen Program Annual Merit Review Meeting, pp. 2 (2008)

Chapter 9
Emerging Trends in Water Photoelectrolysis

Scott C. Warren

9.1 Introduction

Manipulating the structure of matter and the distribution of optical energy at the nanoscale will be crucial to future progress in water photoelectrolysis. Fortunately, the rapid pace of developments in nanomaterials and nanophotonics provides ample inspiration as scientists continue to confront the long-standing challenge to invent an efficient and inexpensive way to store solar energy in chemical bonds of a fuel such as hydrogen.

Preceding chapters have outlined the challenging and often competing criteria that semiconductors must meet to achieve efficient and economical water photoelectrolysis. For a device based on a single photoelectrode to be successful, these criteria must include the following:

- A bandgap between approximately 1.8 and 2.4 eV
- Aqueous stability
- Band edges that straddle the water redox potentials
- Small ohmic losses for charge carrier transport
- Minority carrier mobility and lifetime that permit efficient charge transport to the semiconductor/solution interface
- Water oxidation/reduction kinetics that are faster than surface-mediated recombination

A few semiconductors have come close to meet most of these criteria, but none has met all. It seems that strategies that go beyond compositional variation are warranted. This chapter examines strategies for nanostructuring semiconductors as well as applying plasmonic effects to enhance device efficiency.

S.C. Warren (✉)
Laboratory of Photonics and Interfaces, Swiss Federal Institute of Technology, Lausanne, Switzerland
e-mail: scott.warren@epfl.ch

R. van de Krol and M. Grätzel (eds.), *Photoelectrochemical Hydrogen Production*, Electronic Materials: Science & Technology 102, DOI 10.1007/978-1-4614-1380-6_9,

9.2 Nanomaterials

In recent years, interest in the use of metal oxides as photoelectrodes has greatly increased as significant gains in efficiency have been demonstrated. The interest in metal oxides also arises from challenges facing the aqueous stability of nonoxide semiconductors. Some classes of metal oxides that have inspired recent interest include spinels and delafossites, both of which offer highly tunable electronic and optical properties. But complex metal oxides often have low mobility and short carrier lifetimes, which has so far limited performance in devices based on these materials. Thus, nanostructuring is being increasingly explored as a route to minimize recombination.

In this discussion of nanomaterials, conditions are identified for building efficient devices from semiconductors with poor mobility and short carrier lifetimes. This approach is then analyzed within the context of anodized nanopore and nanotube arrays and directions are suggested for further improvement for this class of photoelectrodes.

9.2.1 Access to Small Length Scales

As new metal oxide semiconductors are identified and explored for water photoelectrolysis, a persistent problem that arises is the low drift mobility of electrons and holes and the short lifetimes of photogenerated carriers. Many materials that seem promising on the basis of bandgap and stability are found inadequate for photoelectrolysis because the low mobility of minority carriers prevents rapid charge transport to the aqueous interface. Metal oxides are also prone to nonstoichiometry, which can result in traps that promote recombination and shorten carrier lifetimes. When short lifetimes and low mobility arise in the same material, device efficiency can drop rapidly and photocurrents are far below those that might be otherwise expected.

Within the last decade, new nanostructuring approaches have emerged that suggest ways to solve these problems. This section reviews the major challenges that semiconductors face with low mobility and short carrier lifetimes and highlights the emerging strategies for addressing these shortcomings.

A central idea in the discussion of charge transport is that of the polaron. It is easiest to understand polarons – of which the two most common types are either large or small – by imagining an uncharged lattice with atoms located at their equilibrium positions. In this thought experiment, an electron or hole is added to the lattice and the lattice responds to this charge by delocalizing it over many atoms (a large polaron) or by localizing it over just a few atoms (a small polaron), as shown in Fig. 9.1. Because the additional charge fills or empties states with bonding or antibonding character, the charge influences bond lengths and angles. If the charge is delocalized the influence is small because it is spread over many atoms, while a highly localized charge

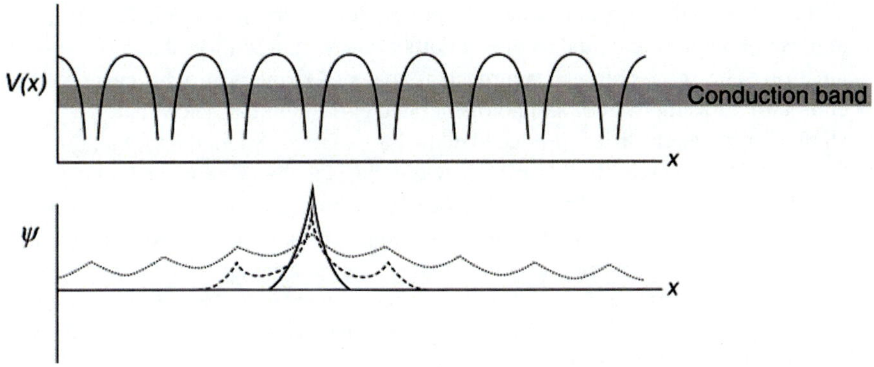

Fig. 9.1 Illustration of polaron size. The periodic potential, $V(x)$, is plotted with the range of conduction band potentials (energies) depicted in *gray*. The wavefunctions, ψ, of a small (*solid line*), intermediate (*dashed line*), and large (*dotted line*) polaron are plotted as a function of a single spatial dimension, x. Illustration is based on [5]

dramatically alters bond lengths and angles because it induces comparatively large changes in electron density. The concept of polarons, then, is used to indicate that the charge carrier and accompanying lattice distortion are coupled together and must therefore be considered as a single unit or quasi-particle.

Holstein has argued that the concept of self-trapping is probably the most important idea arising from the theory of polarons [1]. When a charge carrier is localized as a small polaron, the lattice shifts to new equilibrium positions within a period of a few lattice vibrations [2]. This increases the stability of the state, creating a deeper potential well that must be overcome to transfer the polaron to a neighboring atom. In the context of Marcus theory [3], it is said that there is a large reorganization energy associated with small polarons. It has also been argued that it is more appropriate to distinguish small and large polarons by self-trapping, rather than the extent of lattice deformation [4].

While the phenomena that distinguish small polarons from large polarons has been qualitatively outlined, there are also quantitative measures. Perhaps the most intuitive is electron or hole drift mobility. Because of the larger activation energy associated with transport, small polarons have lower drift mobility than large polarons. Bosman and van Daal have derived estimates of the limit for small polarons by assuming that self-trapping occurs when the persistence of a polaron on a site is much longer than the time period of a lattice vibration [2]. This allows an upper limit for small polaron mobility to be estimated as 0.1 cm^2/V s. A lower limit for large polaron mobility was estimated as 0.8 cm^2/V s, although the value was sensitive to the model's assumptions. Other estimates have placed the lower limit for large polarons at 10 cm^2/V s and the upper limit for small polarons at around 10^{-2} cm^2/V s [4]. Another distinguishing characteristic between small and large polarons is that the mobility of small polarons increases with temperature because transport is assisted by lattice vibrations; large polarons show decreasing mobility with temperature because lattice vibrations scatter carriers [4].

To consider the impact of low mobility on device efficiency, it is first necessary to present the principal models of a Schottky barrier in a semiconductor–electrolyte junction. The most common approach is that of Gärtner [6], who calculated the photocurrent as the sum of all photogenerated carriers in the space charge layer and all photogenerated carriers produced in the bulk of the semiconductor that diffuse to the space charge layer. In this approach, emphasis is therefore placed on determining the minority carrier diffusion distance L_m (which is a function of mobility), the absorption coefficient α, and the space charge layer width w [7]. The Gärtner model assumes that no recombination occurs in the space charge layer, but this may not be suitable for materials with low mobility, where lowering the mobility will increase the time required for a minority carrier to cross the space charge layer and therefore increase the likelihood of recombination.

Seeing this limitation, Jarrett proposed a revised model that allows for recombination in the space charge layer [8]. Jarrett considered the effects of low mobility on device efficiency by first examining the influence of mobility on the transit time, τ_t, the time required for minority carriers to cross the space charge layer. The transit time is calculated as

$$\tau_t = \frac{\varepsilon\varepsilon_0}{eN_i\mu}, \tag{9.1}$$

where ε is the static dielectric constant, ε_0 is the vacuum permittivity, e is the electric charge, N_i is the ionized impurity level concentration, and μ is the mobility [8]. This is plotted in Fig. 9.2 for both small and large polaron transport for varying values of N_i with $\varepsilon = 100$, a typical value for a metal oxide semiconductors. For semiconductors with small polaron transport to have τ_t below 10 ps, ionized impurity concentrations must be above 10^{19}, which implies doping densities in excess of 0.1 at.%. An interesting feature is that τ_t is independent of band bending, V_B, because the higher carrier velocity that accompanies increased V_B is exactly countered by the greater space charge layer width.

There should be some caution in broadly applying (9.1) to all types of carrier transport at interfaces. For example, the relationship does not accurately model the transit time of ballistic transport because the calculation of τ_t depends on the mobility, which is only accurate in so far as it measures a diffusive process, i.e., one that involves multiple scattering events [9]. Because the small polaron conductors have transport mediated by lattice vibrations, numerous scattering events will occur as the carriers cross the space charge layer. Therefore, the transit times as calculated by (9.1) should be representative of the behavior for this class of materials [10].

In calculations of efficiency, τ_t is to be compared with the carrier lifetime in the space charge layer, τ_R. From a qualitative standpoint, if $\tau_t > \tau_R$, a sizable proportion of minority carriers generated inside space charge layer will recombine before making it to the semiconductor/electrolyte interface. Taking a quantitative approach, Jarrett solved the diffusion equation for an illuminated planar semiconductor/electrolyte interface and calculated incident photon-to-current efficiency (IPCE) for systems with varying τ_t/τ_R [8]. One series of calculations used as

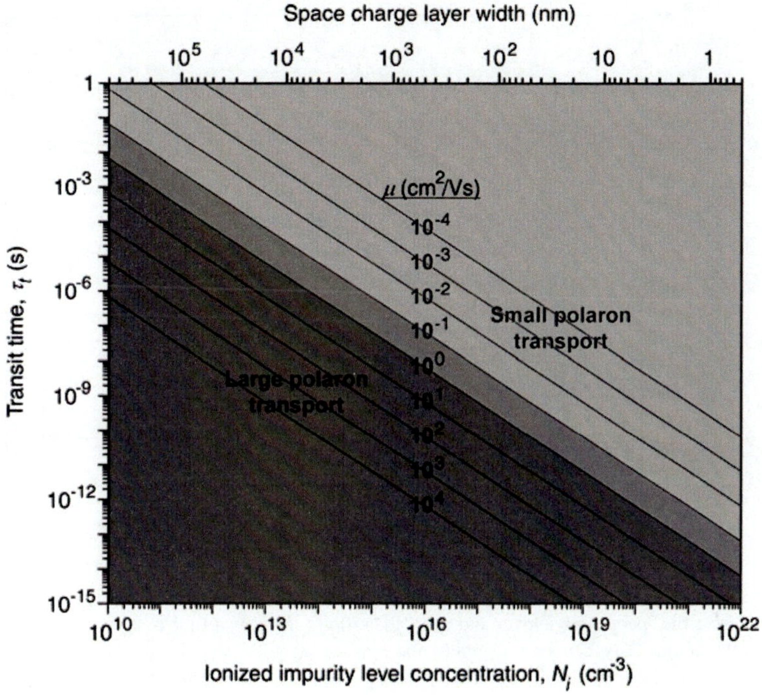

Fig. 9.2 Calculated transit times, τ_t, for the movement of an electron or hole across the space charge layer as a function of ionized impurity level concentration, N_i, and mobility, μ. *Light/dark gray* indicates regions of small/large polaron transport. The space charge layer width is calculated from (9.2) for $\varepsilon = 100$, $V_B = 0.25$ V with the assumption that $N_i = N$

input the conditions that $\alpha L = 1$ and $s/k = 1$, where α is the absorption coefficient, L is the root-mean-square distance traveled in time τ_t in the absence of a field, s is the rate of transfer of minority carriers from the semiconductor surface to the electrolyte and k is the rate of transfer of carriers from within the semiconductor to its surface. These conditions state that most light is absorbed in the space charge layer of the semiconductor (because L must be less than w) and that every minority carrier that makes it to the semiconductor/electrolyte interface contributes to the photocurrent, i.e., there is no surface recombination. Fig. 9.3 shows the results of these calculations, which indicate that the IPCE is dramatically influenced by band bending and τ_t/τ_R. In every circumstance, low efficiencies are obtained for very small degrees of band bending because there is an insufficient field to support charge separation. But with modest band bending, high IPCEs (over 80%) can be obtained only when the transit time is similar to the recombination time.

From the foregoing arguments, it becomes clear that small polaron-type semiconductors with τ_R in the picosecond range need high doping levels ($>10^{19}$ cm^{-3}) to achieve high efficiencies. This implies a narrow space charge layer width, w, as calculated by

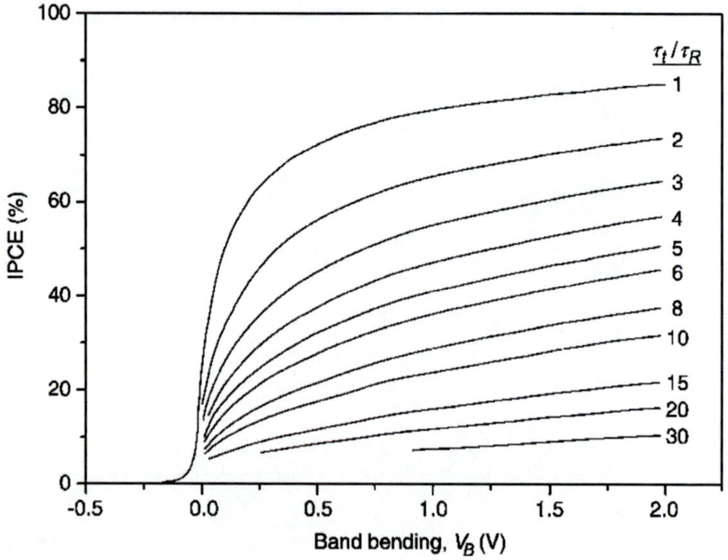

Fig. 9.3 Calculation of incident photon-to-current efficiency (IPCE) as a function of potential drop across the space charge layer and τ_t/τ_R, the ratio of the transit time to the recombination time. The rate of minority carrier transfer to the surface is set to equal the rate of transfer to the electrolyte (i.e., no surface recombination occurs) and $\alpha L = 1$. Modified from [8]

$$w = \sqrt{\frac{2\varepsilon\varepsilon_0}{eN}\left(V_B - \frac{kT}{q}\right)}, \tag{9.2}$$

where N is the carrier concentration. The relationship between w and N is included in Fig. 9.2 for a semiconductor with $\varepsilon = 100$ and $V_B = 0.25$ V, the latter is a typical value for an operating photoelectrochemical device. In Fig. 9.2, it is seen that small space charge layer widths are necessary for semiconductors with small polaron transport if small transit times are desired. For example, under the stated conditions with $N = 10^{19}$ cm^{-3}, the space charge layer width is 16 nm; for $N = 10^{20}$ cm^{-3}, the space charge layer width is 5 nm.

With such small space charge layer widths, it is likely that a significant proportion of light is not absorbed within the space charge layer. Additionally, within the Jarrett model, the absorption coefficient α is to be compared with L, such that high efficiencies are only achievable when αL is equal to or greater than one. But by increasing the doping level to decrease τ_t, L also decreases. For example, a small polaron semiconductor with a mobility of 0.1 cm^2/V s and a doping level of 10^{19} cm^{-3} has a transit time of 55 ps. If this time is similar in magnitude to the lifetime in the space charge layer, then αL is the relevant parameter (not αw) and it must be much less than one since L is just 1.1 nm. Jarrett calculates that in going from $\alpha L = 1$ to $\alpha L = 0.1$, the photocurrent efficiency lowers by a factor of 2.5.

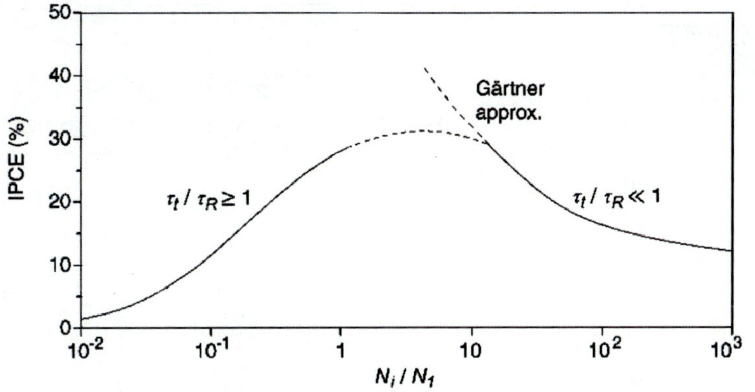

Fig. 9.4 Calculation of the IPCE for a series of systems with varying doping density, where N_1 is defined as the doping density where $\tau_t = \tau_R$. Band bending is 1.0 V and $\alpha L_m = 0.1$. Modified from [8]

Something interesting happens, however, when the doping level increases further. If, through increased doping, τ_t becomes significantly less than τ_R, recombination in the depletion layer can be neglected and consequently the Jarrett model simplifies to the Gärtner model. This is illustrated in Fig. 9.4, where the IPCE is calculated for varying doping levels. It is seen that at low doping levels ($\tau_t < \tau_R$) the Jarrett model applies and predicts low IPCEs because of the significant recombination in the depletion layer. At intermediate doping levels (τ_t is near τ_R) IPCEs reach their peak value because recombination in the space charge layer is diminished. Finally, for high doping levels ($\tau_t < \tau_R$), the Gärtner model applies because recombination in the space charge layer is negligible. But as doping increases, IPCEs decrease because an increasing proportion of photons are absorbed outside of the narrowing space charge layer. This would suggest that there is an optimum doping density where IPCE is maximized, as seen in Fig. 9.4.

The foregoing calculations by Jarrett and Gärtner are modeled on planar semiconductor–electrolyte interfaces and, as a consequence, low IPCEs are predicted for semiconductors with low mobility and short lifetimes. But if nonplanar geometries are explored – specifically, nanostructures in which the semiconductor fits entirely within the space charge layer – it should be possible to make efficient devices. Specifically, the successful approach will involve doping the semiconductor to the point that $\tau_t < \tau_R$ and the Gärtner assumption of no space charge layer recombination applies. If the material can be designed such that all photons are absorbed in the space charge layer – a difficult geometric constraint, considering the small widths as seen in Fig. 9.2 – IPCEs approaching 100% are feasible. Of course, this also assumes that the semiconductor is optically thick such that all above-bandgap photons are absorbed.

This discussion has assumed that τ_R is independent of the doping concentration and the position of the carrier within the depletion layer. In fact, both of these parameters influence τ_R, because the recombination rate depends on the

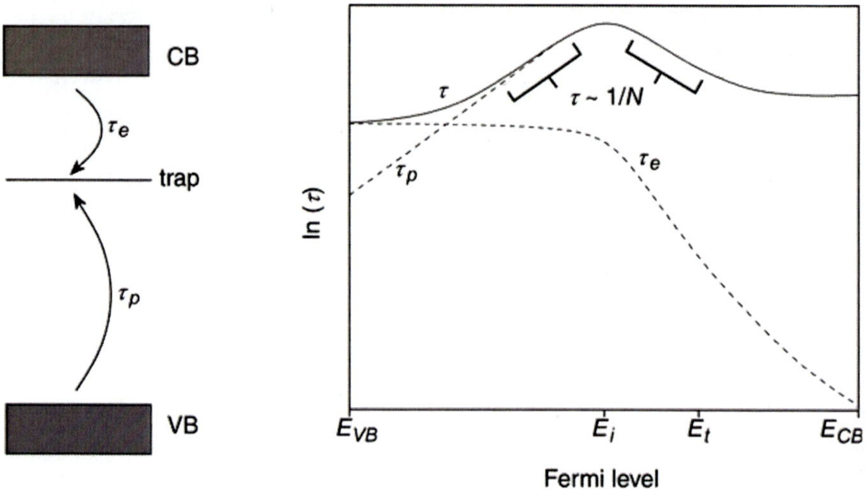

Fig. 9.5 The drawing (*left*) shows trap-mediated recombination with the electron (τ_e) and hole (τ_p) lifetimes; the total lifetime (τ) for movement of a carrier from the conduction band (CB) to the valence band (VB) is the sum of τ_e and τ_p. The graph (*right*) plots lifetime as a function of Fermi level – i.e., doping – for the case in which the energy level of the trap (E_t) is between the middle of the bandgap (E_i) and the conduction band (E_{CB}). The ranges of Fermi levels in which $\tau \sim 1/N$ are indicated within *brackets*. The graph is modified from [11]

concentration of charge carriers in the bands and the relative occupation of electrons and holes in traps [11]. In general, band-to-band recombination is dominant in direct bandgap semiconductors while trap-mediated recombination is dominant in indirect bandgap semiconductors [9]. Minority carrier lifetime for band-to-band recombination is inversely proportional to the majority carrier concentration N. In this case, increasing the doping density does not significantly alter the ratio of τ_t to τ_R since both have the same functional relationship to N. On the other hand, band bending transports excess electrons and holes away from each other, which minimizes direct band-to-band recombination within the space charge layer.

The situation is more complicated for trap-mediated recombination. When traps are located near the middle of the bandgap and light intensities are low (the situation in which there are fewer photogenerated carriers than majority carriers), minority carrier lifetimes are independent of N and inversely proportional to the trap concentration [9]. The case for traps closer to a band edge is more complex, as shown in Fig. 9.5. Shockley and Read discussed four regimes of recombination behavior under steady-state conditions for this type of system [11]. First, for highly doped p-type semiconductors, the rate-determining step is electron trapping because the very high concentration of holes leads to a short τ_p. In this regime, τ_e and therefore τ is relatively insensitive to changes in doping. Second, for semiconductors with modest levels of p-type dopants, the rate-determining step is now the trapping of holes because of their lowered concentration. Here, τ_p (and τ) change with doping as $1/N$. Third, for modestly doped n-type samples, traps are still mostly unoccupied and the even lower concentration of holes makes their removal the rate-limiting step. But in this regime, increasing doping fills the traps which decreases the hole lifetime; as a consequence τ changes with doping as $1/N$. Fourth, with all the traps filled by electrons, the rate-determining step is hole trapping. Lifetime is now independent of doping because increasing the doping level further does not increase the concentration of electrons in traps. As shown in Fig. 9.5, only in the range of modest doping does lifetime scale as $1/N$.

These four regimes were derived based on the assumption of low light intensities. As long as the majority carrier concentration does not change significantly upon light irradiation, the light intensity is considered low. For the regimes of high doping, this condition should hold for typical light intensities. It should also be noted that lifetime was calculated assuming the absence of a field. In the presence of a field, however, lifetimes should be further increased because the field separates photogenerated carriers. With the effect of the field accounted for, τ can be equated to τ_R.

Two additional effects can influence device efficiency – in these cases, by altering τ_t. In general, carrier mobility decreases with increasing impurity concentration because of impurity scattering [9], although the relative importance of impurity scattering in materials of low mobility – which already have mean free paths on the length scale of atomic distances – is probably minimal. It is also found that mobility changes under strong electric fields, which is a condition that arises within the space charge layer at high doping levels. For a potential drop of 0.25 V across a 25-nm space charge layer, the average electric field is 100 kV/cm. Semiconductors with high mobility typically have reduced mobility under high fields, although the situation is not straightforward for materials with low mobility. For example, if the average time required for a carrier to hop between centers becomes comparable to the lattice relaxation time, mobility can increase because self-trapping is diminished [12]. In fact, in some materials carriers can shift between small and large polaron transport [13]. It is therefore difficult to make any general predictions and the behavior of mobility under high fields must be determined experimentally for each case.

To summarize the foregoing arguments, semiconductors with small polaron transport and short carrier lifetimes can be used to make efficient devices only if several design criteria are met. First, doping levels must be sufficiently high such that $\tau_t < \tau_R$. This will eliminate most recombination within the depletion layer. Second, light absorption should be confined to the space charge layer. Based on the arguments presented, this condition is unlikely to be met for electrodes with planar geometries. This can be achieved, however, if the semiconductor is suitably nanostructured. By designing the material such that the electrolyte interface is never more than 5–10 nm from any place where minority carriers are generated in the semiconductor, high IPCEs are possible.

In the discussion that follows, a synthetic method that exploits nanostructuring concepts is presented, allowing these design criteria to be met. Several compelling synthetic strategies have emerged in recent years and further progress may require the application of these strategies to new metal oxide materials. The discussion

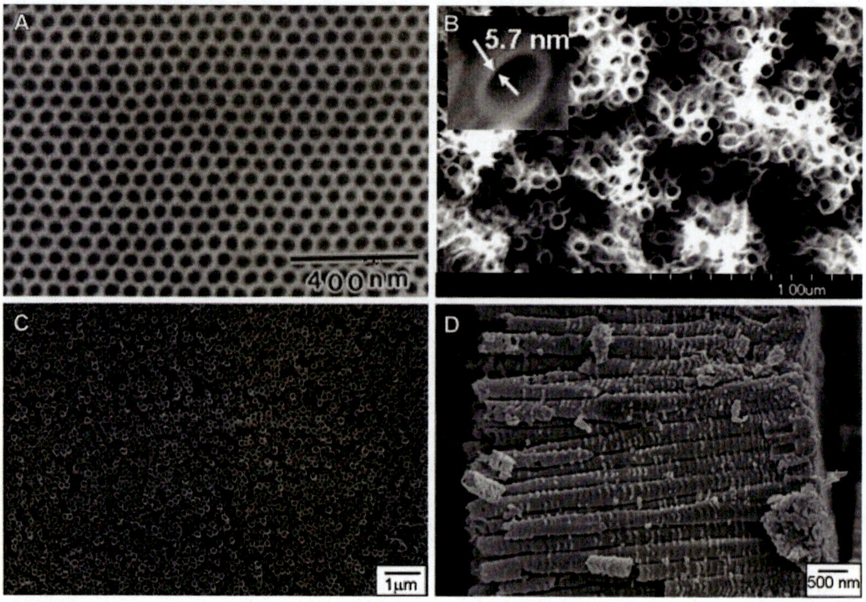

Fig. 9.6 SEM images of the anodization products. (**a**) Large-scale ordered aluminum oxide made by anodization in sulfuric acid at 65 V. From [29], with permission. (**b**) Hematite (α-Fe_2O_3) nanotube arrays prepared by sonoelectrochemical anodization in ethylene glycol with NH_4F and water. The *inset* shows a single nanotube. Reprinted with permission from [28], copyright 2009 Americal Chemical Society. (**c**) Titania nanotube arrays prepared by anodization in an aqueous electrolyte with KF. (**d**) Side view of same tubes from (**c**). (**c**) and (**d**) are reproduced from [30], with permission of American Scientific Publishers

addresses anodization of metals to produce nanotube arrays. It has already been seen that the application of these synthetic techniques to photoelectrochemical devices leads to improved performance.

9.2.2 Nanopore and Nanotube Arrays by anodization

It has long been known that the anodization of certain metals leads to the production of porous metal oxides and hydroxides. These metals include Al [14–16], Ti [17, 18], Ta [19], Cd [20], Nb [21], Mg alloys [22], W [23], Sn [24], Fe [12, 25], Ag [26], and Si [27]. Only recently conditions have been identified that allow the formation of well-controlled, uniform structures. For example, in 1995 an aluminum anodization process to develop hexagonally packed pores with ordered domains was developed [16]. In 2001, it was discovered that anodization of titanium foils led to the production of TiO_2 nanotube arrays [18]. More recently, a anodization process to form nanoporous [25] and nanotubular [28] iron oxides was developed. Examples of these materials are shown in Fig. 9.6. Pore sizes in the range of 20–100 nm are

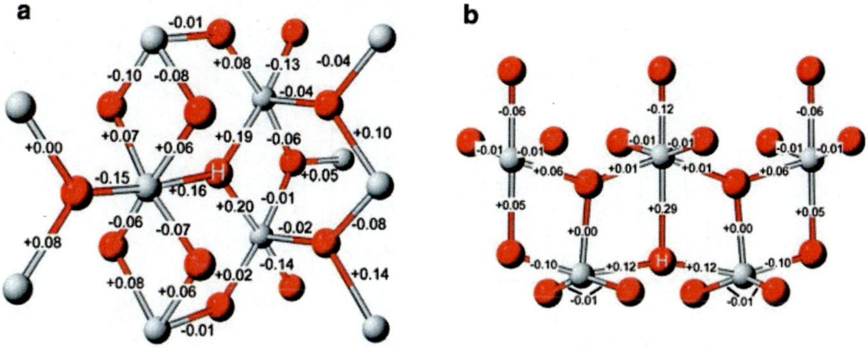

Fig. 9.7 Distortions in bond lengths, in Ångströms, surrounding a hole "H" on an oxygen atom in (**a**) rutile and (**b**) anatase. The images show just part of the cluster that was used for the calculation. Reproduced with permission from [36]. Copyright 2009 American Chemical Society

typical for this class of materials, although some materials, such as aluminum, have achieved even greater variations [31]. In the case of the nanotubular materials such as TiO_2 and Fe_2O_3, wall thicknesses in the order of 5–30 nm are typical.

In the context of water splitting, TiO_2 nanotube arrays have attracted the greatest attention. TiO_2 is an excellent candidate for nanostructuring because of its low room-temperature mobility and modest carrier lifetime. TiO_2 is generally n-type and therefore the hole mobility is the relevant parameter in determining τ_t. Only a few experimental [32–34] and theoretical [35, 36] determinations of hole mobility have been performed. Recent calculations of room temperature hole mobility in rutile provide values of 5×10^{-3} to 4×10^{-5} cm^2/V s for the most favorable (adiabatic) modes of charge transfer, while in anatase only one adiabatic mode was found, yielding a mobility of 2×10^{-3} cm^2/V s [36]. Where these calculations could be compared with experiment – high temperature conductivity measurements on rutile [33] – good agreement was found. Models of lattice distortion arising from the hole polarons were also calculated, as shown in Fig. 9.7. These showed that although the distortion influenced bond lengths up to several atoms away from the hole, these long range distortions were small in magnitude and had little influence on the hole structure and hopping energies. The large reorganization energies in both anatase and rutile (mostly between 2.0 and 2.5 eV) suggested that the hole transport be classified as a small polaron [4].

Hole lifetime in TiO_2, as in all materials, is sensitive to impurity concentration and identity, grain boundaries, defects, surface states, electric fields, and the concentration of photogenerated carriers. For this reason, determination of hole lifetime under one set of conditions may vary substantially from another set. Studies of TiO_2 nanoparticles (of diameter ~2 nm) showed electron–hole lifetimes (τ) of 30 ns [37, 38]. With the cautious approximation that τ_R of TiO_2 nanotube arrays is also 30 ns – probably an underestimate of τ_R – (9.1) shows that a doping level of approximately 1×10^{18} cm^{-3} is necessary to have τ_t equal to τ_R. This corresponds to a space charge layer width of 55 nm. As suggested by the trends discussed above, decreasing τ_t/τ_R further diminishes

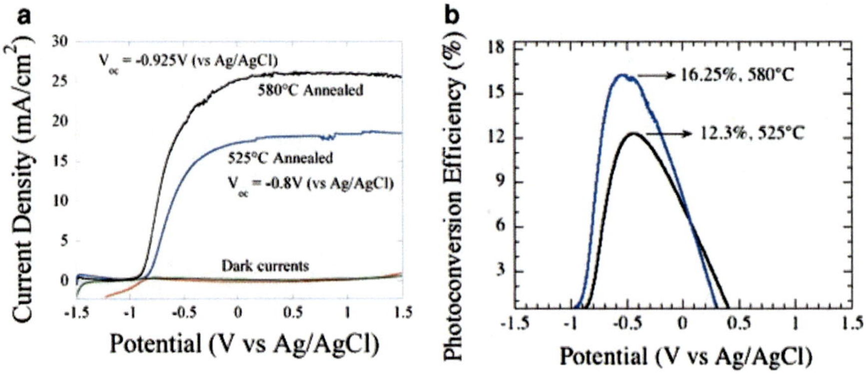

Fig. 9.8 Current densities and photoconversion efficiencies for TiO_2 nanotube calcined at either 525 or 580°C. Reproduced from [41] with permission of IOP Publishing Ltd

recombination in the space charge layer. This speaks, therefore, to the opportunities in nanostructuring TiO_2. In practice, because the lifetimes in TiO_2 are relatively long, much of the work on TiO_2 does not seek to explicitly dope the material for the sake of improving charge separation. Rather, the material is doped only modestly as an n-type semiconductor through the formation of oxygen vacancies and the incorporation of contaminants from either the titanium foil (the typical impurity content of titanium foil used to make the nanotubes is 0.1–0.3%) or the electrolyte.

In 2004, Grimes reported the application of TiO_2 nanotube arrays to water photoelectrolysis [39] and in 2005 provided an improved synthetic method that produced 224-nm long anatase nanotubes with a wall thickness of 34 nm. This led to an IPCE of over 90% at an applied bias of 1 V when irradiated at 337 nm [40]. In the later work, it was suggested that a key aspect of the improved performance was that the space charge layer would fit entirely within the nanotube walls, thereby implying that electric field-mediated charge separation was important to achieve high efficiencies in these nanostructures. The best performing materials synthesized using TiO_2 nanotubes converted light energy to hydrogen energy at an efficiency of 16.25%, as shown in Fig. 9.8, although it should be emphasized that ultraviolet light and not the full solar spectrum was used [41].

Considering the large bandgap of TiO_2, materials exposed to sunlight produce relatively low photocurrents. When AM 1.5 is used to photoelectrolyze water on titanium dioxide nanotubes, the maximum current densities are 1.1 mA/cm^2 with conversion efficiencies of 0.3% [42]. Numerous efforts have been made to dope TiO_2 to increase its response to visible light [43–46]. While visible light absorption can be induced in these materials and this visible light reaction can be used to drive photochemical reactions at the nanoparticle surface, efficiencies from visible light absorption are usually low and efficiencies from ultraviolet light typically decrease as well. The decreased efficiencies often arise from the creation of localized states within the bandgap that act to trap electrons or holes.

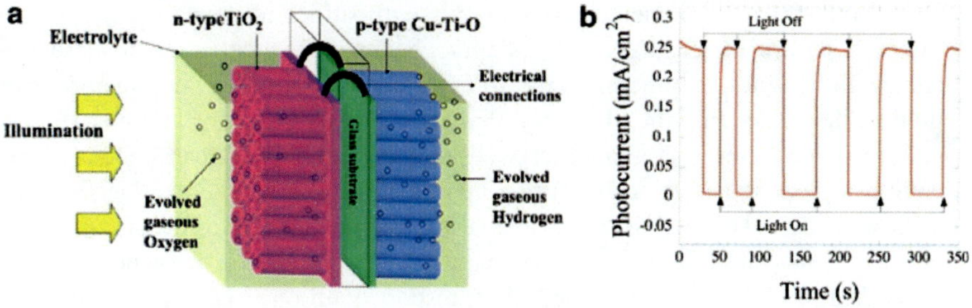

Fig. 9.9 Illustration of a self-biased photoelectrochemical pn-junction diode (**a**) and the corresponding photocurrent produced from the device (**b**). Reproduced with permission from [47]. Copyright 2008 Americal Chemical Society.

An approach that has been successfully implemented to improve light absorption in TiO_2-type nanotubes has been the incorporation of either iron or copper. A thin metallic film is sputtered that contains a mixture of titanium and iron, which is then anodized. This leads to the growth of nanotubular metal oxides with compositions that depend on the starting Ti:Fe or Ti:Cu ratio. In the case of iron, incorporating 6% into titanium provided the best results, where visible light photocurrents were obtained. Compared to the Ti-only materials, however, the sample had to be biased to a much greater degree to see similar photocurrents: onset potentials were nearly 1 V more positive. Nevertheless, a plateau photocurrent of 2.0 mA/cm^2 was achieved under AM 1.5, nearly twice that of the TiO_2-only materials. The large bias most probably resulted from the fact that the iron provides trap sites well below the TiO_2 conduction band edge, implying that large biases would be needed to free these trapped electrons.

A similar approach was used to incorporate copper into TiO_2-based films. Metallic Cu–Ti films with Cu contents of 60–74 at.% were anodized and subsequently calcined. This led to the production of nanotubular arrays composed of CuO, anatase, and rutile. The films exhibited p-type behavior, arising from CuO. This allowed a two-photoelectrode device to be built comprising a p-type CuO–TiO_2 nanotubular array and an n-type TiO_2 nanotubular array, as illustrated in Fig. 9.9. As the device employed two photoelectrodes, sufficient bias was established to drive unassisted water photoelectrolysis (Fig. 9.9b). Photocurrents of approximately 0.25 mA/cm^2 were established.

More efficient devices will require the improved utilization of visible light at both photoelectrodes. As suggested in the foregoing discussion, doping TiO_2 provides a compelling route to achieve this. Yet, it is well established that the doping also leads to an increased degree of recombination depending on conditions (dopant concentration, identity, redox behavior). The seminal studies of Hoffmann indicated, however, that the use of certain dopants in extremely small particles can, additionally, prolong lifetimes and facilitate interfacial charge transfer [43]. This effect was lost when particle sizes were increased. This therefore suggests that doping TiO_2 strategies that can decrease τ_t can also be used to enhance light absorption and charge transfer if implemented in materials with sufficiently small particle sizes. These sizes are increasingly being approached with synthetic methods such as anodization, which may permit TiO_2 to both absorb increasing amounts of visible light while minimizing the likelihood that holes recombine with electrons.

Iron oxide (α-Fe_2O_3) is another well-known material that can benefit from a combined doping and nanostructuring approach, as outlined above. Hematite has been variously classified as either a small polaron conductor or a borderline case between small and large polaron behavior [2]. Some of the discrepancy probably arises from its highly anisotropic hole transport, with one experimental study on heavily doped p-type hematite showing that mobility within (0001) being somewhere between a factor of 10 and 10,000 higher than in [0001] [48]. A theoretical study has predicted a 1,000-fold difference in conductivity [49]. Other studies have shown that the anisotropy is highly dependent on dopant identity and concentration [50]. The most careful measurements of hole and electron transport in poly-crystalline hematite have indicated a mobility of 0.2 cm^2/V s at 1,400 K and 0.1 cm^2/V s at 1,200 K, respectively [2]. Theoretical studies, in particular, have suggested that the transport mechanism is based on small polaron in spite of bond distances being distorted over significant distances [51].

Lifetimes in hematite are much shorter than TiO_2. Carrier relaxation to the conduction or valence band edge occurs within a few hundred fs, and recombination or trapping occurs within 3–5 ps. To have $\tau_t = \tau_R$, the doping density should be 10^{20}, or approximately 0.25 at.%, assuming that these values for lifetime are similar to τ_R. This corresponds to a space charge width of 5 nm for a potential drop of 0.25 V. Therefore, it would be desirable to synthesize highly doped hematite nanostructures with dimensions of less than 10 nm.

Investigations of iron anodization have indicated that it is possible to synthesize porous, amorphous iron oxide with a nanotubular morphology (Fig. 9.6) [25, 28]. Heat treatment crystallizes the iron oxide to form hematite. Recently, anodization conditions were found which lead to the formation of nanotubular hematite with wall thickness of approximately 6 nm [28]. The best performing nanotubes were formed under a reducing environment, which should create a high concentration of oxygen vacancies. It has been shown that oxygen vacancies create a donor level about 0.20 eV below the conduction band [2], thereby serving as an electron trap when empty. The resulting material exhibits photocurrents of 1.41 mA/cm^2 at 1.5 V vs. RHE, although it should be noted that the sample was only illuminated with 87 mW/cm^2 at AM 1.5. There are better dopants for hematite than oxygen vacancies since the vacancies also act as traps, and achieving a controlled doping density would further lead to improved control over τ_t. Nevertheless, the initial studies with α-Fe_2O_3 represents promising initial investigations of a compelling and versatile nanostructuring technique. An interesting example of this combined doping and nanostructuring approach has been demonstrated for Si-doped hematite prepared by chemical vapor deposition [52]; if applied to the nanotube structures significant performance gains may be possible.

To apply this as a fabrication technique for new metal oxide compositions, some understanding of the formation mechanism must be had. There are numerous studies that outline the mechanism of the formation of such anodized films and these studies have revealed that there are many similarities in spite of differing compositions [17, 53]. The anodization process occurs as follows:

1. Prior to anodization, there is an oxide layer of varying thickness on the metal surface. Application of an anodizing potential causes electric fields to be concentrated in thinner regions of the oxide layer.
2. The anodizing potential causes the reaction $M^0 \rightarrow M^{n+} + ne^-$ to occur at the metal/oxide interface, with the reaction occurring at the greatest rate where electric fields are most concentrated. The high-electric field drives the metal ion to the metal oxide/electrolyte interface.
3. The anodizing potential drives water decomposition ($H_2O \rightarrow 2\,H^+ + O^{2-}$) and, possibly, electrolyte decomposition at the electrode/solution interface. The oxygen anions diffuse through the oxide layer to form the metal oxide. Some proportion of anions from the electrolyte are also incorporated into the oxidized material. This metal oxide layer through which metal and oxygen ions diffuse is called the barrier layer.
4. The water decomposition reaction increases the concentration of electrolyte species that etch the metal oxide – usually this is H^+ or F^-. Because these species are most concentrated in the vicinity of the concentrated electric field, etching occurs most rapidly in these locations, which leads to pore formation.
5. A self-reinforcing process develops in which the electric field becomes more concentrated in the pore bottoms because this is where the barrier layer is the thinnest. This promotes steps 2, 3, and 4, which allows the anodization and pore formation to continue.
6. The formation of an ordered pore structure arises only when a careful balance of oxide growth and dissolution are achieved [54, 55]. As the metal is oxidized, it also expands, which induces stresses. Modest stresses force the growing pores to organize in a hexagonal array so as to provide the most efficient packing of the expanding oxide [54, 56, 57]. Temperature, electrolyte, and voltage are the parameters typically used to control growth and dissolution rates.

Figure 9.10 outlines these steps for the anodization of nanoporous and nanotubular morphologies. The nanotubular structures can be seen as containing two sets of pores: one large pore inside each tube and a second pore that surrounds the outside of the tube. In contrast, the nanoporous structures only develop the single large pore. It has been proposed [17] that variations in ion mobility in different metal oxides determine whether nanotubular or nanoporous morphologies form. For example, Ti and O ions have low mobility, which implies that the oxide layer at the bottom of the pores will be relatively thin under steady-state conditions. Because of the relatively thin oxide layer, some Ti will remain unoxidized within the pore walls. The electric field then becomes sufficiently concentrated in these unoxidized regions for electric field-assisted dissolution to occur. This is a self-reinforcing process that continues throughout the anodization and leads to nanotube formation.

An important point is taken from the work on Cu–Ti systems [47]. While copper has not yet been found to be suitable for anodization as a pure metal, it has been

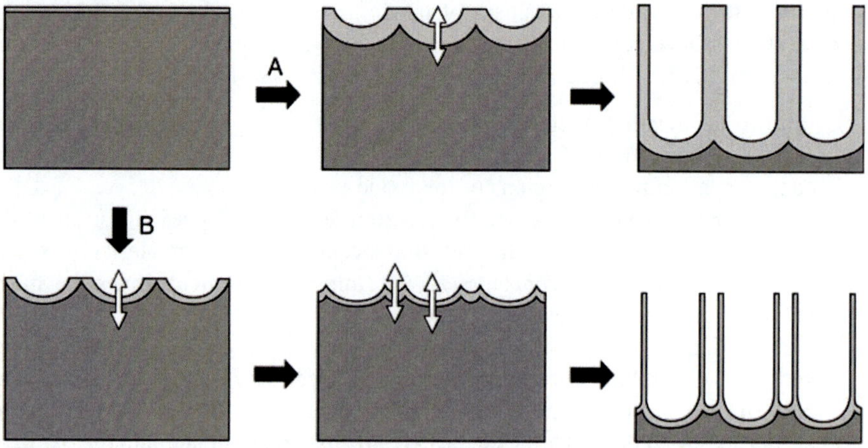

Fig. 9.10 Cross-sectional view of the anodization process of (**a**) ordered porous metal oxides such as Al_2O_3 and (**b**) nanotube arrays such as TiO_2 and Fe_2O_3. *Dark gray* is metal, *light gray* is metal oxide. *White arrows* indicate the locations where metal and oxygen anions diffuse through the metal oxide layer. The barrier layer is much thinner in (**b**) than in (**a**), which promotes nanotube formation

successfully implemented in the form of Cu–Ti alloys with Cu compositions up to 74 at.%. This would therefore suggest that mesoporous complex oxides can be fabricated by anodization as long as one component of the starting alloy can be anodizied. Because the resulting oxide is most often amorphous [58], heat treatment that leads to a crystalline, homogeneous complex oxide is critical. Based on the extensive experience in heat treatments of TiO_2 nanotubes, annealing temperatures should take into consideration the possibility of structural collapse.

The controlled incorporation of dopants is another important direction for further improvement in anodized materials. Suitable methods may include direct incorporation of a metal ion dopant during preparation of the metal film or by incorporation of anions from the electrolyte during anodization.

With such material properties suitably controlled, it may be possible to fabricate efficient devices made from complex oxides with poor mobility and short carrier lifetimes.

9.3 Nanophotonics

In addition to improved methods for controlling the structure and composition of semiconductors, there is emerging work in the field of nanophotonics that may be favorably exploited for use in water photoelectrolysis. Much of the work in nanophotonics seeks to manipulate photons using the properties of nanostructured materials in ways that cannot be achieved with macroscopic optical components.

There are several emerging areas of nanophotonics, which include plasmonics, multiple exciton generation, photonic crystals, and upconversion. While these concepts have been applied to the design of solar cells, the exploration of these concepts in the context of water photoelectrolysis devices is limited. It is therefore of interest to consider the ways in which advances in nanophotonics could influence the direction of future work in water photoelectrolysis. This section emphasizes developments in plasmonics that may be useful if successfully applied to water photoelectrolysis.

9.3.1 Plasmonics

Much interest has been stimulated in plasmonics, which arises from the useful property of certain nanostructured materials to confine and manipulate information encoded within a signal of photons. This opens the prospect for building microprocessors capable of computing with optical signals rather than electrical ones, which could lead to breakthroughs in speed and efficiency of computing devices [59]. Other demonstrations within plasmonics include the ability to create lenses with resolution well below the diffraction limit of light, thus paving the way to optical microscopes with unparalleled resolution [60–62]. The underlying physics that are exploited in plasmonics are also of interest in solar energy conversion devices, since the ability to confine and scatter photons should improve photon absorption and utilization in water photoelectrolysis devices.

Already, several plasmonic effects have been explored within the contexts of thin-film silicon, dye-sensitized, and organic solar cells. The phenomena that have been observed and exploited – principally near-field absorption and light scattering – are described below. Together, these findings suggest strategies for further efficiency enhancements in water photoelectrolysis.

When a metal nanoparticle is subjected to an oscillating electric field such as the field intrinsic to light, the free electrons in the nanoparticle will respond to the field by oscillating. The response to the field is quantified by the polarizability of the metal. The polarizability α is given by

$$\alpha = 4\pi a^3 \frac{\varepsilon - \varepsilon_m}{\varepsilon + 2\varepsilon_m}, \tag{9.3}$$

where a is the particle radius, ε is the wavelength-dependent dielectric function of the metal, and ε_m is the dielectric constant of the surrounding medium. The equation indicates that the polarizability is greatest when the real part of ε is equal or close to $-2\varepsilon_m$. This provides the condition for a strong resonance of the free electrons, which is called the dipole surface plasmon. Although most metals have resonances in the ultraviolet, gold and silver are most polarizable in the visible or infrared.

The interaction of light with the metal nanoparticle is defined by the extinction cross-section C_{ext}, which is the sum of the absorption cross-section C_{abs} and the scattering cross-section C_{sca}:

$$C_{\text{abs}} = k\text{Im}[\alpha] = 4\pi k a^3 \text{Im}\left|\frac{\varepsilon - \varepsilon_{\text{m}}}{\varepsilon + 2\varepsilon_{\text{m}}}\right|, \tag{9.4}$$

$$C_{\text{sca}} = \frac{k^4}{6\pi}|\alpha|^2 = \frac{8\pi}{3}k^4 a^6 \left|\frac{\varepsilon - \varepsilon_{\text{m}}}{\varepsilon + 2\varepsilon_{\text{m}}}\right|^2. \tag{9.5}$$

For small particle sizes (<25 nm diameter for gold), C_{abs} is the dominant term and C_{sca} can be neglected [63]. But with increasing particle size, C_{sca} becomes larger, owing to the a^6 dependence on radius. Therefore, the behavior of the plasmon can be tuned with particle size.

When C_{abs} is the dominant term, the plasmon – i.e., the collective oscillation of electrons – persists on the metal nanoparticle for between 5 and 10 fs before decaying as heat [64]. The nonrandom displacement of electrons produces an electric field that extends beyond the nanoparticle's surface. When this electric field extends into a region of space that contains a semiconductor, an electron–hole pair can be generated in the semiconductor if the energy of the plasmon is greater than the semiconductor's bandgap. The electric field decays with distance d from the metal particle as d^{-3} so that the probability of absorption in the semiconductor decreases with increasing distance from the metal nanoparticle. Thus, the plasmon effectively "focuses" photons to increase photon absorption in regions of the semiconductor that are close to a metal nanoparticle; this is termed near-field absorption.

Plasmonic near-field absorption has been modeled and exploited in silicon, organic, and dye-sensitized solar cells. For example, an organic solar cell achieved a 2.7× increase in photocurrent after incorporating copper nanoparticles [65]. A tandem organic cell exploiting the plasmon resonance of silver nanoparticles achieved a photocurrent of 2.5 mA/cm^2, while the identical device without the plasmon resonance was expected to achieve 1.9 mA/cm^2 [66]. Two examples of the modeling of the electric field induced by the surface plasmon are shown in Fig. 9.11. At the resonance wavelength, significant field enhancements are observed and the effect extends over 50 nm from the silver, albeit with decreasing field.

When C_{sca} is the dominant term, the oscillating dipole primarily causes scattering of the incident light. In general, scattering with metal nanoparticles has been exploited to increase the optical path of photons through the absorbing layer or to couple light into the waveguiding modes of thin films. This was elegantly demonstrated in a thin film silicon solar cell in which 12 or 16-nm silver nanoparticles scattered light into waveguiding modes, leading to a 16-fold enhancement in absorption at certain wavelengths in the infrared [67]. Using a similar thin-film silicon geometry, the waveguiding modes were modeled, as shown in Fig. 9.12a. A strong enhancement of the field is observed, which is localized in

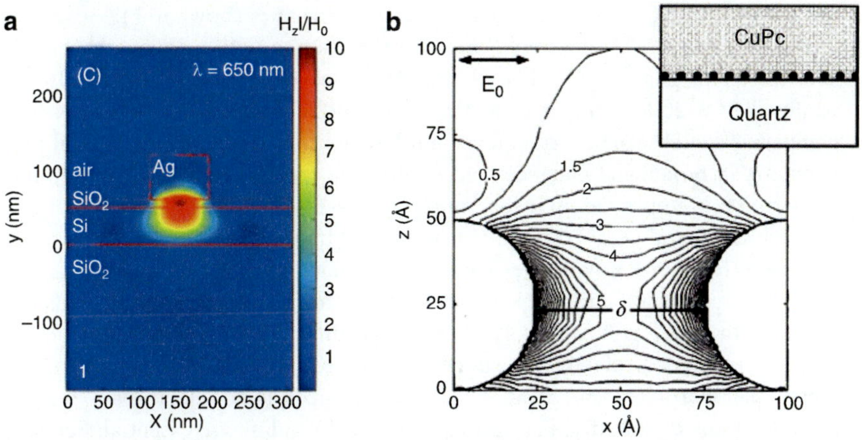

Fig. 9.11 Plots of field intensity for two types of solar cells. (**a**) A cross-sectional view of a 50-nm thick silicon solar cell separated by a 10 nm silica layer from a 100-nm wide silver nanowire. As compared to the incident field, a nearly tenfold increase is observed in regions of the silicon close to the silver. Reproduced from [68], with permission. (**b**) A cross-sectional view of 5-nm diameter silver nanowires sitting on a quartz substrate and in contact with copper phthalocyanine (CuPc). Nanoparticles are spaced 5-nm apart (δ), which induces coupling between the dipoles and creates an increased field between adjacent nanoparticles. Contour labels represent the absorption enhancement factor as compared to the incident field. Reprinted with permission from [66]. Copyright 2004, American Institute of Physics

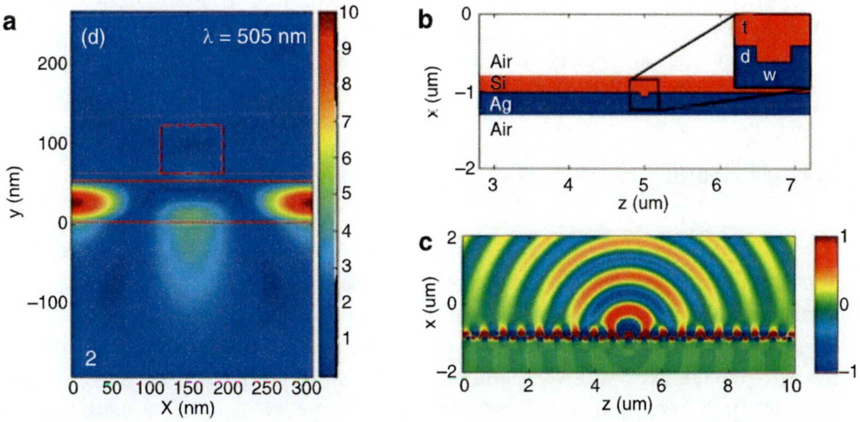

Fig. 9.12 Models of scattering from metal nanostructures. (**a**) Same geometry as Fig. 9.11a showing a waveguiding mode. Reproduced from [68], with permission. (**b**) A model of a Si/Ag device with a 100-nm wide and 50-nm deep groove and (**c**) a simulation of the field scattered from the groove. Reproduced with permission from [69]. Copyright 2008 Americal Chemical Society

the silicon layer [68]. An alternate scattering geometry is shown in Fig. 9.12b, with a thin-film silicon cell deposited on a silver substrate with a groove in the silver substrate. This geometry leads to a high degree of lateral scattering, mostly being confined to a 200-nm layer from the silver surface, as shown in Fig. 9.12c. In comparing the absorbance of silicon in the scattering geometry to a typical planar geometry, a broadband enhancement in absorbance was observed. At 700 nm the scattering film improved light absorption by a factor of 2, at 900 nm a factor of 1.2 was achieved, and at 1,100 nm it was improved by a factor of 2.4.

The efforts to enhance efficiency in solar cells through use of metal nanoparticles should provide a general guide for the application of plasmonic concepts to water photoelectrolysis, although differences in device design, such as the presence of an electrolyte, suggest that there may be additional criteria. When metal nanoparticles are located at the semiconductor/electrolyte interface, considerations of stability, Fermi level, and band bending are essential. Secondary considerations may include the influence of catalytic effects and the possible formation of trap states at the metal/semiconductor interface.

Near-field absorption enhancement can be used to confine light absorption to the proximity of metal nanoparticles. This could prove useful in a number of circumstances, including those described in Sect. 9.2, where the transport of minority carriers across the space charge layer competes with recombination. In such a case, the placement of nanoparticles at the semiconductor/electrolyte interface would lead to carrier generation at or near the interface, thereby limiting recombination losses. Alternatively, particle placement at the back interface could assist semiconductors with poor transport of majority carriers. In the case of scattering from metal structures, either of the designs that are illustrated in Fig. 9.12 may have promise, although variations in the dielectric function of the semiconductor can dramatically influence the efficacy of such structures [69]. In general, however, the possibility of confining light absorption to a decreased volume of semiconductor, as such designs allow, could allow a larger photovoltage to develop while potentially decreasing material costs.

9.4 Outlook

Future directions in water photoelectrolysis are myriad and there are many opportunities to discover cost-effective methods for the conversion of solar energy to chemical energy. Although an emerging trend in water photoelectrolysis is the development of nanostructured photoelectrodes, the full potential of nanostructures can only be achieved when the limitations of a specific material are accounted for, such as high recombination rates or poor charge transport. In particular, the directions suggested here provide a rationale for the development of highly doped nanostructures in which the feature size of the structure is commensurate with twice the space charge width. This strategy is expected to lead to significant improvements in efficiency for semiconductors with small polaron transport and short lifetimes. With this design approach, many semiconductors – among the most promising are complex oxides – can realistically be considered as candidates for water photoelectrolysis.

The application of plasmonic metal nanostructures to water photoelectrolysis is an intriguing and potentially rich area for future exploration. By providing several general mechanisms to alter the traditional $e^{-\alpha x}$ absorption profile in semiconductors, the construction of new design rules for water photoelectrolysis devices is possible.

Acknowledgments S. C. W. thanks Michael Grätzel and Hen Dotan for fruitful discussions and the Swiss Federal Office of Energy (PEChouse, project number 102326) and the European Commission's Framework Project 7 (NanoPEC, Project 227179) for support.

References

1. Holstein, T.: Studies of polaron motion. Part I. The molecular-crystal model. Ann. Phys. **8**, 325–342 (1959)
2. Bosman, A.J., van Daal, H.J.: Small-polaron versus band conduction in some transition-metal oxides. Adv. Phys. **19**, 1–117 (1970)
3. Marcus, R.A.: Chemical and electrochemical electron-transfer theory. Annu. Rev. Phys. Chem. **15**, 155–196 (1964)
4. Shluger, A.L., Stoneham, A.M.: Small polarons in real crystals: concepts and problems. J. Phys. D Condens. Mat. **5**, 3049–3086 (1993)
5. Austin, I.G., Mott, N.F.: Polarons in crystalline and non-crystalline materials. Adv. Phys. **18**, 41–102 (1969)
6. Gärtner, W.W.: Depletion-layer photoeffects in semiconductors. Phys. Rev. **116**, 84–87 (1959)
7. Butler, M.A.: Photoelectrolysis and physical properties of the semiconducting electrode WO_3. J. Appl. Phys. **48**, 1914–1920 (1977)
8. Jarrett, H.S.: Photocurrent conversion efficiency in a schottky barrier. J. Appl. Phys. **52**, 4681–4689 (1981)
9. Sze, S.M., Ng, K.K.: Physics of Semiconductor Devices. Wiley, Hoboken (2007)
10. Miller, R.J.D., Memming, R.: Fundamentals in photoelectrochemistry. In: Archer, M.D., Nozik, A.J. (eds.) Nanostructured and Photoelectrochemical Systems For Solar Photon Conversion, pp. 760. Imperial College Press, London (2008)
11. Shockley, W., Read, W.T.: Statistics of the recombinations of holes and electrons. Phys. Rev. **87**, 835 (1952)
12. Emin, D.: Lattice relaxation and small-polaron hopping motion. Phys. Rev. B **4**, 3639–3651 (1971)
13. Emin, D., Holstein, T.: Adiabatic theory of an electron in a deformable continuum. Phys. Rev. Lett. **36**, 323 (1976)
14. Dell'Oca, C.J., Fleming, P.J.: Initial stages of oxide growth and pore initiation in the porous anodization of aluminum. J. Electrochem. Soc. **123**, 1487–1493 (1976)
15. Parkhutik, V.P., Shershulsky, V.I.: Theoretical modeling of porous oxide growth on aluminum. J. Phys. D Appl. Phys. **25**, 1258–1263 (1992)
16. Masuda, H., Fukuda, K.: Ordered metal nanohole arrays made by a two-step replication of honeycomb structures of anodic alumina. Science **268**, 1466–1468 (1995)
17. Mor, G.K., Varghese, O.K., Paulose, M., Shankar, K., Grimes, C.A.: A review on highly ordered, vertically oriented TiO_2 nanotube arrays: fabrication, materials properties, and solar energy applications. Solar Energ. Mater. Solar Cells **90**, 2011–2075 (2006)

18. Gong, D., Grimes, C.A., Varghese, O.K., Hu, W., Singh, R.S., Chen, Z., Dickey, E.C.: Titanium oxide nanotube arrays prepared by anodic oxidation. J. Mater. Res. **16**, 3331–3334 (2001)
19. Sieber, I., Kannan, B., Schmuki, P.: Self-assembled porous tantalum oxide prepared in H_2SO_4/HF electrolytes. Electrochem. Solid-State Lett. **8**, J10–J12 (2005)
20. Vijayavalli, R., Vasudeva Rao, P.V., Udupa, H.V.K.: Effect of ac superimposition on dc in the cathodic polarization of anodized cadmium in alkaline solution. Electrochim. Acta **16**, 1197–1200 (1971)
21. Sieber, I., Hildebrand, H., Friedrich, A., Schmuki, P.: Formation of self-organized niobium porous oxide on niobium. Electrochem. Comm. **7**, 97–100 (2005)
22. Hsiao, H.-Y., Tsai, W.-T.: Characterization of anodic films formed on AZ91D magnesium alloy. Surf. Coat. Technol. **190**, 299–308 (2005)
23. Tsuchiya, H., Macak, J.M., Sieber, I., Taveira, L., Ghicov, A., Sirotna, K., Schmuki, P.: Self-organized porous WO_3 formed in NaF electrolytes. Electrochem. Commun. **7**, 295–298 (2005)
24. Metikos-Hukovic, M., Reseti'c, A., Gvozdic, V.: Behaviour of tin as a valve metal. Electrochim. Acta **40**, 1777–1779 (1995)
25. Prakasam, H.E., Varghese, O.K., Paulose, M., Mor, G.K., Grimes, C.A.: Synthesis and photoelectrochemical properties of nanoporous iron (III) oxide by potentiostatic anodization. Nanotechnology **17**, 4285–4291 (2006)
26. Wales, C.P., Burbank, J.: Oxides on the silver electrode. J. Electrochem. Soc. **106**, 885–890 (1959)
27. Föll, H., Christophersen, M., Carstensen, J., Hasse, G.: Formation and application of porous silicon. Mater. Sci. Eng. R **39**, 93–141 (2002)
28. Mohapatra, S.K., John, S.E., Banerjee, S., Misra, M.: Water photooxidation by smooth and ultrathin α-Fe_2O_3 nanotube arrays. Chem. Mater. **21**, 3048–3055 (2009)
29. Masuda, H., Hasegwa, F., Ono, S.: Self-ordering of cell arrangement of anodic porous alumina formed in sulfuric acid solution. J. Electrochem. Soc. **144**, L127–L130 (1997)
30. Varghese, O.K., Paulose, M., Shankar, K., Mor, G.K., Grimes, C.A.: Water-photolysis properties of micron-length highly-ordered titania nanotube-arrays. J. Nanosci. Nanotechnol. **5**, 1158–1165 (2005)
31. Lee, W., Ji, R., Gosele, U., Nielsch, K.: Fast fabrication of long-range ordered porous alumina membranes by hard anodization. Nat. Mater. **5**, 741–747 (2006)
32. Odier, P., Baumard, J.F., Panis, D., Anthony, A.M.: Thermal emission, electrical conductivity, and hall effect for defects study at high temperature ($T \geq 1250$ K) in refractory oxides (Y_2O_3, TiO_2). J. Solid State Chem. **12**, 324–328 (1975)
33. Bak, T., Nowotny, M.K., Sheppard, L.R., Nowotny, J.: Mobility of electronic charge carriers in titanium dioxide. J. Phys. Chem. C **112**, 12981–12987 (2008)
34. Bak, T., Nowotny, J., Rekas, M., Sorrell, C.C.: Defect chemistry and semiconducting properties of titanium dioxide: III. Mobility of electronic charge carriers. J. Phys. Chem. Solids **64**, 1069–1087 (2003)
35. Kerisit, S., Deskins, N.A., Rosso, K.M., Dupuis, M.: A shell model for atomistic simulation of charge transfer in titania. J. Phys. Chem. C **112**, 7678–7688 (2008)
36. Deskins, N.A., Dupuis, M.: Intrinsic hole migration rates in TiO_2 from density functional theory. J. Phys. Chem. C **113**, 346–358 (2008)
37. Rothenberger, G., Moser, J., Grätzel, M., Serpone, N., Sharma, D.K.: Charge carrier trapping and recombination dynamics in small semiconductor particles. J. Am. Chem. Soc. **107**, 8054–8059 (1985)
38. Bahnemann, D.W., Hilgendorff, M., Memming, R.: Charge carrier dynamics at TiO_2 particles: reactivity of free and trapped holes. J. Phys. Chem. B **101**, 4265–4275 (1997)
39. Mor, G.K., Shankar, K., Varghese, O.K., Grimes, C.A.: Photoelectrochemical properties of titania nanotubes. J. Mater. Res. **19**, 2989–2996 (2004)
40. Mor, G.K., Shankar, K., Paulose, M., Varghese, O.K., Grimes, C.A.: Enhanced photocleavage of water using titania nanotube arrays. Nano Lett. **5**, 191–195 (2005)

41. Shankar, K., Mor, G.K., Prakasam, H.E., Yoriya, S., Paulose, M., Varghese, O.K., Grimes, C.A.: Highly-ordered TiO_2 nanotube arrays up to 220 μm in length: use in water photoelectrolysis and dye-sensitized solar cells. Nanotechnology **18**, 065707 (2007)
42. Paulose, M., Mor, G.K., Varghese, O.K., Shankar, K., Grimes, C.A.: Visible light photoelectrochemical and water-photoelectrolysis properties of titania nanotube arrays. J. Photochem. Photobiol. A **178**, 8–15 (2006)
43. Choi, W., Termin, A., Hoffmann, M.R.: The role of metal ion dopants in quantum-sized TiO_2: correlation between photoreactivity and charge carrier recombination dynamics. J. Phys. Chem. **98**, 13669–13679 (1994)
44. Irie, H., Watanabe, Y., Hashimoto, K.: Nitrogen-concentration dependence on photocatalytic activity of $TiO_{2-x}N_x$ powders. J. Phys. Chem. B **107**, 5483–5486 (2003)
45. Asahi, R., Morikawa, T., Ohwaki, T., Aoki, K., Taga, Y.: Visible-light photocatalysis in nitrogen-doped titanium oxides. Science **293**, 269–271 (2001)
46. Burda, C., Lou, Y., Chen, X., Samia, A.C.S., Stout, J., Gole, J.L.: Enhanced nitrogen doping in TiO_2 nanoparticles. Nano Lett. **3**, 1049–1051 (2003)
47. Mor, G.K., Varghese, O.K., Wilke, R.H.T., Sharma, S., Shankar, K., Latempa, T.J., Choi, K.-S., Grimes, C.A.: P-type Cu–Ti–O nanotube arrays and their use in self-biased heterojunction photoelectrochemical diodes for hydrogen generation. Nano Lett. **8**, 1906–1911 (2008)
48. Nakau, T.: Electrical conductivity of α-Fe_2O_3. J. Phys. Soc. Jpn. **15**, 727 (1960)
49. Iordanova, N., Dupuis, M., Rosso, K.M.: Charge transport in metal oxides: a theoretical study of hematite α-Fe_2O_3. J. Chem. Phys. **122**, 144305–144310 (2005)
50. Benjelloun, D., Bonnet, J.-P., Dordor, P., Launay, J.-C., Onillon, M., Hagenmuller, P.: Anisotropie des propriétés électriques de monocristaux de Fe_2O_3-α dopés au nickel. Rev. Chim. Miner. **21**, 721–731 (1984)
51. Rosso, K.M., Dupuis, M.: Reorganization energy associated with small polaron mobility in iron oxide. J. Chem. Phys. **120**, 7050–7054 (2004)
52. Kay, A., Cesar, I., Grätzel, M.: New benchmark for water photooxidation by nanostructured α-Fe_2O_3 films. J. Am. Chem. Soc. **128**, 15714–15721 (2006)
53. Li, F., Zhang, L., Metzger, R.M.: On the growth of highly ordered pores in anodized aluminum oxide. Chem. Mater. **10**, 2470–2480 (1998)
54. Nielsch, K., Choi, J., Schwirn, K., Wehrspohn, R.B., Gosele, U.: Self-ordering regimes of porous alumina: The 10% porosity rule. Nano Lett. **2**, 677–680 (2002)
55. Macák, J.M., Tsuchiya, H., Schmuki, P.: High-aspect-ratio TiO_2 nanotubes by anodization of titanium. Angew. Chem. Int. Ed. **44**, 2100–2102 (2005)
56. Jessensky, O., Muller, F., Gosele, U.: Self-organized formation of hexagonal pore arrays in anodic alumina. Appl. Phys. Lett. **72**, 1173–1175 (1998)
57. Li, A.P., Muller, F., Birner, A., Nielsch, K., Gosele, U.: Hexagonal pore arrays with a 50–420 nm interpore distance formed by self-organization in anodic alumina. J. Appl. Phys. **84**, 6023–6026 (1998)
58. Allam, N.K., Shankar, K., Grimes, C.A.: A general method for the anodic formation of crystalline metal oxide nanotube arrays without the use of thermal annealing. Adv. Mater. **20**, 3942–3946 (2008)
59. Barnes, W.L., Dereux, A., Ebbesen, T.W.: Surface plasmon subwavelength optics. Nature **424**, 824–830 (2003)
60. Fang, N., Lee, H., Sun, C., Zhang, X.: Sub-diffraction-limited optical imaging with a silver superlens. Science **308**, 534–537 (2005)
61. Lee, H., Xiong, Y., Fang, N., Srituravanich, W., Durant, S., Ambati, M., Sun, C., Zhang, X.: Realization of optical superlens imaging below the diffraction limit. New J. Phys. **7**, 255–255 (2005)
62. Melville, D., Blaikie, R.: Super-resolution imaging through a planar silver layer. Opt. Express **13**, 2127–2134 (2005)
63. Kreibig, U., Vollmer, M.: Optical Properties of Metal Clusters. Springer, Berlin (1995)

64. Link, S., El-Sayed, M.A.: Size and temperature dependence of the plasmon absorption of colloidal gold nanoparticles. J. Phys. Chem. B **103**, 4212–4217 (1999)
65. Stenzel, O., Stendal, A., Voigtsberger, K., von Borczyskowski, C.: Enhancement of the photovoltaic conversion efficiency of copper phthalocyanine thin film devices by incorporation of metal clusters. Solar Energ. Mater. Solar Cells **37**, 337–348 (1995)
66. Rand, B.P., Peumans, P., Forrest, S.R.: Long-range absorption enhancement in organic tandem thin-film solar cells containing silver nanoclusters. J. Appl. Phys. **96**, 7519–7526 (2004)
67. Pillai, S., Catchpole, K.R., Trupke, T., Green, M.A.: Surface plasmon enhanced silicon solar cells. J. Appl. Phys. **101**, 093105–093108 (2007)
68. Pala, R.A., White, J., Barnard, E., Liu, J., Brongersma, M.L.: Design of plasmonic thin-film solar cells with broadband absorption enhancements. Adv. Mater. **21**, 3504–3509 (2009)
69. Ferry, V.E., Sweatlock, L.A., Pacifici, D., Atwater, H.A.: Plasmonic nanostructure design for efficient light coupling into solar cells. Nano Lett. **8**, 4391–4397 (2008)

Index

R. van de Krol and M. Grätzel (eds.), *Photoelectrochemical Hydrogen Production*, Electronic Materials: Science & Technology 102, DOI 10.1007/978-1-4614-1380-6,

N

O

P

Q

R